I0036883

Advances in Computer Simulation Studies on Crystal Growth

Advances in Computer Simulation Studies on Crystal Growth

Special Issue Editor

Hiroki Nada

MDPI • Basel • Beijing • Wuhan • Barcelona • Belgrade

MDPI

Special Issue Editor
Hiroki Nada
Environmental Management Research Institute,
National Institute of Advanced Industrial Science and Technology (AIST)
Japan

Editorial Office
MDPI
St. Alban-Anlage 66
4052 Basel, Switzerland

This is a reprint of articles from the Special Issue published online in the open access journal *Crystals* (ISSN 2073-4352) from 2016 to 2018 (available at: https://www.mdpi.com/journal/crystals/special_issues/advances_in_computer_simulation_studies)

For citation purposes, cite each article independently as indicated on the article page online and as indicated below:

LastName, A.A.; LastName, B.B.; LastName, C.C. Article Title. *Journal Name* **Year**, *Article Number*, Page Range.

ISBN 978-3-03897-356-0 (Pbk)
ISBN 978-3-03897-357-7 (PDF)

Articles in this volume are Open Access and distributed under the Creative Commons Attribution (CC BY) license, which allows users to download, copy and build upon published articles even for commercial purposes, as long as the author and publisher are properly credited, which ensures maximum dissemination and a wider impact of our publications. The book taken as a whole is © 2018 MDPI, Basel, Switzerland, distributed under the terms and conditions of the Creative Commons license CC BY-NC-ND (http://creativecommons.org/licenses/by-nc-nd/4.0/).

Contents

About the Special Issue Editor

Hiroki Nada, Senior Researcher. Hiroki Nada received his Bachelor's degree in Applied Physics and his Master's degree in Geophysics from Hokkaido University in 1991 and 1993, respectively. He earned a Ph.D. from Hokkaido University under the supervision of Prof. Yoshinori Furukawa in 1995. Since 2001, he has been a senior researcher at the National Institute of Advanced Industrial Science and Technology (AIST). From October 2000 to September 2001, he was a visiting researcher at Debye Institute, Utrecht University. From April 2013 to March 2016, he was a visiting professor at the Institute of Low Temperature Science, Hokkaido University. He received a JSAP Award B in 1996 and a JACG Award in 2008. His research interests include computational science, data science, crystal growth, environmental earth science, biomolecules, ice, water, minerals, and functional materials.

crystals

MDPI

Editorial

Computer Simulations: Essential Tools for Crystal Growth Studies

Hiroki Nada

National Institute of Advanced Industrial Science and Technology (AIST), 16-1 Onogawa, Tsukuba 305-8569, Japan; hiroki.nada@aist.go.jp

Received: 31 July 2018; Accepted: 2 August 2018; Published: 4 August 2018

Abstract: This special issue discusses recent advances in computer simulation studies of crystal growth. Crystal growth is a key to innovation in science and technology. Owing to recent progress in computer performance, computer simulation studies of crystal growth have become increasingly important. This special issue covers a variety of simulation methods, including the Monte Carlo, molecular dynamics, first-principles, multiscale, and continuum simulation methods, which are used for studies on the fundamentals and applications of crystal growth and related phenomena for different materials, such as hard-sphere systems, ice, organic crystals, semiconductors, and graphene.

Keywords: molecular dynamics (MD); Monte Carlo (MC); first-principles (FP) simulation; continuum simulation; multiscale simulation

1. Introduction

Crystals are ubiquitous in daily life and technology. Many kinds of crystalline products, such as salt, sugar, and fat, are used in cooking, and electronic devices are made from semiconductor crystals. Crystals also play an important role in life and the global environment. Living organisms produce mineral crystals to maintain biogenic activity, and snow and ice crystals play a crucial role in climate change. For most topics related to crystals, crystal growth is an important research area.

Owing to recent progress in computer performance, computer simulation studies of crystal growth have become increasingly important. Computer simulations can be used to analyze and predict various aspects of the crystal growth process, such as growth and nucleation mechanisms, as well as the structures and dynamics of surfaces and interfaces, and pattern formation.

This special issue discusses recent advances in computer simulation studies of crystal growth. We present 10 papers, covering fundamental studies and applications of crystal growth or related phenomena. A variety of simulation methodologies are used in the studies.

2. Methodologies of Crystal Growth Computer Simulations

2.1. Molecular Simulation

Molecular simulations, such as molecular dynamics (MD) and Monte Carlo (MC) simulations, are powerful tools for investigating the growth mechanisms and interface structures of crystals at the molecular scale [1–3]. MD simulations analyze the structure, dynamics, mechanical properties, electrical properties, and optical properties of a condensed phase by solving the Newtonian equations of motion for each atom or molecule. MC simulations generate new states of the atomic or molecular arrangement stochastically according to a Boltzmann probability distribution. In addition to single condensed phases, crystal–liquid interfaces or crystal–vapor interfaces can be examined by MD or MC simulations. Therefore, we can analyze the atomic- or molecular-scale mechanisms of crystal growth by these simulation methods.

In this special issue, Mori [4], Qiu and Molinero [5], Barcaro et al. [6], Elts et al. [7], Hagiwara et al. [8], and Y.-P. Liu et al. [9] used MD simulations. Ito and Akiyama [10] and Akutsu [11] used MC simulations. Elts et al. also used a kinetic MC method, which is a simulation method for the mesoscale growth or dissolution of a crystal [7].

2.2. First-Principles Simulation

First-principles (FP) methods, such as density functional theory (DFT) [12], can reproduce the atomic-scale structure and energetic state of a real material precisely. In the FP method, the electronic structures of atoms and molecules in a material are computed by solving the Schrödinger wave equation. The FP method can be combined with the MD simulation method (FP-MD method) [13]. In principle, the FP-MD method can provide precise information on the crystal growth mechanism and interface structure of a real material. However, FP-MD simulations of crystal growth for a large system are too time-consuming. Thus, for many cases, the FP method can be used effectively in parts of computer simulation studies of crystal growth or related phenomena.

In this special issue, Ito and Akiyama used the DFT method for calculating chemical potential precisely in their simulation studies [10]. Barcaro et al. used the DFT method to obtain energetic and structural information about small Si clusters, which was then used to optimize the reactive force-field parameters [6].

2.3. Continuum Simulations

Continuum simulations can be used to study mesoscale or macroscale phenomena related to crystal growth, such as crystal morphology and fluid dynamics in crystal growth. There are a variety of continuum simulation methods. The phase-field method is a popular continuum simulation method for studies of crystal growth kinetics and crystal growth morphology [3,14], although this issue does not include studies using the phase-field method.

In this special issue, Elts et al. used continuum simulations in their multiscale simulation studies of crystal growth [7]. Related to continuum simulations, Ruan used a morphology evolution model and the MC method for simulations of polymer crystallization in a shear flow [15]. Z. Liu et al. used a particle level set method [16] for simulations of polymer crystallization under isothermal and temperature gradient conditions [17].

2.4. Multiscale Simulations

Recently, multiscale simulations have attracted a great deal of attention in the field of crystal growth. In this special issue, Elts et al. review the recent progress made by their group in multiscale simulations, including MD simulations, kinetic MC simulations, and continuum simulations for crystal growth [7]. Using multiscale simulations, they predicted the macroscopic morphology and growth or dissolution rate of a crystal from the molecular structure.

3. Materials for Crystal Growth Computer Simulations

3.1. Hard-Sphere System

The hard-sphere system is the simplest off-lattice model used for computer simulation studies of crystal growth. Crystallization of a hard-sphere system is often used as a model for crystallization of a colloidal system. In this special issue, Mori reviews computer simulation studies of crystal growth using a hard-sphere model for crystal-fluid coexistence in the equilibrium state and hard-sphere systems in gravity [4].

3.2. Organic Molecules

Crystal growth of organic molecules, such as carbohydrates, amino acids, peptides, proteins, and polymers, is related to many phenomena in nature and in industries such as the biology, pharmaceutical, nutraceutical, food, and cosmetic industries. Computer simulations have been used for studies of the growth mechanism, growth morphology, and equilibrium morphology of organic molecule crystals.

In this special issue, Qiu and Molinero report MD simulations of the crystal growth of alkanes [5]. They found that the strength of the alkane–fluid attractive interaction controls the interfacial orientation of liquid alkanes and their crystallization. Elts et al. performed multiscale simulations of the growth and dissolution of aspirin crystals [7]. They predicted the aspirin dissolution rates, which agreed well with the experimental rates. Ruan performed MC simulations of polymers crystallizing in a shear flow [15], and provided simulation results for the growth kinetics, morphology, and rheology of the polymer crystals that agreed well with earlier experimental and theoretical studies. Z. Liu et al. performed simulations of polymer crystallization using a particle level set method [17]. They clarified the development of crystallinity during crystallization under quiescent isothermal conditions, and their results were consistent with theory.

3.3. Ice

Ice is a familiar material in daily life and studies of its crystal growth are important, both scientifically and practically, in connection with topics such as the freezing of water in biological systems, pattern formation of snow crystals, artificial snow, cryopreservation of tissues, and food processing. In this special issue, Hagiwara et al. studied the structural and dynamic properties of an aqueous solution including a winter flounder antifreeze protein and salt ions near the secondary prismatic and pyramidal planes of ice [8]. Their MD simulation indicated that hydrogen bonding between water molecules in the solution is inhibited, which may be related to the fact that the antifreeze activity of the protein is enhanced if salt ions are present.

3.4. Functional Materials

Controlling the growth, size, and morphology of crystals is essential for developing functional materials, which can be used for applications including devices, solar cells, and optical materials. Computer simulations have been used for studies of the growth of various crystals of functional materials, such as semiconductors and graphene.

In this special issue, Ito and Akiyama review recent progress in computational materials science in the area of semiconductor epitaxial growth [10]. They present their computer simulation studies of the heteroepitaxial growth of InAs on GaAs and the formation of InP nanowires with their ab initio approach. Barcaro et al. performed a computer simulation study of the nucleation and growth of Si nanoclusters [6]. They proposed a theoretical approach that can be used to model the nucleation and growth of small particles for which experimental studies are difficult to perform. Akutsu studied the surface tension, growth rate, and size of macrosteps on the surface of 4H-SiC crystals using the restricted solid-on-solid model [11]. The effects of the driving force on the size of a faceted macrostep and on the growth rate of the vicinal surface were discussed. Y.-P. Liu et al. studied the growth of graphene sheets embedded with single-wall carbon nanocones (SWCNCs) and suggested conditions suitable for SWCNCs growing on a Cu substrate [9].

4. Conclusions

This special issue presents advances in computer simulation studies of crystal growth. Crystal growth is important in many fields of science and technology. Because the performance of computers is still improving, computer simulations will continue to be essential tools. By covering various types of computer simulation studies of crystal growth and related phenomena from fundamental research to practical applications, this special issue provides helpful information for future simulation studies.

Acknowledgments: I thank all the authors who contributed to this special issue for preparing interesting papers. I also thank Ms. Sweater Shi for her kind editorial assistance during the publication of this special issue. In this special issue, Mori [4], Qiu and Molinero [5], Barcaro et al. [6], Elts et al. [7], and Ito and Akiyama [10] contributed "Feature Papers".

Crystals **2018**, *8*, 314

Conflicts of Interest: The authors declare no conflict of interest.

References

1. Frenkel, D.; Smit, B. *Understanding Molecular Simulation: From Algorithms to Applications*; Frenkel, D., Klein, M., Parrinello, M., Smit, B., Eds.; Academic Press, A division of Harcourt, Inc.: San Diego, CA, USA, 1996.
2. Allen, M.P.; Tildesley, D.J. *Computer Simulation of Liquids*; Oxford University Press: Oxford, UK, 1987.
3. Nada, H.; Miura, H.; Kawano, J.; Irisawa, T. Observing Crystal Growth Processes in Computer Simulations. *Prog. Cryst. Growth Charact.* **2016**, *62*, 404–407. [CrossRef]
4. Mori, A. Computer Simulations of Crystal Growth using a Hard-Sphere Model. *Crystals* **2017**, *7*, 102. [CrossRef]
5. Qiu, Y.; Molinero, V. Strength of Alkane-Fluid Attraction Determines the Interfacial Orientation of Liquid Alkanes and Their Crystallization through Heterogeneous or Homogeneous Mechanisms. *Crystals* **2017**, *7*, 86. [CrossRef]
6. Barcaro, G.; Monti, S.; Sementa, L.; Carravetta, V. Atomistic Modelling of Si Nanoparticles Synthesis. *Crystals* **2017**, *7*, 54. [CrossRef]
7. Elts, E.; Greiner, M.; Briesen, H. In Silico Prediction of Growth and Dissolution Rates for Organic Molecular Crystals: A Multiscale Approach. *Crystals* **2017**, *7*, 288. [CrossRef]
8. Yasui, T.; Kaijuma, T.; Nishio, K.; Hagiwara, Y. Molecular Dynamics Analysis of Synergistic Effects of Ions and Winter Flounder Antifreeze Protein Adjacent to Ice-Solution Surfaces. *Crystals* **2018**, *8*, 302. [CrossRef]
9. Liu, Y.P.; Li, J.T.; Song, Q.; Zhuang, J.; Ning, X.J. A Scheme for the Growth of Graphene Sheets Embedded with Nanocones. *Crystals* **2017**, *7*, 35. [CrossRef]
10. Ito, T.; Akiyama, T. Recent Progress in Computational Materials Science for Semiconductor Epitaxial Growth. *Crystals* **2017**, *7*, 46. [CrossRef]
11. Akutsu, N. Disassembly of Faceted Macrosteps in the Step Droplet Zone in Non-Equilibrium Steady State. *Crystals* **2017**, *7*, 42. [CrossRef]
12. Kohn, W.; Becke, A.D.; Parr, R.G. Density Functional Theory of Electronic Structure. *J. Phys. Chem.* **1996**, *100*, 12974–12980. [CrossRef]
13. Car, R.; Parrinello, M. Unified Approach for Molecular Dynamics and Density Functional Theory. *Phys. Rev. Lett.* **1985**, *55*, 2471–2474. [CrossRef] [PubMed]
14. Sekerka, R.F. Fundamentals of Phase Field Theory. In *Advances in Crystal Growth Research*; Saito, K., Furukawa, Y., Nakajima, K., Eds.; Elsevier Science B.V: Amsterdam, The Netherlands, 2001; Chapter 2; p. 21.
15. Ruan, C. Kinetics and Morphology of Flow Induced Polymer Crystallization in 3D Shear Flow Investigated by Monte Carlo Simulation. *Crystals* **2017**, *7*, 51. [CrossRef]
16. Enright, D.; Fedkiw, R.; Ferziger, J.; Mitchell, I. A Hybrid Particle Level Set Method for Improved Interface Capturing. *J. Comput. Phys.* **2002**, *183*, 83–116. [CrossRef]
17. Liu, Z.; Ouyang, J.; Ruan, C.; Liu, Q. Simulation of Polymer Crystallization under Isothermal and Temperature Gradient Conditions Using Particle Level Set Method. *Crystals* **2016**, *6*, 90. [CrossRef]

© 2018 by the author. Licensee MDPI, Basel, Switzerland. This article is an open access article distributed under the terms and conditions of the Creative Commons Attribution (CC BY) license (http://creativecommons.org/licenses/by/4.0/).

crystals

MDPI

Article

Simulation of Polymer Crystallization under Isothermal and Temperature Gradient Conditions Using Particle Level Set Method

Zhijun Liu [1], Jie Ouyang [1],*, Chunlei Ruan [2] and Qingsheng Liu [1]

[1] Department of Applied Mathematics, Northwestern Polytechnical University, Xi'an 710129, China;
mpingke@mail.nwpu.edu.cn (Z.L.); qingsheng408@163.com (Q.L.)
[2] Department of Computational Mathematics, Henan University of Science and Technology,
Luoyang 471003, China; ruanchunlei622@mail.nwpu.edu.cn
* Correspondence: jieouyang@nwpu.edu.cn; Tel.: +86-29-8849-5234

Academic Editor: Hiroki Nada
Received: 28 June 2016; Accepted: 1 August 2016; Published: 8 August 2016

Abstract: Morphological models for polymer crystallization under isothermal and temperature gradient conditions with a particle level set method are proposed. In these models, the particle level set method is used to improve the accuracy in studying crystal interaction. The predicted development of crystallinity during crystallization under quiescent isothermal condition by our model is reanalyzed with the Avrami model, and good agreement between the predicted and theoretical values is observed. In the temperature gradient, the computer simulation results with our model are consistent with the experiment results in the literature.

Keywords: kinetics; microstructure; crystallization; level set

1. Introduction

The final properties of a product produced from semi-crystalline polymer are to a great extent determined by the final internal microstructure [1,2]. This final internal microstructure, in turn, is determined by the crystallization/processing conditions. Therefore, it is very important to accurately model the solidification process and predict the final microstructure formed under different processing conditions.

To date, a number of investigators are interested in predicting the morphological development of polymer crystallization and many research studies have been carried out on this topic [2–12]. In order to obtain the internal microstructure of polymer products, different approaches have been proposed for morphological modeling of polymer crystallization by researchers [7–12]. Charbon and Swaminarayan [7,8] presented front-tracking methods to predict the evolution of microstructures during spherulitic crystallization under realistic crystallization conditions. Raabe and Godara [9] studied the topology of spherulite growth during crystallization of isotactic polypropylene (iPP) by using a cellular automata method. Xu and Bellehumeur [10,11] proposed a modified phase-field method to capture the spatiotemporal morphology development with the crystallization behavior of ethylene copolymers in the rotational-molding process. Ruan et al. [12] investigated the evolution of morphology of crystallization in the short fiber reinforced system using a pixel coloring method.

We presented a level set method for simulating the solidification structure of polymer crystallization during cooling stage in [13] to reduce the computation complexity in studying crystal interaction. In that method, each crystal can be distinguished from others by its assigned color, the problem of evolving multiple crystal interfaces is reduced to tracking one level set variable (signed distance function) and determining the color of a newly solidification node point. That method is easy to be implemented and it is also applicable for any system that displays nucleation and growth. However, just like other Eulerian methods, level set methods have the main drawback that they are not conservative [14–16] (see Figure 1a). To fix this problem, several attempts to improve mass conservation of level set methods have been done, such as the improved level set methods [16–18] and the particle level set methods [14,19]. In the particle level set method, particles that are distributed within both an interior and exterior band of the interface are used to preserve volume so as to maintain the interface. Literature [14] indicates the particle level set method compares favorably with volume of fluid methods in the conservation of mass and purely Lagrangian schemes for interface resolution.

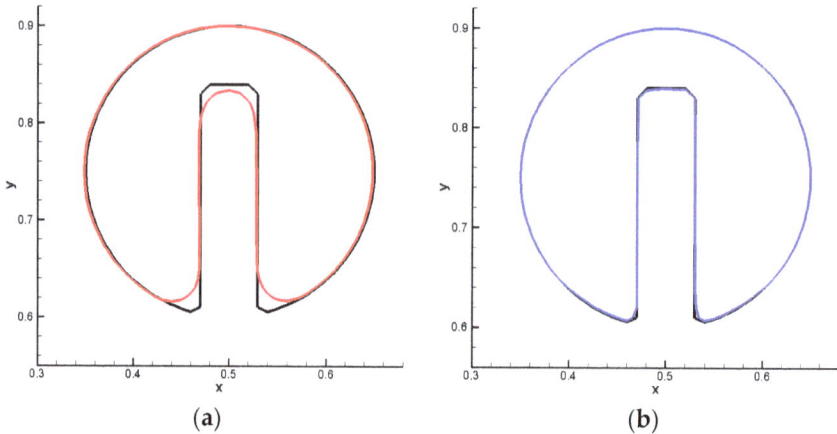

Figure 1. Comparison of the level set solution (red), particle level set solution (blue), and theory (black) after one revolution of rigid body rotation of Zalesak's disk in a constant vorticity velocity field (more information about Zalesak's disk can be found in [20]): (**a**) level set method; and (**b**) particle level set method.

In this paper, we aim to develop a more accurate interface tracking method for studying morphology of polymer crystallization, and further apply the proposed method to morphological models of iPP crystallization under isothermal and temperature gradient conditions. To achieve our goal, firstly, based on the method described earlier [13], we use a particle level set method instead of the level set method to correct any volume loss that resulted from advecting the level set. Secondly, because there already exist many particles in the particle level set method and the accuracy of Lagrangian advection, we use some of these particles to help to color the node points instead of using the semi-Lagrangian method that was used in [13]. Finally, for different temperature conditions, we use different method to deal with the problems in determining morphology of polymer crystallization.

The outline of this paper is as follows: Section 2 presents the particle level set method for polycrystals growth. In Section 3, the morphological modeling using the particle level set method for polymer crystallization under isothermal conditions is introduced. In Section 4, the extended algorithm for polymer crystallization in uniaxial linear temperature field is established. Section 5 gives the numerical results and discussions. Finally the conclusions are drawn in Section 6.

2. Particle Level Set for Polycrystals Growth

2.1. Level Set Method

The level set method was originally designed by Osher and Sethian [21,22] in 1988, and then it has been manipulated in moving interfaces of fluid mechanics, combustion, computer animation, image processing and some other interfaces of evolution problems. According to this method, the interface whose motion is recast as a time-dependent Eulerian initial value partial differential equation is denoted implicitly by the zero set of a continuous function.

A level set function $\varphi(x,t)$ is defined as:

$$\varphi(x,t) = \begin{cases} +d(x,t) & x \in \Omega_{interior} & \text{(the solid regions)} \\ 0 & x \in \partial\Omega = \Gamma(t) & \text{(the melt-solid interfce)} \\ -d(x,t) & x \in \Omega_{exterior} & \text{(the melt regions)} \end{cases} \tag{1}$$

where $d(x,t)$ is set as the smallest distance between a given point in the domain Ω and the interface $\Gamma(t)$:

$$d(x,t) = \min_{\substack{x_\Gamma \in \Gamma \\ x \in \Omega}} (|x - x_\Gamma|) \tag{2}$$

Additionally, the level set function has the following feature:

$$|\nabla\varphi| = 1 \tag{3}$$

The instantaneous interface associates with the contour $\varphi(x,t) = 0$, i.e.,

$$\Gamma(t) = (x \in \Omega : \varphi(x,t) = 0) \tag{4}$$

The normal unit vector on the interface is expressed as:

$$n = \left.\frac{\nabla\varphi}{|\nabla\varphi|}\right|_{\varphi=0} \tag{5}$$

The equation for the evolution of φ corresponding to the motion of the interface is given by:

$$\varphi_t + u \cdot \nabla\varphi = 0 \tag{6}$$

where u represents velocity. With an evolution of the interface, the re-initialization is often necessarily due to a generally deviation of φ from its initialized value which represents signed distance. We apply the re-initialization until φ reach steady-state, i.e., the following equation is iterated:

$$\varphi_t = \frac{\varphi_0}{\sqrt{\varphi_0^2 + \varepsilon^2}}(1 - |\nabla\varphi|) \tag{7}$$

where φ_0 is the initial level set value to be re-initialized. When φ reaches steady-state, it satisfies the condition $|\nabla\varphi| = 1$, i.e., φ is a signed distance. It is imperative for the formulation to remain well-posed as $\varphi \to 0$ if the parameter ε in Equation (7) is assigned some small value.

2.2. Particle Level Set Method

The main problem that the level set method suffers from is numerical dissipation. The particle level set method merges the best aspects of Eulerian front-capturing schemes and Lagrangian front-tracking methods for improved mass conservation. In the particle level set method, two sets of massless particles, positive and negative particles, which are placed within a band across the interface, are used to correct the level set function.

The particle correction procedures in the particle level set method are summarized as follows:

(i) Particle initialization. When the initial surface is defined, the particles need to be placed within three cells of the interface. Each particle stores its position and radius, which is used to perform error correction on the level set function. The radius is set so that the boundary is just touching the interface:

$$r_p = \begin{cases} r_{max} & if\ s_p\varphi(x_p) > r_{max} \\ s_p\varphi(x_p) & if\ r_{min} \leqslant s_p\varphi(x_p) \leqslant r_{max} \\ r_{min} & if\ s_p\varphi(x_p) < r_{min} \end{cases} \tag{8}$$

where $r_{min} = 0.1\min(\Delta x, \Delta y)$, $r_{max} = 0.5\min(\Delta x, \Delta y)$, and s_p is the sign of the particle, set to +1 if $\varphi(x_p) > 0$ and -1 if $\varphi(x_p) < 0$. In [14], they recommend that 16 particles be placed in each cell in 2D.

(ii) Particle update: The positions of the particles are updated using a second order Runga Kutta (RK2) time integration:

$$x_p(t+1) = x_p(t) + dtu_t(x_p(t)) \tag{9}$$

$$x_p^*(t+1) = x_p(t+1) + dtu_t(x_p(t+1)) \tag{10}$$

$$x_p(t+1) = \frac{x_p^*(t+1) + x_p(t)}{2} \tag{11}$$

Error correction: Whenever a particle escapes the interface by more than its radius, it will be used to perform error correction on the interface. To enable error correction, a local level set value for each corner of the escaped particle is defined as follows:

$$\varphi_p(x) = s_p(r_p - |x - x_p|) \tag{12}$$

Error correction is performed using the positive particles to create a temporary grid φ^+ and the negative particles to a temporary grid φ^-. For all of the escaped positive particles, the φ_p values on cell corners containing the escaped particles are calculated by Equation (12), the value for each corner is then set to

$$\varphi^+ = \max(\varphi_p, \varphi^+) \tag{13}$$

Similarly, for all the escaped negative particles, the value for each corner is set to

$$\varphi^- = \min(\varphi_p, \varphi^-) \tag{14}$$

Then, for each grid node, the minimum absolute value is chosen as the final correction for φ

$$\varphi = \begin{cases} \varphi^+ & if\ |\varphi^+| \leqslant |\varphi^-| \\ \varphi^- & if\ |\varphi^+| > |\varphi^-| \end{cases} \tag{15}$$

(iii) Particle reseeding: With the interface stretching and tearing, regions that lack a sufficient number of particles in the computational domain will form. Reseeding is carried out to delete the particles that are superfluous or far away from the interface and distribute a new set of particles to ensure that there is a uniform distribution of particles near the interface. It is important to note that if the simulation does not cause the particles to be unevenly distributed, there is no reason to reseed.

Further details of the particle level set method can be found in Reference [14]. The accuracy and efficiency of this method are shown in Figure 1b.

2.3. Particle Level Set Method for Polycrystals Growth

Typically, many crystals grow from individual seeds, and each of them will grow until it collides with other crystals and forms grain boundaries. Determining the contact boundaries is a difficult task. It is the reason why we employ a single signed distance to implicitly denote the interface of crystals and we also allot each crystal a "color" (respectively, a number) to distinguish different crystals. Because there already exists many particles in the particle level set method and the accuracy of Lagrangian advection, we detect the contact boundaries of crystals by these particles instead of the semi-Lagrangian method used in [13]. As demonstrated in Figure 2, the interfaces of two crystals are represented by the dotted lines, which are captured by the particle level set method at time t_{n-1}. The two crystals are differentiated by the colors of the nodes (the big circles), which lie in the two crystals respectively. To be used to color the nodes inside the two crystals at the next time step, the particles (the small circles) which are distributed inside the interfaces in the particle level set method are dyed in the same colors of the nodes near them (see Figure 2a). After one time step, the two crystals contact and we can capture their interfaces (denote with the solid lines) by the particle level set method, but the contacted boundary of the two crystals is not yet determined. Meanwhile, the colored particles move to the new positions, the undyed nodes inside the solid line and outside the dotted line are in the crystalline phase (Figure 2b). Here we need to determine which crystal these nodes belong to, i.e., what color should the undyed nodes be colored? Noting that the dyed particles that belong to the same crystal have the same color, we thus color each undyed node with the color of the nearest dyed particle, then the boundary of the two crystals is determined (Figure 2c). Additionally, we have to point out that firstly, the node coloring procedure can also applied to the growth of crystal before contact. Secondly, once particle reseeding is imperative, attention must be paid to color the particles that are placed into the cells where there exist different colored particles. The boundaries lie in these cells, so error boundaries can be resulted from any inappropriate coloring method. Maybe there is a better way to cope with this problem. In this study, we only use a simple technique: no coloring to these particles (Figure 3). This is the reason: on the one hand, it can ensure a sufficient number of particles in the computational domain; on the other hand, it would not lead to error results for the particles that are uncolored. More details of the particle level set method for polycrystals growth can be found in Section 3.

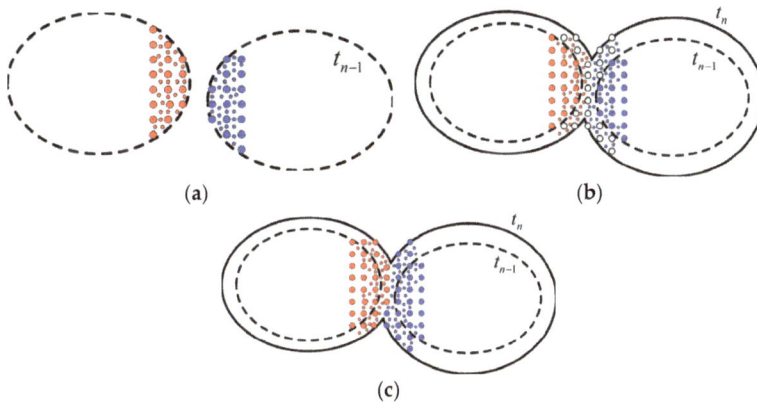

Figure 2. Schematic representation of the particle level set method for polycrystals growth: (**a**) two crystals are distinguished; (**b**) Undyed nodes in the crystalline colors of their nodes at time t_{n-1} phase are captured by particle level set method at time t_n. The dyed particles move to new positions; (**c**) Color the undyed nodes by the colors of the nearest particles at time t_n, determine the boundary of the two crystals.

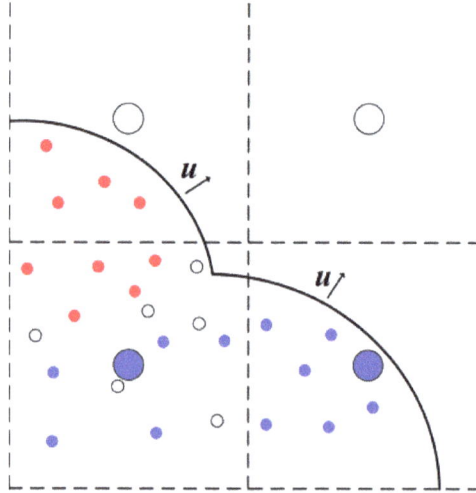

Figure 3. Schematic representations of the uncolored and reseeding particles in the lower-left cell where there exists different colored particles.

3. Morphological Model for Isothermal Crystallization of Polymer

Crystallization is a mechanism of phase change in semicrystalline polymeric materials. An isothermal crystallization process generally includes three steps, namely, nucleation, growth and impingement. Under the favorable thermodynamic condition, nuclei appear randomly in the polymer melt, and then the formed nuclei act as seeds for spherulites to grow with the same rate. Each spherulite grows until it impinges on adjacent spherulites and stops growing. Impingement takes place and continues until all possible material is transformed.

3.1. Nucleation

In semicrystalline polymers, the number of nuclei per unit volume or so-called nucleus density depends on the temperature and supercooling. Because the nucleation of semicrystalline polymers is typically heterogeneous, it is difficult to apply the theoretical models of the nucleation [4]. As a result, empirical approaches are adopted to solve this problem and nucleation laws are presented to represent the empirical relations between the nucleus density and the temperature. In the simulations, we use the following relation of the nucleation density [8,23]:

$$N = 1.458 \left\{ \exp[111.265 - 0.2544\varphi(T + 273.15)] \right\}^{2/3} \tag{16}$$

where the unit of N is mm^{-2}, and the unit of T is $^\circ C$.

3.2. Growth and Impingement

Growth rate is an important factor that affects the development of morphology. Generally, for each polymer, the rate of crystal growth is a function of the crystallization temperature and can be considered a constant when the crystallization is considered under an isothermal condition. For a spherulite, the radial growth rate G can be calculated by applying the Hoffman–Lauritzen theory [24]:

$$G = G_0 \exp\left[-\frac{U}{R_g(T - T_\infty)}\right] \exp\left[-\frac{K_g}{T\Delta T}\right] \tag{17}$$

where G_0 and K_g are constants, U is the activation energy of motion, R_g is the gas constant, T_∞ is a temperature typically 30 K below the glass transition, and $\Delta T = T_m^0 - T$ is the degree of undercooling. It should be noted that, in the level set method, the velocity μ should be defined on the whole domain or on a narrow band near the interface. Therefore, we should extend the interface velocity away from the interface (solid/liquid boundary). There are many techniques to construct the extension velocity, details of which can be found in [25,26].

During the crystallization process, various stages of spherulite growth occur. In the early stages, probably no spherulites impinge. As the spherulites continue to grow, more and more of them will impinge each other. It is impingement that makes the grain boundaries form and the growth of spherulites stop. In fact, the shapes of the grains can directly influence the final properties of the polymers. In this study, the particle level set method for polycrystals growth is used to simulate the growth and impingement of spherulites.

3.3. Algorithm for Polymer Crystallization under Isothermal Conditions

Under isothermal conditions, we use the stochastic method utilized in [13] to place the nucleation sites in the nucleation process. In this method, a node is chosen randomly in the computational domain when a new nucleus appears. Then a new color is allotted to this node and the signed distance field is updated with the following expression:

$$\varphi(y) = \max(\varphi^0(y), ||x - y|| - R_0), \forall y \in \Omega \tag{18}$$

where x is the nucleation site, φ^0 is the signed distance before the potential nucleation site is nucleated at x, R_0 is the size of the initial crystal seed at location x, and y is the location of a node. It should be noted that, unlike in [13], in the nucleation process in this paper, the particles inside each crystal also need to be colored by the same color of the crystal. It is the colored particles that provide an effective way to distinguish different crystals after growth under different conditions.

In the crystal growth process, the particles inside the crystals are first used to correct the volume loss that resulted from advecting the level set in the particle level set method. Then they are utilized to color the uncolored nodes in the crystalline phase by their colors. If new particles need to be added to the computational zone, the way to color them is introduced in Section 2.3.

In our scheme, the relative crystallinity can be calculated by [13]:

$$\alpha = \frac{number\ of\ nodes\ that\ have\ been\ colored}{total\ number\ of\ nodes} \tag{19}$$

In addition, the mean of the maximum size of spherulites is defined as:

$$\overline{R}_{max} = (\frac{A}{N\pi})^{1/2} \tag{20}$$

where A is the volume of the spherulites which can be calculated with the number of notes occupied by the spherulites and the volume of a cell, and N is the number of the spherulites.

Figure 4 shows the algorithm for polymer crystallization under isothermal conditions using the particle level set method.

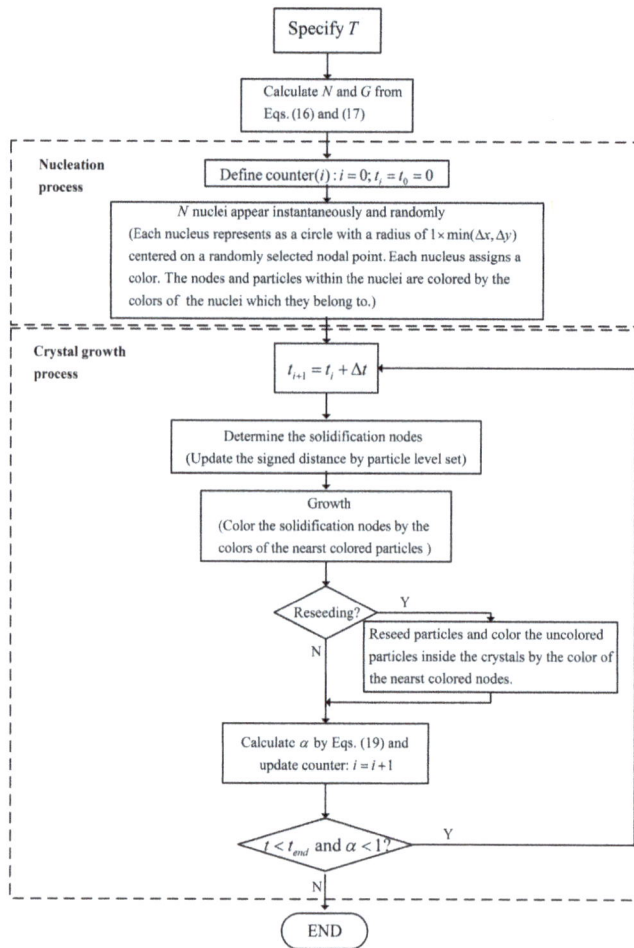

Figure 4. Algorithm for polymer crystallization under isothermal conditions using the particle level set method.

4. Morphological Model for Polymer Crystallization in a Temperature Gradient

In this paper, the polymer crystallization in a temperature gradient is modeled in a uniaxial linear temperature field: $T = T_0 + \Lambda x$. It is obvious when $\Lambda = 0$, $T = T_0$, the temperature is constant, i.e., it is isothermal. Thus, we need to extend the algorithm presented in Figure 4 to the uniaxial linear temperature field. To do this, on the one hand, we divide the computational zone into a series of small enough rectangles with respect to the x coordinate, so as to place the nucleation sites into the computational zone. In each small rectangle, the temperature is considered equal to that at the middle line of the vertical side on calculating the number of nuclei by Equation (16), and also the positions of spherulite nuclei are randomly chosen. On the other hand, $G(T)$ needs to be recalculated on the x coordinate by $G(T_0 + \Lambda x)$ each time step. With these preparations, the extended algorithm for polymer crystallization in the uniaxial linear temperature field using the particle level set method is shown as follows (Figure 5).

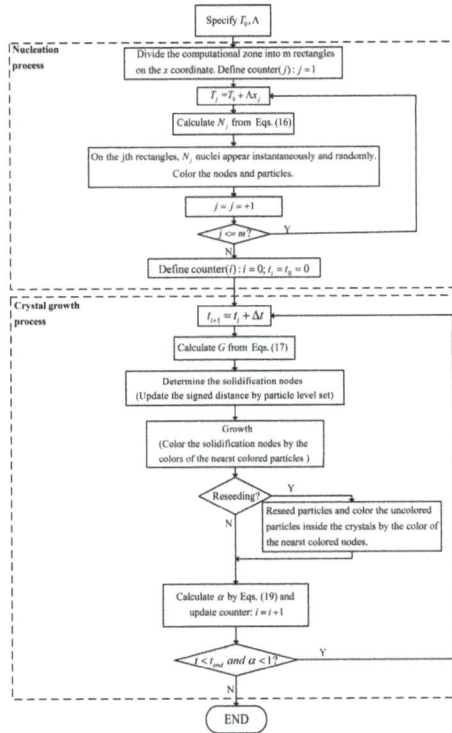

Figure 5. Algorithm for polymer crystallization in the uniaxial linear temperature field by the particle level set method.

5. Results and Discussion

5.1. Problem Formulation

We consider the graphical simulation of the crystallization process on square samples. The maximum size of the samples is 600 μm × 600 μm. In order to ensure each unit cell is no more than the real space of 1 μm² a regular mesh of 601 × 601 nodes is selected to solve this problem. It should be pointed out that the finer the unite cell the more accurate of the computational results. However, the finer the unit cell the more CPU time is needed. The material used in this simulation is an iPP and parameters used are listed in Table 1 [23].

Table 1. Numerical values for the parameters of iPP used in the models.

Parameter	Physical Meaning	Value	
U (cal·mol^{-1})	Activation energy of motion	1500	
G_0 (cm·s^{-1})	Parameter in Hoffman–Lauritzen expression for the laminar growth rate	$\begin{cases} 0.3359 & T \geqslant 136\,C \\ 3249 & T < 136\,C \end{cases}$	
T_∞ (°C)	A temperature typically 30 K below the glass transition	−41.95	
K_g (K^2)	Parameter in Hoffman–Lauritzen expression for the growth rate	$\begin{cases} 1.47 \times 10^5 & T \geqslant 136\,C \\ 3.30 \times 10^5 & T < 136\,C \end{cases}$	
T_m^0 (°C)	Melting temperature	185.05	
R_g (J (K·mol)$^{-1}$)	Gas constant	8.314472	

5.2. Isothermal Case

5.2.1. Morphological Development

Figure 6 shows the morphological development of iPP at $T = 106.85$ °C acquired from the algorithm for polymer crystallization under isothermal conditions by the particle level set method, the size of the square is 160 μm × 160 μm. As is illustrated in the Figure 6, in the process of evolutionary, the white region is polymer melt and the other colored region is crystals. Figure 6a illustrates iPP morphology at $t = 0$ s when heterogeneous nucleation takes place. Then each spherulite grows individually without impingement (Figure 6b) until $t = 6.8$ s, as time goes by, adjacent crystals touch and intercrystalline impingement boundaries arise (Figure 6c,d). Finally, we depict the final morphology (Figure 6e). We notice that the structures and characteristics of crystals obtained in this study are in accordance with the literature [27]. Figure 7 shows the evolution of crystal morphology with different temperatures and a square size of 500 μm × 500 μm. It is obvious that the lower the temperature the faster the crystallization time, the more the number of spherulites and the smaller size of the final grain. Figure 8 is plotted to demonstrate the mean of the maximum size of spherulites against time with different temperatures, and we observe that the mean of the maximum size of spherulites rises rapidly as time increases and reaches a plateau finally. Moreover, we perceive that the crystallization time becomes longer as the temperature rises. Apparently, we obtain many of the same conclusions analyzing Figures 7 and 8.

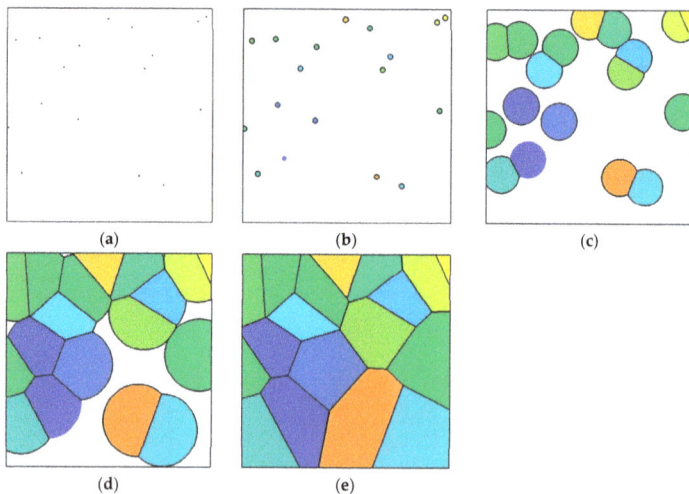

(a) (b) (c)

(d) (e)

Figure 6. Examples of spherulitic morphology obtained from the stochastic simulation scheme at $T = 106.85$ °C: (a) $t = 0$ s; (b) $t = 6.8$ s; (c) $t = 50.3$ s; (d) $t = 98.7$ s; and (e) final morphology.

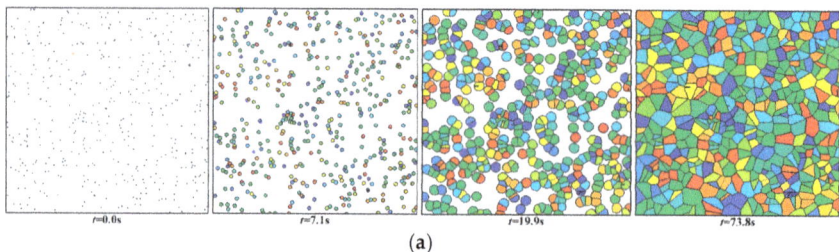

$t=0.0$s $t=7.1$s $t=19.9$s $t=73.8$s

(a)

Figure 7. *Cont.*

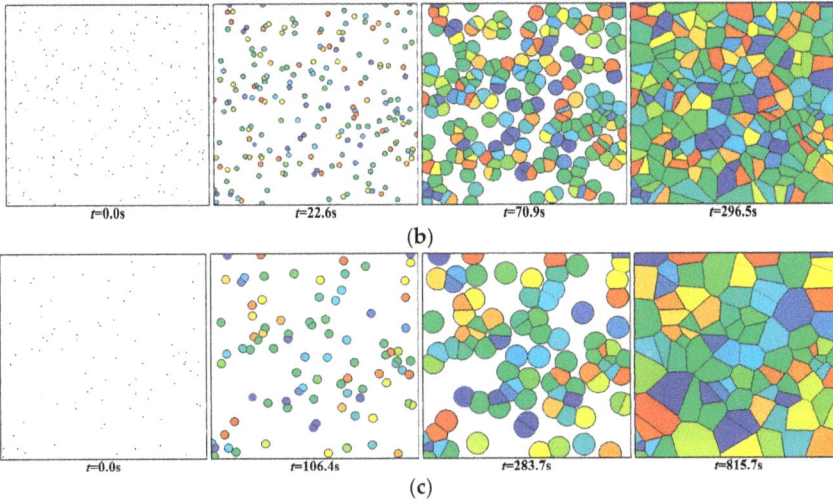

Figure 7. Evolution of crystal morphology under different temperatures: (a) $T = 121.85\ ^\circ$C; (b) $T = 126.85\ ^\circ$C; and (c) $T = 131.85\ ^\circ$C.

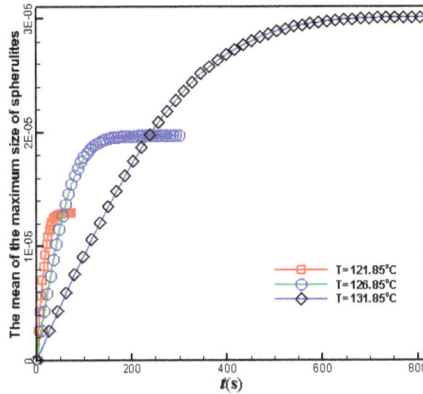

Figure 8. The mean of the maximum size of spherulites versus time under different temperatures.

5.2.2. Overall Crystallization Kinetics

The most generally used approach for the description of the isothermal polymer crystallization kinetics is the Avrami model [28,29]. In the Avrami model, the relative crystallinity $\alpha(t)$ is usually written in the following form:

$$\alpha(t) = 1 - \exp(-k_a t^{n_a}) \tag{21}$$

where k_a and n_a are the isothermal crystallization rate constant and the Avrami index (crystal geometry information), respectively. In our study, heterogeneous nucleation and spherulitic morphology are assumed. Following [30], we take $n_a = 2$, $k_a = \pi N(T)[G(T)]^2$. Thus, Equation (21) can be rewritten as:

$$\alpha(t) = 1 - \exp\left\{-\pi N(T)[G(T)]^2 t^2\right\} \tag{22}$$

Figure 9 displays the relative crystallinity evolution under different temperatures in isothermal condition. Symbols represent the predicted results, and lines represent the analytical solutions

15

according to Avrami model. It is clear that the predicted results show good agreement with the theoretical values.

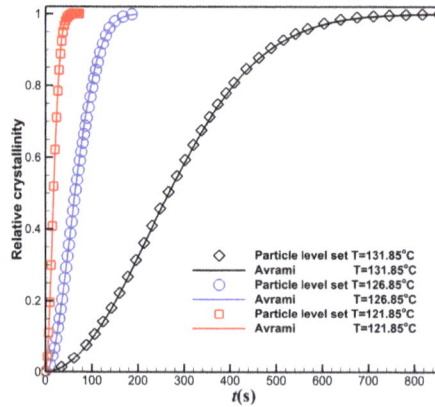

Figure 9. Relative crystallinity evolution under different temperatures in isothermal condition.

5.3. Temperature Gradient Case

In our simulations, the temperature gradient in the sample is parallel to the x axis, and the temperature, T, increases from the left to the right.

5.3.1. Effects of Temperature Gradient

In order to better illustrate the effects of temperature gradient, we assume that there is no other spherulite except the one(s) given in the computational domain in this section. The reasons for this are listed as follows. First, the fewer the number of spherulites in the computational domain, the more obvious the effect of the temperature gradient becomes. Second, different polymer material has different nucleation density at the same temperature. Therefore, even if we limit the numbers of nuclei, we can also obtain meaningful simulation results by selecting sizes of the samples appropriately.

Figure 10 shows the spherulite shapes predicted by the particle level set method for different temperature gradient. The sizes of these samples are 400 µm × 400 µm. In Figure 10a, the temperature gradient, Λ, equal to 50 °C·mm^{-1}; and in Figure 10b, Λ = 100 °C·mm^{-1}. As we can see in these two figures, the temperature gradient results in the spherulite shape anisotropy, which increases with time. Besides, in spite of the centers of the two spherulites are all locate at 130 °C, the shapes of them are quite different at the same moment (see Figure 10c). The main reason for this is that different temperature gradient makes the temperature distributions different on both sides of the center, and different temperature leads to different growth rate (see Figure 11). Figure 12 shows the shapes of interspherulitic boundaries in different temperature gradients of samples with sample size: 200 µm × 200 µm. We can see that, the three spherulites nucleate at the same sites in the three samples and the temperatures on the left side of the samples are all 110 °C, while the shapes of interspherulitic boundaries are very different. It is clear when Λ = 0 °C·mm^{-1}, the boundaries are straight lines, but as the temperature gradient is higher, the boundaries bend toward higher T, and the curvature of the boundaries also increases.

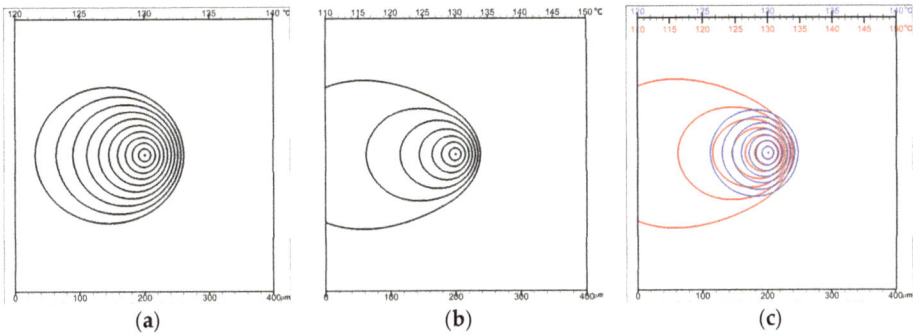

Figure 10. Spherulite shapes predicted by the particle level set method for different temperature gradient. The positions of spherulite growth fronts are plotted in 1-min intervals: (**a**) $\Lambda = 50\,°C·mm^{-1}$; (**b**) $\Lambda = 100\,°C·mm^{-1}$; and (**c**) shapes of the spherulites are different at the same moment.

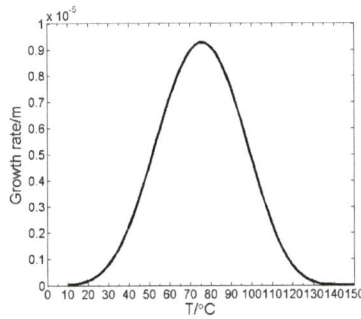

Figure 11. Growth rate versus temperature.

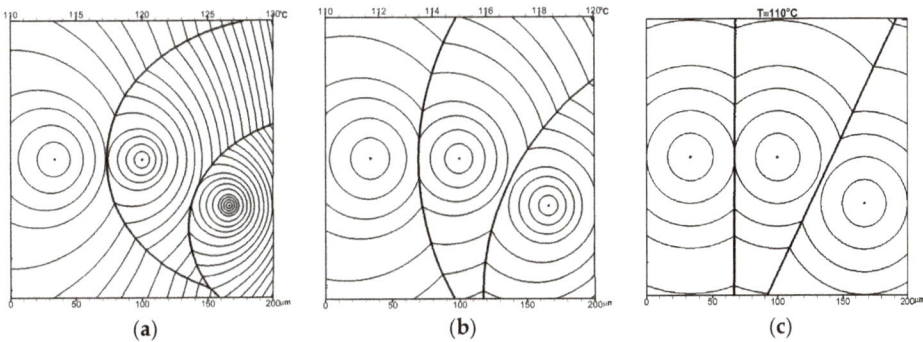

Figure 12. Shapes of interspherulitic boundaries predicted by the particle level set method for different temperature gradients: (**a**) $\Lambda = 100\,°C·mm^{-1}$; (**b**) $\Lambda = 50\,°C·mm^{-1}$; and (**c**) $\Lambda = 0\,°C·mm^{-1}$.

5.3.2. Morphological Development

In order to validate our model using the particle level set method for polymer crystallization in a temperature gradient, we simulate iPP film crystallization in the temperature gradient of $35\,°C·mm^{-1}$. To compare our results to those in [23,31], all parameters utilized in the literature are also used in this simulation (see Table 1). The size of this sample is $600\,\mu m \times 600\,\mu m$. As we can see from Figure 13, the temperature gradient affects not only the nucleation density but also the spherulitic pattern.

17

Figure 13a shows the spherulites only nucleate at the low-temperature side. Figure 13b,c demonstrates that the spherulites grow faster toward the low-temperature side and soon impinge on each other. The spherulites can still grow toward the high-temperature side, as shown in Figure 13d–f. We can also see that, in Figure 13f, the spherulites in the high-temperature side are elongated in shape, the collision boundaries of them are almost parallel to the temperature gradient, but the joint growth front of them is perpendicular to the temperature gradient. Figure 14 shows the computer-simulated positions of the crystallization front in the iPP film during crystallization in the uniaxial linear temperature field. The simulated results are in good agreement with the experiment and calculated results in the literature [23,31]. It indicates the correctness of our model.

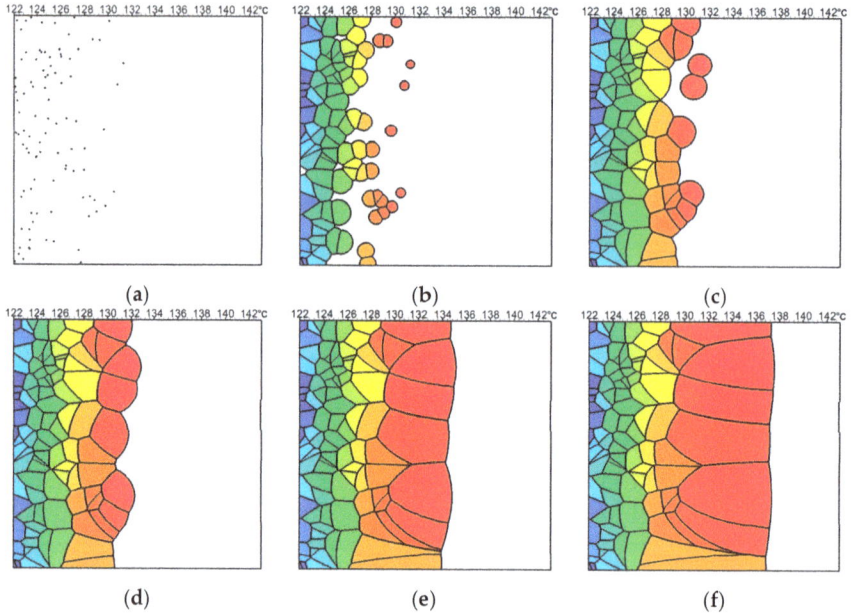

Figure 13. Evolution of crystal morphology in iPP film during crystallization in the uniaxial linear temperature field: (**a**) $t = 0$ s; (**b**) $t = 78$ s; (**c**) $t = 234$ s; (**d**) $t = 468$ s; (**e**) $t = 1560$ s; (**f**) $t = 3633$ s.

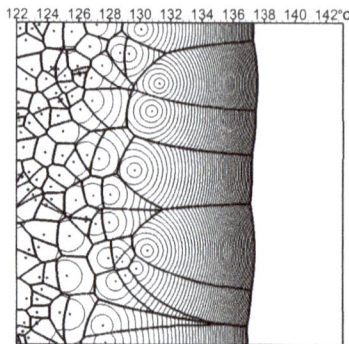

Figure 14. Positions of the crystallization front in iPP film during crystallization in the uniaxial linear temperature field.

6. Conclusions

We have presented an efficient particle level set method for polycrystals growth. In this method, the particles placed within an interior band of the interface in the particle level set method are used not only to correct any volume loss that resulted from advecting the level set and but also to determine the boundaries of crystals. With this method, an algorithm for polymer crystallization under quiescent isothermal condition has been developed. By the simulation model, for iPP, not only have we predicted the crystal morphological development and its distributions, but we have also obtained the relative crystalline and the mean of the maximum size of spherulites at different temperature *T*. The predicted development of crystallinity during crystallization has been reanalyzed with the Avrami model, and good agreement between the predicted and theoretical values has been observed.

We have also extended the algorithm from isothermal to temperature gradient conditions. In the uniaxial linear temperature field, we have presented a new method to place the nuclei in the computational zone. Numerical experiments have been used to analyze the effects of the temperature gradient. We also have simulated iPP film crystallization in the temperature gradient. The computer simulation results are consistent with the experiment results in the literature.

It should also be pointed out that the method used in this paper has two advantages compared with that in [13]: First, the particle level set method preserves volume better than the level set method. Thus, to the crystals with sharp edges, such as dendrites, more accurate crystallinity can be acquired. Second, as more particles are used to color the nodes, more precise boundaries of the crystals can be achieved. However, there are always two sides to everything, even if the particles utilized to color the nodes are also used in the original particle level set method, that more particles are used in this method increases the computational cost. In addition, the method cannot be used to determine the trajectory of the fibrils. However, this problem might be solved if we consider the trajectory of the particles, which are used in this method.

All in all, the morphological models for polymer crystallization under different conditions proposed in this study are valid. The calculated results with them give us another way to observe the microstructure of polymer products.

Acknowledgments: This work was financially supported by the National Basic Research Program of China (973 Program, contract Grant Number: 2012CB025903), the Major Research plan of the National Natural Science Foundation of China (Contract Number: 91434201), and the Natural Sciences Foundation of China (Contract Number: 11402078), which are gratefully acknowledged.

Author Contributions: Zhijun Liu and Jie Ouyang conceived the idea, developed the models and wrote the paper. Chunlei Ruan and Qingsheng Liu performed the computations and analyzed the results.

Conflicts of Interest: The authors declare no conflict of interest.

References

1. Zhou, H.M. *Computer Modeling for Injection Molding: Simulation, Optimization, and Control*; John Wiley & Son: Hoboken, NJ, USA, 2012.
2. Zheng, R.; Tanner, R.; Fan, X.J. *Injection Molding: Integration of Theory and Modeling Methods*; Springer: Berlin, Germany, 2011.
3. Spina, R.; Spekowius, M.; Hopmann, C. Multiphysics simulation of thermoplastic polymer crystallization. *Mater. Des.* **2016**, *95*, 455–469. [CrossRef]
4. Spina, R.; Spekowius, M.; Dahlmann, R.; Hopmann, C. Analysis of polymer crystallization and residual stresses in injection molded parts. *Int. J. Precis. Eng. Manuf.* **2014**, *15*, 89–96. [CrossRef]
5. Roozemond, P.C.; van Erp, T.B.; Peters, G.W.M. Flow-induced crystallization of isotactic polypropylene: Modeling formation of multiple crystal phases and morphologies. *Polymer* **2016**, *89*, 69–80. [CrossRef]
6. Xu, K.L.; Guo, B.H.; Reiter, R.; Xu, J. Simulation of secondary nucleation of polymer crystallization via a model of microscopic kinetics. *Chin. Chem. Lett.* **2015**, *26*, 1105–1108. [CrossRef]
7. Swaminarayan, S.; Charbon, C. A multiscale model for polymer crystallization. I: Growth of individual spherulites. *Polym. Eng. Sci.* **1998**, *38*, 634–643. [CrossRef]

8. Charbon, C.; Swaminarayan, S. A multiscale model for polymer crystallization. II: Solidification of a macroscopic part. *Polym. Eng. Sci.* **1998**, *38*, 644–656. [CrossRef]
9. Raabe, D.; Godara, A. Mesoscale simulation of the kinetics and topology of spherulite growth during crystallization of isotactic polypropylene (iPP) by using a cellular automaton. *Model. Simul. Mater. Sci. Eng.* **2005**, *13*, 733–751. [CrossRef]
10. Xu, H.; Bellehumeur, C.T. Modeling the morphology development of ethylene copolymers in rotational molding. *J. Appl. Polym. Sci.* **2006**, *102*, 5903–5917. [CrossRef]
11. Xu, H.; Bellehumeur, C.T. Morphology development for single-site ethylene copolymers in rotational molding. *J. Appl. Polym. Sci.* **2008**, *107*, 236–245. [CrossRef]
12. Ruan, C.; Ouyang, J.; Liu, S. Multi-scale modeling and simulation of crystallization during cooling in short fiber reinforced composites. *Int. J. Heat Mass Transf.* **2012**, *55*, 1911–1921. [CrossRef]
13. Liu, Z.J.; Ouyang, J.; Zhou, W.; Wang, X.D. Numerical simulation of the polymer crystallization during cooling stage by using level set method. *Comput. Mater. Sci.* **2015**, *97*, 245–253. [CrossRef]
14. Enright, D.; Fedkiw, R.; Ferziger, J.; Mitchell, I. A hybrid particle level set method for improved interface capturing. *J. Comput. Phys.* **2002**, *183*, 83–116. [CrossRef]
15. Osher, S.; Fedkiw, R. *Level Set Methods and Dynamics Implicit Surfaces*; Springer: New York, NY, USA, 2006.
16. Sussman, M.; Fatemi, E.; Smereka, P.; Osher, S. An improved level set method for incompressible two-phase flows. *Comput. Fluids* **1998**, *27*, 663–680. [CrossRef]
17. Sussman, M.; Puckett, E.G. A coupled level set and volume-of-fluid method for computing 3D and axisymmetric incompressible two-phase flows. *J. Comput. Phys.* **2000**, *162*, 301–337. [CrossRef]
18. Bourlioux, A. A coupled level-set volume-of-fluid algorithm for tracking material interfaces. In Proceedings of the 6th International Symposium on Computational Fluid Dynamics, Lake Tahoe, CA, USA, 4–8 September 1995.
19. Enright, D.; Losasso, F.; Fedkiw, R. A fast and accurate semi-Lagrangian particle level set method. *Comput. Struct.* **2005**, *83*, 479–490. [CrossRef]
20. Zalesak, S.T. Fully multidimensional flux-corrected transport algorithms for fluids. *J. Comput. Phys.* **1979**, *31*, 335–362. [CrossRef]
21. Osher, S.; Sethian, J.A. Fronts propagating with curvature-dependent speed: algorithms based on Hamilton-Jacobi formulations. *J. Comput. Phys.* **1988**, *79*, 12–49. [CrossRef]
22. Sethian, J.A. *Level Set Methods and Fast Marching Methods: Evolving Interfaces in Computational Geometry, Fluid Mechanics, Computer Vision and Materials Science*; Cambridge University Press: Cambridge, UK, 1999.
23. Piorkowska, E. Modeling of polymer crystallization in a temperature gradient. *J. Appl. Polym. Sci.* **2002**, *86*, 1351–1362. [CrossRef]
24. Lauritzen, S.I.; Hoffman, J.D. Theory of formation of polymer crystals with folded chains in dilute solution. *J. Res. Natl. Bur. Stand.* **1960**, *64A*, 73–102. [CrossRef]
25. Tan, L.J. Multiscale Modeling of Solidification of Multi-Component Alloys. Ph.D. Thesis, Cornell University, Ithaca, NY, USA, 2007.
26. Adalsteinsson, D.; Sethian, J.A. The fast construction of extension velocities in level set methods. *J. Comput. Phys.* **1999**, *148*, 2–22. [CrossRef]
27. Anantawaraskul, S.; Ketdee, S.; Supaphol, P. Stochastic simulation for morphological development during the isothermal crystallization of semicrystalline polymers: A case study of syndiotactic polypropylene. *J. Appl. Polym. Sci.* **2009**, *111*, 2260–2268. [CrossRef]
28. Avrami, M. Kinetics of phase change I General theory. *J. Chem. Phys.* **1939**, *7*, 1103–1112.
29. Avrami, M. Transformation-time relations for random distribution of nuclei. *J. Chem. Phys.* **1940**, *8*, 212–224. [CrossRef]
30. Piorkowska, E.; Gregory, C.R. *Handbook of Polymer Crystallization*; Wiley: Hoboken, NJ, USA, 2013.
31. Pawlak, A.; Piorkowska, E. Crystallization of isotactic polypropylene in a temperature gradient. *Colloid Polym. Sci.* **2001**, *279*, 939–946. [CrossRef]

© 2016 by the authors. Licensee MDPI, Basel, Switzerland. This article is an open access article distributed under the terms and conditions of the Creative Commons Attribution (CC BY) license (http://creativecommons.org/licenses/by/4.0/).

crystals

MDPI

Article

Disassembly of Faceted Macrosteps in the Step Droplet Zone in Non-Equilibrium Steady State

Noriko Akutsu

Faculty of Engineering, Osaka Electro-Communication University, Hatsu-cho, Neyagawa,
Osaka 572-8530, Japan; nori@phys.osakac.ac.jp; Tel.: +81-72-824-1131

Academic Editor: Hiroki Nada
Received: 23 December 2016; Accepted: 2 February 2017; Published: 8 February 2017

Abstract: A Wulff figure—the polar graph of the surface tension of a crystal—with a discontinuity was calculated by applying the density matrix renormalization group method to the p-RSOS model, a restricted solid-on-solid model with a point-contact-type step–step attraction. In the step droplet zone in this model, the surface tension is discontinuous around the (111) surface and continuous around the (001) surface. The vicinal surface of 4H-SiC crystal in a Si–Cr–C solution is thought to be in the step droplet zone. The dependence of the vicinal surface growth rate and the macrostep size $\langle n \rangle$ on the driving force $\Delta\mu$ for a typical state in the step droplet zone in non-equilibrium steady state was calculated using the Monte Carlo method. In contrast to the known step bunching phenomenon, the size of the macrostep was found to decrease with increasing driving force. The detachment of elementary steps from a macrostep was investigated, and it was found that $\langle n \rangle$ satisfies a scaling function. Moreover, kinetic roughening was observed for $|\Delta\mu| > \Delta\mu_R$, where $\Delta\mu_R$ is the crossover driving force above which the macrostep disappears.

Keywords: Monte Carlo simulation; crystal growth; surface and interface; density matrix renormalization group calculation; surface tension; kinetic roughening; surface free energy; step–step attraction

PACS: 81.10.Aj; 68.35.Ct; 05.70.Np; 68.35.Md

---

1. Introduction

The faceted macrosteps on a crystal surface are known to degrade the grown crystal [1]. Although studies have investigated methods of dispersing faceted macrosteps, an effective method has not yet been established. For example, in the case of solution growth for 4H-SiC, the faceted macrosteps remain near equilibrium. This formation of faceted macrosteps near equilibrium has been considered to be due to the effects of anomalous surface tension. To control the dynamics of macrosteps, the fundamentals of the phenomena with macrosteps must be clarified.

The connection between the surface tension and the instability with respect to macrostep formation is also scientifically interesting. In 1963, Cabrera and Coleman [2] phenomenologically studied anomalous surface tension and studied its effect on a vicinal surface near equilibrium. They assumed several anomalous surface tensions, and then discussed the possible equilibrium crystal shapes (ECSs). They also declared the instability with respect to the macrostep formation to be the result of anomalous surface tension [3]. However, at that time, the microscopic model used to determine the anomalous surface tension was not provided.

Jayaprakash et al. [4] studied the faceting transition of the ECS using a terrace–step–kink (TSK) model with the long-range step–step interaction expressed by the equation $\sum_{i \neq j} \sum_{\tilde{y}} g_0 / [\tilde{x}_i(\tilde{y}) - \tilde{x}_j(\tilde{y})]^2$ [5–7], where g_0 is the coupling constant of the elastic repulsion, $\tilde{x}_i$ is the location of the ith step on a vicinal surface normal to the mean running direction of steps, and $\tilde{y}$ is the location along the steps. They showed that the step–step interaction affects the coefficient of the $O(\rho^3)$ term in the surface free energy, which is given by

$$f(\rho) = f(0) + \gamma\rho + B\rho^3 + O(\rho^4), \tag{1}$$

where ρ is the step density, γ is the step tension, and B is step interaction coefficient. This ρ-expanded form of the free energy excluding the quadratic term with respect to ρ is called the Gruber–Mullins–Pokrovsky–Talapov (GMPT) [8–11] universal form. It is well known that the ECS can be obtained using the Landau–Andreev method [12,13] or from the polar graph of the surface tension using the Wulff theorem [14–17]. When the long-range step–step interaction is attractive, B becomes negative at low temperatures, and the slope of the surface on the ECS has a jump at the facet edge. This jump in the surface slope is referred to as a first-order shape transition [4,18] after the fashion of a phase transition.

The step bunching for Si near equilibrium in ultrahigh vacuum has been studied for Si(111) [19–24] and Si(113) [25–28]. Williams et al. [19,20] have experimentally demonstrated that the step bunching of Si(111) is caused by the competition between the polymorphic surface free energies for the Si(111) (1×1) and (7×7) structures based on the GMPT surface free energy (Equation (1)). Additionally, Song et al. obtained the temperature–slope phase diagram for the step bunching on Si(113) by analyzing the results of X-ray diffraction, and have reported that an anomalous surface free energy caused by a step–step attraction may explain the step bunching phenomenon. Lössig [29] and Bhattacharjee [30] have stated that the TSK model with a short-range step–step attraction and with the long-range step–step repulsion can represent the step bunching phenomenon. By analyzing the scanning tunneling microscopy results on a vicinal surface of Si(113) tilted toward the $[1\bar{1}0]$ direction, van Dijken et al. [27] demonstrated that there is a step–step attraction to condense steps and a large long-range step–step repulsion. Shenoy et al. [31,32] have shown that the TSK model with a short-range step–step attraction and with the long-range step–step repulsion causes the periodic array of the n-merged steps using a renormalization group method. Einstein et al. [33] introduced the idea of the random matrix to the terrace width distribution to assist in the determination of the strength of the long-range step–step repulsion.

Step bunching or step faceting near equilibrium occurs without long-range step–step repulsion [18,34–45]. Rottman and Wortis [18] studied an ECS using a three-dimensional ferromagnetic Ising model with both nearest-neighbor (NN) and next-nearest-neighbor (NNN) interactions. They calculated ECS using mean-field theory. When the NNN interaction is negative, they showed that the ECS has the first-order shape transition at the (001) facet edge at low temperatures. Using a lattice model with a point-contact-type step–step attraction [38–45], Akutsu showed that a faceted macrostep self-assembles at equilibrium in association with the morphological change resulting from the discontinuous surface tension. The lattice model was a restricted solid-on-solid (RSOS) model with point-contact-type step–step attraction (p-RSOS model, Figure 1). The term "restricted" means that the height difference between NN sites is restricted to $\{0, \pm1\}$. It was considered that the origin of the point-contact-type step–step attraction is the orbital overlap of the dangling bonds at the meeting point of the neighboring steps. The energy gained by forming the bonding state is regarded as the attractive energy between steps. The surface tension of the model was numerically calculated using the density matrix renormalization group (DMRG) method [46–51], and it was demonstrated that the surface tension of the vicinal surface tilted toward the $\langle 111 \rangle$ direction is discontinuous at low temperatures because of the point-contact-type step–step attraction.

This connectivity of the surface tension for the p-RSOS model is directly linked to the faceting diagram (Figure 2). There are two transition temperatures in the p-RSOS model—$T_{f,1}$ and $T_{f,2}$. At temperatures of $T < T_{f,1}$, the surface tension around the (111) surface is discontinuous; for $T < T_{f,2}$ $(< T_{f,1})$, the surface tension around the (001) surface is discontinuous. Based on the connectivity of the surface tension, the temperature region $T < T_{f,2}$ is called the step-faceting zone, the region $T_{f,2} \leq T < T_{f,1}$ is called the step droplet zone, and the region $T_{f,1} \leq T$ is called the GMPT [8,9] zone. Moreover, the plot of the roughening transition temperature of the (001) surface divides the step droplet and GMPT zones into step droplet zones I and II and GMPT zones I and II, respectively [43,44].

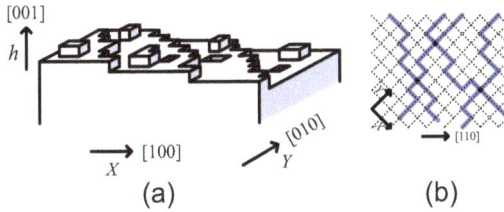

Figure 1. (**a**) Perspective view of the restricted solid-on-solid (RSOS) model tilted toward the ⟨110⟩ direction; (**b**) Top-down view of the RSOS model. Thick blue lines represent surface steps. This figure was reproduced from Akutsu [45].

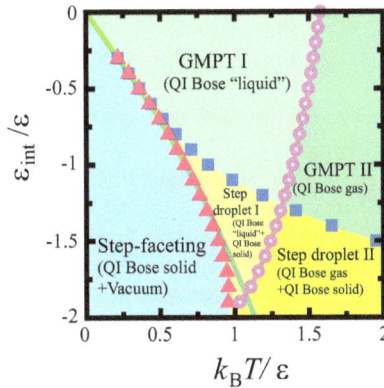

Figure 2. Faceting diagram of the p-RSOS model (RSOS model with point-contact-type step–step attraction) for a vicinal surface obtained using the density matrix renormalization group (DMRG) method. Squares: calculated values of $T_{f,1}$. Triangles: calculated values of $T_{f,2}$. Open circles: calculated roughening transition temperatures of the (001) surface. Solid line: zone boundary line calculated using the two-dimensional Ising model. For the definitions of and details about the QI Bose solid, liquid, and gas, please refer to Akutsu [43]. This figure was reproduced from Akutsu [43]. GMPT: Gruber–Mullins–Pokrovsky–Talapov.

In our previous study [45], the height profile of a faceted macrostep at *equilibrium* was investigated on the p-RSOS model, and it was demonstrated that the characteristics of the height profile of a macrostep can be classified by the connectivity of the surface tension. The characteristics of the height profile of a macrostep are irrelevant to the details of the crystal structure. Hence, the height profile of a macrostep can be used to determine in which zone the surface exists. For example, the height profile of the macrostep in the case of 4H-SiC [1] is similar to the profile in the step droplet zone [45]. This suggests that the surface tension of 4H-SiC around the faceted side surface is discontinuous.

In the present work, under a driving force $\Delta\mu$, the disassembly of the faceted macrostep in the step droplet zone to form a kinetically roughened homogeneous surface (Figure 3) in *non-equilibrium steady state* was studied using the Monte Carlo method on the p-RSOS model. Step droplet zone I was the focus of this study; the step-faceting zone will be studied in future work. To focus on the dynamics affected by the surface tension, surface diffusion [52–55], elastic effects [5–7,20,21], and polymorphic effects [19,20,28] were excluded.

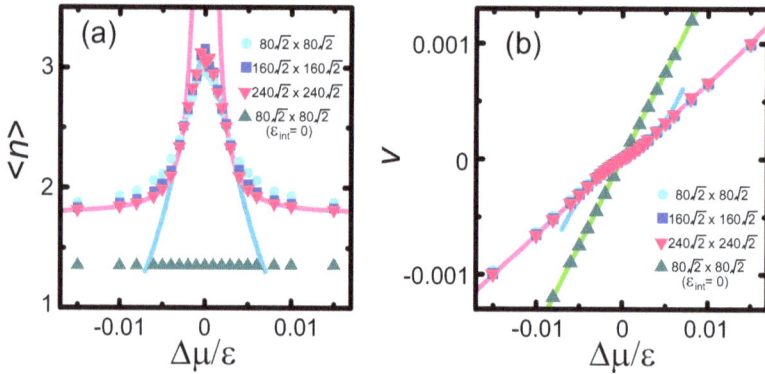

Figure 3. (**a**) Macrostep size. Blue line: $\langle n \rangle = 1/(z + 2/N_{\text{step}})$ with $z = 0.307 + 17.1|\Delta\mu|/\epsilon + 6.85 \times 10^3(|\Delta\mu|/\epsilon)^2$. Pink lines: $\langle n \rangle = 1.79 + 9.16 \times 10^{-6}(|\Delta\mu|/\epsilon)^{-1.89}$; (**b**) Vicinal surface growth rate. Pink line: $v = 0.0855 \, \text{sign}(\Delta\mu)(|\Delta\mu|/\epsilon)^{1.06}$, where $\text{sign}(x) = 1, 0$, and -1 for $x > 0$, $x = 0$, and $x < 0$, respectively. Blue line: $v = p_1 v_1 = 0.132 p_1 \Delta\mu/\epsilon$ with $p_1 = 0.332 + 15.6|\Delta\mu|/\epsilon + 4.43 \times 10^3(|\Delta\mu|/\epsilon)^2$. Green line: $v = 0.151\Delta\mu/\epsilon$. $\epsilon_{\text{int}}/\epsilon = -0.9$ and $k_B T/\epsilon = 0.63$ (step droplet zone I). The plotted values were obtained as averages over 2×10^8 Monte Carlo steps per site (MCS/site).

This paper is organized as follows. In §2, the model and the discontinuous surface tension are briefly explained. In §3, the results obtained using the Monte Carlo method are presented. The process of detaching the steps from a macrostep is discussed in §4 through the analysis of the results in the case of small $|\Delta\mu|$. In §5, the case of large $|\Delta\mu|$ is considered; for this case, the size of a macrostep is modeled using a scaling function, and a crossover point $\Delta\mu_R$ associated with kinetic roughening is introduced. The implications of the results are discussed in §6, and §7 concludes the paper.

2. Restricted Solid-on-Solid Model with Point-Contact-Type Step–Step Attraction

2.1. Restricted Solid-on-Solid Model with Point-Contact-Type Step–Step Attraction

The microscopic model considered in this study is the p-RSOS model (Figure 1). The Hamiltonian of the (001) surface can be written as

$$\mathcal{H}_{\text{p-RSOS}} = \mathcal{N}\epsilon_{\text{surf}} + \sum_{n,m} \epsilon[|h(n+1,m) - h(n,m)| + |h(n,m+1) - h(n,m)|]$$
$$+ \sum_{n,m} \epsilon_{\text{int}}[\delta(|h(n+1,m+1) - h(n,m)|, 2) + \delta(|h(n+1,m-1) - h(n,m)|, 2)], \quad (2)$$

where $\mathcal{N}$ is the total number of lattice points, ϵ_{surf} is the surface energy per unit cell on the planar (001) surface, ϵ is the microscopic step energy, $\delta(a, b)$ is the Kronecker delta, and ϵ_{int} is the microscopic step–step interaction energy. The summation with respect to (n, m) is taken over all sites on the square lattice. The RSOS condition is required implicitly. When ϵ_{int} is negative, the step–step interaction becomes attractive (sticky steps).

It should be noted that the p-RSOS model (Equation (2)) automatically includes the "entropic step–step repulsion". Since the p-RSOS model is an RSOS model, the overhang structures with respect to the height of the surface are inhibited by the geometrical restriction. This is the microscopic origin of the entropic step–step repulsion.

2.2. Discontinuous Surface Tension

The polar graphs of the surface tension and the surface free energy were calculated using the DMRG method [46–51], and are shown in Figure 4. In low-dimensional cases, more precision is required than can be provided by a mean field calculation of the partition function [56]. Hence, to obtain reliable results, the DMRG method—which was developed for one-dimensional (1D) quantum spin systems [46]—was adopted. The transfer matrix version of the DMRG method—known as the product wave function renormalization group (PWFRG) method [49–51]—was used in this study. Details of the calculation method for the surface tension and the surface free energy are given in Appendix A. In Figure 4, the angle θ is the tilt angle from the (001) surface toward the $\langle 111 \rangle$ direction. The surface gradient p is related to θ as $|p| = \pm \tan \theta$. The surface tension was calculated from the surface free energy $f(p)$ as

$$\gamma_{\text{surf}}(p_x, p_y) = \frac{f(p_x, p_y)}{\sqrt{1 + p_x^2 + p_y^2}}. \tag{3}$$

The calculated surface tension and the surface free energy at $k_B T / \epsilon = 0.63$ and $\epsilon_{\text{int}} / \epsilon = -0.9$ are shown in Figure 4.

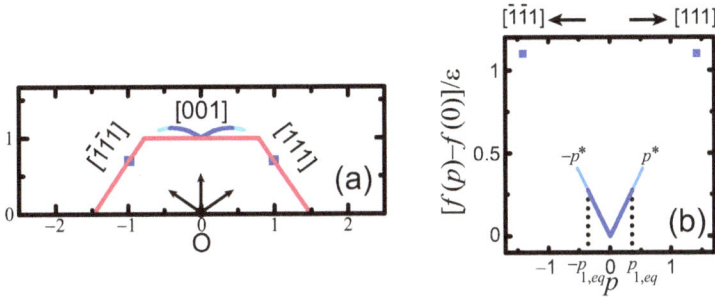

Figure 4. (**a**) Polar graph of the surface tension (Wulff figure) and Andreev free energy (equilibrium crystal shape, ECS) calculated using the DMRG method. This figure was reproduced from Akutsu [45]; (**b**) Surface free energy. $k_B T / \epsilon = 0.63$, $\epsilon_{\text{int}} / \epsilon = -0.9$ (step droplet zone I). Red lines: $Z(R)$ calculated using the DMRG method. Blue lines and squares: polar plots of (**a**) the surface tension $\gamma_{\text{surf}}(\theta) / \epsilon$ ($-54.74° < \theta < 54.74°$), where θ is the tilt angle of the vicinal surface from the (001) surface toward the $\langle 111 \rangle$ direction; and (**b**) the surface free energy. Pale blue lines: values for the metastable surfaces for (**a**) the surface tension; and (**b**) the surface free energy. End points of the pale blue lines (p^* and $-p^*$): approximate spinodal points. Point O: Wulff point. ϵ_{surf} was assumed to equal ϵ. An enlarged figure of the ECS near the facet edges is shown in Appendix A.

The surface tension contains discontinuities near the (001) and (111) surfaces in the case of $\epsilon_{\text{int}} / \epsilon = -0.9$, $k_B T_{f,2} / \epsilon = 0.613 \pm 0.02$, and $k_B T_{f,1} / \epsilon = 0.709 \pm 0.02$. At $k_B T / \epsilon = 0.63$, the surface is in the step droplet zone (Figure 2), where the surface tension is continuous around the (001) surface but discontinuous around the (111) surface.

The ECS calculated using the DMRG method (Appendix A) is plotted in red in Figure 4a. This ECS result agrees well with the ECS obtained using the Wulff theorem [14–17]. In step droplet zone I, the (001) surface meets the curved area without a discontinuity in its slope point P in Figure A1) (The Gauss curvature, which is the determinant of the curvature tensor, jumps at the facet edge of the (001) surface), whereas the (111) surfaces meet the curved areas with a discontinuity in their slopes (point Q in Figure A1).

At the zone boundary lines in Figure 2, the following conditions are satisfied [40,43]:

$$f^{(111)}(q) = f^{(111)}(0) + \gamma_{q,1}(T)|q| + C_{q,\text{eff}}(T)|q|^4 + \mathcal{O}(|q|^5),$$

$$B_{q,\text{eff}}(T) = 0, \ C_{q,\text{eff}}(T) > 0, \ (\text{at } T = T_{f,1}) \tag{4}$$

$$\gamma_{q,1}(T) = \lim_{n \to \infty} \gamma_{q,n}(T)/n \quad (\text{at } T = T_{f,2}), \tag{5}$$

where $q(T)$ is the surface gradient on the vicinal surface near the (111) surface, $f^{(111)}(q)$ is the surface free energy of the vicinal surface near the (111) surface, $\gamma_{q,n}(T)$ is the step tension of an n-merged (negative) step [43], and $B_{q,\text{eff}}(T)$ and $C_{q,\text{eff}}(T)$ are coefficients. In GMPT zones I and II, since $B_{q,\text{eff}}(T) > 0$, the surface free energy $f^{(111)}(q)$ has a form similar to Equation (1). As the temperature decreases, $B_{q,\text{eff}}(T)$ decreases and $C_{q,\text{eff}}(T)$ increases. For $T < T_{f,1}$, where the vicinal surface exists in the step droplet zone, $B_{q,\text{eff}}(T)$ becomes negative and the first-order transition occurs. Hence, the upper zone boundary line of $T = T_{f,1}$ is a critical curve.

The key points to obtain Equations (4) and (5) are the meeting of neighboring steps and the inhomogeneity of the vicinal surface [40,43]. For the vicinal surface tilting toward the $\langle 111 \rangle$ direction, two neighboring steps can occupy one site at the same time, and no more than three steps can occupy one site at a time because of the geometrical restrictions of the RSOS model. Hence, the surface cannot be mapped to a 1D fermion model [57–59]. The double occupancy of a site gives rise to the point-contact-type step–step interaction. When the interaction is repulsive, the term $C_{q,\text{eff}}(T)|q|^4$ is present [43,60,61], whereas in the case of attractive interaction, the vicinal surface becomes inhomogeneous and can be expressed as a mixture of the various n-merged steps (macrosteps) [40]. Since the population of the n-merged steps depends on the surface slope, $B_{q,\text{eff}}(T)|q|^3$ is affected by the point-contact-type step–step attraction through the slope dependence of the size of the macrosteps.

3. Monte Carlo Results

3.1. Monte Carlo Method

To study the non-equilibrium steady state with macrosteps, the vicinal surface of the following Hamiltonian with a fixed number N_{step} of steps was investigated using the Monte Carlo method with the Metropolis algorithm:

$$\mathcal{H}_{\text{noneq}} = \mathcal{H}_{\text{p–RSOS}} - \Delta\mu \sum_{n,m} [h(n, m, t + 1) - h(n, m, t)], \tag{6}$$

where $\Delta\mu = \mu_{\text{ambient}} - \mu_{\text{crystal}}$ is the driving force, μ_{crystal} is the chemical potential of the bulk crystal, and μ_{ambient} is the chemical potential of the ambient phase. The explicit form of $\Delta\mu$ is given in Markov [62]. When $\Delta\mu > 0$, the crystal grows because its chemical potential is lower than that of the ambient phase, whereas when $\Delta\mu < 0$, the crystal recedes (evaporate, dissociates, or melts).

The explicit procedure of the application of the Monte Carlo method in this study is as follows. At the initial time, the steps are positioned at equal distances. Then, the lattice site to be updated is randomly selected. The surface structure is updated non-conservatively using the Monte Carlo method with the Metropolis algorithm. With the RSOS restriction taken into consideration, the structure is updated with probability 1 when $\Delta E \leq 0$ and with probability $\exp(-\Delta E/k_B T)$ when $\Delta E > 0$, where $\Delta E = E_f - E_i$, E_i is the energy of the present configuration and E_f is the energy of the updated configuration. The energy is calculated from the Hamiltonian (Equation (6)). A periodic boundary condition was imposed in the direction parallel to the steps. In the direction normal to the steps, the lowest side of the structure was connected to the uppermost side by adding a height with a number N_{step} of steps.

Snapshots of the simulated surfaces after 2×10^8 Monte Carlo steps per site (MCS/site) are shown in Figure 5. The height profiles of the surfaces for the cross section along the bottom of the top-down views are also shown as side views.

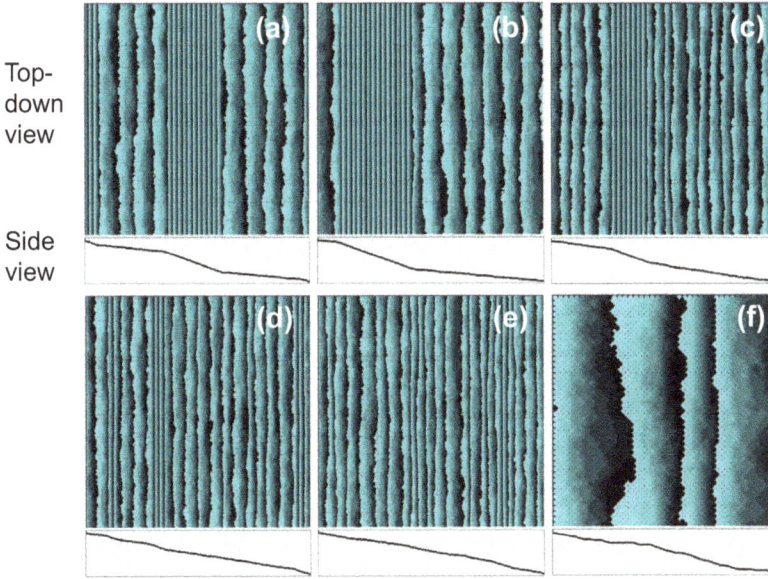

Figure 5. Snapshots at 2×10^8 MCS/site. $\Delta\mu/\epsilon =$ (**a**) 0; (**b**) 0.0005; (**c**) 0.003; (**d**) 0.004; and (**e**) 0.006. $k_B T/\epsilon = 0.63$, $N_{step} = 240$, system size: $240\sqrt{2} \times 240\sqrt{2}$; (**f**) $\Delta\mu/\epsilon = 0.015$, system size: $40\sqrt{2} \times 40\sqrt{2}$, $N_{step} = 40$, $\epsilon_{int}/\epsilon = -0.9$. The surface height is represented by brightness with 10 gradations, with brighter regions indicating a larger height. The darkest areas next to the brightest ones represent terraces that are actually higher by a value of unity, because of the finite gradation. The height profiles for the cross section along the bottom of each surface map are shown below each map.

3.2. Macrostep Size and Surface Growth Rate

At equilibrium ($\Delta\mu = 0$), the vicinal surface showed a homogeneous stepped surface for a small mean surface slope $\bar{p}$ ($|\bar{p}| < p_{1,eq}$, Figure 4b) because the surface tension around the (001) surface is continuous in the step droplet zone. In contrast, when the mean surface slope satisfies $p_{1,eq} < \bar{p} < \sqrt{2}$, homogeneous stepped surfaces are thermodynamically unstable [40]. Then, the surface is realized through two-surface coexistence; the two surfaces are the surface with a slope equal to $p_{1,eq}$ and the (111) surface [40–45]. This is illustrated in Figure 5a. Because the (111) surface is smooth (which means a small number of kinks exist on it), it hardly moves. From the time-dependent Ginzburg–Landau equation of the surface [63,64], the smooth surface does not move because the surface stiffness is divergent [65]. The reason the faceted macrosteps move at equilibrium is the finiteness of the system size [45].

For $\Delta\mu > 0$, the size of a single macrostep decreases as $\Delta\mu$ increases, as shown in Figure 5. Furthermore, for $\Delta\mu/\epsilon \gtrsim 0.006$, the macrostep disassembles to form a homogeneous *rough* surface. Because the temperature was the same for all cases shown in Figure 5, the surfaces with $\Delta\mu/\epsilon \gtrsim 0.006$ roughen kinetically. Interestingly, the change in the size of macrostep is symmetric in the case of $\Delta\mu < 0$, where the steps are receding. The patterns obtained in the case of $\Delta\mu < 0$ (e.g., $\Delta\mu/\epsilon = -0.003$) is quite similar to the patterns obtained in the case of $|\Delta\mu|$ (e.g., $\Delta\mu/\epsilon = 0.003$).

To study the characteristics of the vicinal surface on a mesoscopic scale (approximately 10 nm to 1 μm) in detail, the size of the macrosteps and the growth rate of the surface were measured during the Monte Carlo simulation. To evaluate the size of a macrostep, the number n of elementary steps in a locally merged step [40] was introduced. The average size of the locally merged steps is obtained as

$$\langle n \rangle = \frac{\sum_n N_n n}{\sum_n N_n},$$

(7)

where N_n is the number of n-merged steps on the vicinal surface.

The time evolutions of $\langle n \rangle$ at different values of $\Delta\mu/\epsilon$ are shown in Figure 6a. As shown in Figure 6a, for $|\Delta\mu| \geq 0.003$, $\langle n \rangle$ is constant near 2×10^8 MCS/site. Hence, surfaces with $|\Delta\mu| \geq 0.003$ are in non-equilibrium steady state. The time evolutions of $\langle n \rangle$ in non-equilibrium steady state were also obtained for $|\Delta\mu| = 0.001$ and 0.002, the results of which are not shown in Figure 5. The $\Delta\mu$-dependence of $\langle n \rangle$ at 2×10^8 MCS/site is shown in Figure 3a. For small $|\Delta\mu|$, $\langle n \rangle$ decreases linearly as $|\Delta\mu|$ increases.

To estimate the growth rate v, the average surface height $\bar{h}(t)$ was calculated as $\bar{h}(t) = (1/\mathcal{N}) \sum_{n,m} h(n, m)$. The time evolutions of $\bar{h}(t)$ at different values of $\Delta\mu/\epsilon$ are shown in Figure 6b. As shown in the figure, $\bar{h}(t)$ increases or decreases linearly with increasing t. Hence, v is defined as

$$v = \frac{\bar{h}(t) - \bar{h}(0)}{t}.$$

(8)

The $\Delta\mu$-dependence of v is shown in Figure 3b. $|v|$ increases as $|\Delta\mu|$ increases.

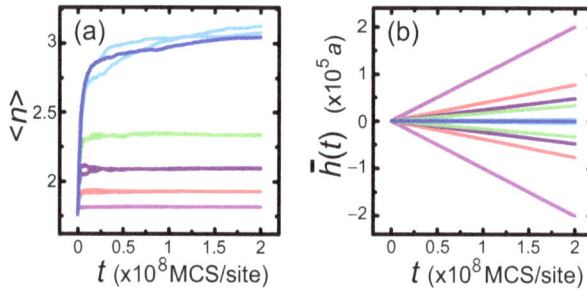

Figure 6. (a) Time evolution of the average number $\langle n \rangle$ of steps in a merged step; (b) Time evolution of the average surface height $\bar{h}$. $k_B T/\epsilon = 0.63$. Dark blue lines: $\Delta\mu = 0$. Light blue lines: $\Delta\mu/\epsilon = \pm 0.0005$. Green lines $\Delta\mu/\epsilon = \pm 0.003$. Purple lines: $\Delta\mu/\epsilon = \pm 0.004$. Coral lines: $\Delta\mu/\epsilon = \pm 0.006$. Pink lines: $\Delta\mu/\epsilon = \pm 0.01$. System size: $240\sqrt{2} \times 240\sqrt{2}$, $N_{step} = 240$, $\epsilon_{int}/\epsilon = -0.9$.

4. Detachment of Steps from Macrosteps

4.1. Size of a Macrostep

As shown in Figure 5b,c, for sufficiently small $|\Delta\mu|$, $\langle n \rangle$ can be approximated as

$$
\begin{aligned}
\langle n \rangle &\approx \frac{N_1 + n_m N_m}{N_1 + N_m} = \frac{N_{step}}{N_1 + N_m} \\
&= \frac{1}{z + N_m/N_{step}}, \\
N_1 &= z N_{step} \\
&= p_1 N_{step} \zeta \left[1 - \frac{(1-z)}{\sqrt{2\zeta}} \right], \quad \zeta = \frac{L}{a N_{step}} = \frac{1}{\bar{p}},
\end{aligned}
$$

(9)

(10)

where n_m is number of elementary steps contained in the most dominant size of the macrosteps, N_m is the number of n_m-merged steps, N_1 is the number of elementary steps outside of the macrostep,

z is the ratio N_1/N_{step}, and p_1 is the slope of the "terrace" surface that is in contact with the (111) surface. From the snapshots, it was assumed that $N_m = 2$.

From the definitions of z and Equation (9), z can be calculated from $\langle n \rangle$ as

$$z = \frac{1}{\langle n \rangle} - \frac{2}{N_{step}}. \tag{11}$$

Thus, the curve of best fit was obtained as $z = 0.307 + 17.1|\Delta\mu|/\epsilon + 6.86 \times 10^3 (|\Delta\mu|/\epsilon)^2$ by applying the method of least squares to the values of z, which was estimated from $\langle n \rangle$ using Equation (11) on a system of size $240\sqrt{2} \times 240\sqrt{2}$. The values of $\langle n \rangle$ reproduced by the best fit equations are plotted as blue lines in Figure 3a. The lines agree well with the values of $\langle n \rangle$ for small $|\Delta\mu|$. The best fit equation reveals the following. In the case of $\bar{p} = 0.707$, 69.0% of all N_{step} steps self-assemble to form a macrostep in the limit of $|\Delta\mu| \to 0$. For small $|\Delta\mu|$, $\langle n \rangle$ decreases linearly as $|\Delta\mu|$ increases. This indicates that $\langle n \rangle$ has a cusp singularity at $|\Delta\mu| = 0$, because the slope of the line in Figure 3 in the limit of $\Delta\mu \to +0$ (from the right-hand side) is different from the slope of the line in the limit of $\Delta\mu \to -0$ (from the left-hand side).

4.2. Growth Rate

Next, the growth rate was investigated. Here, the following model for the time evolution of $\langle n \rangle$ was considered:

$$\frac{\partial \langle n \rangle}{\partial t} = n_+ - n_-, \tag{12}$$

where n_+ is the rate at which elementary steps join the macrostep, and n_- is the rate at which elementary steps detach from the macrostep. In the growth condition ($\Delta\mu > 0$), when $n_+ < n_-$, the macrostep dissociates. In this case, n_+ limits the surface growth rate. In contrast, when $n_+ > n_-$, $\langle n \rangle$ increases up to N_{step}, and n_- limits the surface growth rate. When the surface is in steady state, $n_+ = n_-$.

Because a surface with a slope of p_1 shows step flow growth, n_+ is considered to be $p_1 v_1$ for small $|\Delta\mu|$, where v_1 is the growth rate of a single step. Hence, the growth rate in steady state is expressed as

$$v = n_+ = p_1 v_1. \tag{13}$$

p_1 is obtained from Equation (10) as

$$p_1 = \frac{z}{\zeta - (1-z)/\sqrt{2}}. \tag{14}$$

At equilibrium, the terrace surface with a slope of $p_{1,eq} = 0.3532$, which was calculated using the DMRG method, coexists with the (111) surface, which forms the side surface of the macrostep. For $|\Delta\mu| > 0$, p_1 can be obtained from Equation (14) using the value of z. In the same manner as for z, the curve of best fit was obtained for p_1 on a system of size $240\sqrt{2} \times 240\sqrt{2}$ as $p_1 = 0.332 + 15.6|\Delta\mu|/\epsilon + 4.43 \times 10^3 (|\Delta\mu|/\epsilon)^2$. In this manner, it was found that p_1 increases linearly with $|\Delta\mu|$ for small $|\Delta\mu|$. Equation (13) with $v_1 = 0.132\Delta\mu$ is plotted in Figure 3b as blue lines.

Since $n_+ = n_-$ in steady state, the equation for n_- may be used. To model the mechanism of the detachment of an elementary step from a macrostep, the two-dimensional (2D) nucleation mode was considered. In this case, v should be proportional to $\exp(-const./|\Delta\mu|)$ [66]. However, the Monte Carlo data could not be fit by this equation. The 2D multi-nucleation mode was also considered. In this case, v should be proportional to $|\Delta\mu|^{2/3} \exp(-const./|\Delta\mu|)$ [62,66]. However, the Monte Carlo data could not be fit by this equation either. From this, the detachment of an elementary step from a macrostep was considered to be caused by the "noise" of the attachment and detachment of atoms ("Atoms" in the model correspond to unit cells) from the ambient phase in association with thermal fluctuations.

5. Kinetic Roughening

As shown in Figure 5, macrosteps do not form at large $|\Delta\mu|$ ($|\Delta\mu|/\epsilon \gtrsim 0.006$). This means that the vicinal surface is kinetically roughened [67–69].

Analysis of the results obtained by the Monte Carlo calculations for $\langle n \rangle$ yielded the scaling function $Y(x)$ for $\langle n \rangle$, and the power law behavior for v (Figure 7). The scaling function $Y(x) = L^{-0.375} \ln[\langle n \rangle - \langle n \rangle_\infty]$, where $\langle n \rangle_\infty = 1.79$ is $\langle n \rangle$ at $|\Delta\mu| \to \infty$, is expressed as

$$Y(x) = \begin{cases} 0.037 & x \to -\infty \\ -0.242x - 1.49 & x > x_R, \quad (x_R = -5.3 \text{ in Figure 7}), \end{cases} \tag{15}$$

where $x = \ln|\Delta\mu|/\epsilon$. x_R is determined as the point of intersection of the blue and the pink lines in Figure 7b. The value of x_R corresponds to $\Delta\mu_R/\epsilon = 0.005$. This yields

$$\langle n \rangle = \langle n \rangle_\infty + \exp\left\{ L^{0.38} Y \left(\ln\left[\frac{|\Delta\mu|}{\epsilon}\right] \right) \right\}. \tag{16}$$

This means that $\langle n \rangle - \langle n \rangle_\infty$ shows power law behavior, with the power depending on the system size. The lines based on the power law equation for a system size of $240\sqrt{2} \times 240\sqrt{2}$ are shown as pink lines in Figure 3a. The values obtained using the Monte Carlo method agree well with the power law functions for large $|\Delta\mu|$.

v also shows power law behavior. $\ln|v|$ (Figure 7b) increases linearly with increasing $\ln|\Delta\mu|/\epsilon$ for $|\Delta\mu| > \Delta\mu_R$. Thus, v can be expressed as

$$|v| \approx |\Delta\mu|^\beta, \quad \beta = 1.06 \pm 0.06. \tag{17}$$

Here, the choice of β as the symbol for the exponent is in accordance with Reference [70]. The power law equation given by Equation (17) is plotted as pink lines in Figure 3b. The values obtained using the Monte Carlo method agree well with the lines for large $|\Delta\mu|$.

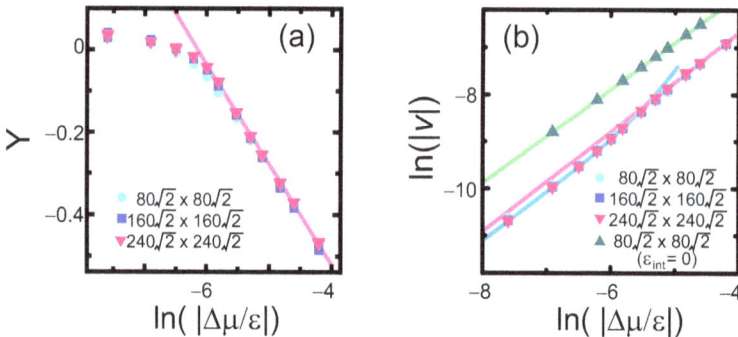

Figure 7. (a) Scaling function for $\langle n \rangle$. $Y = L^{-0.38} \ln[\langle n \rangle_{\Delta\mu} - \langle n \rangle_\infty]$. Pink line: $Y = -0.242 \ln(|\Delta\mu|/\epsilon) - 1.49$; (b) $\ln(|v|)$ plotted against $\ln(|\Delta\mu|/\epsilon)$. Pink line: $\ln(|v|) = 1.06(|\Delta\mu|/\epsilon) - 2.46$. Blue line: $\ln(|v|) = p_1 v_1 = 0.132 p_1 |\Delta\mu|/\epsilon$ with $p_1 = 0.332 + 15.6|\Delta\mu|/\epsilon + 4.43 \times 10^3 (|\Delta\mu|/\epsilon)^2$. Green line: $\ln(|v|) = 0.996(|\Delta\mu|/\epsilon) - 1.92$. All data are averaged over 2×10^8 MCS/site.

As shown in Figure 7, $\Delta\mu_R$ is not a roughening transition point in the Kosterlitz–Thouless universality class [71–74], but a crossover point from the two-surface coexistence state to a homogeneous rough surface. To clearly observe this crossover, the case with $\epsilon_{int}/\epsilon = -1.3$ and $k_B T/\epsilon = 0.83$—which is also in step droplet zone I—was calculated (Figure 8). In this case, $\langle n \rangle$ and v showed behaviors quite similar to the case of $\epsilon_{int}/\epsilon = -0.9$ and $k_B T/\epsilon = 0.63$. In the case of

$\epsilon_{int}/\epsilon = -1.3$, $\Delta\mu_R = 0.0045$ was obtained. The shift from the power law behavior for v at small $|\Delta\mu|$ in the case of $\epsilon_{int}/\epsilon = -1.3$ was more evident than in the case of $\epsilon_{int}/\epsilon = -0.9$.

It should be noted that the surface structure contains locally merged steps to some extent, though the surface is homogeneous at the thermodynamic limit ($L \to \infty$). For $|\Delta\mu| > \Delta\mu_R$, the power of v is similar to that in the case of the original RSOS model ($\epsilon_{int} = 0$, Figure 7). In spite of this, the growth rate remained lower than that in the original RSOS model at large $|\Delta\mu|$ (Figure 3b). To see the structure at a more microscopic scale, the surface was simulated with a small system size, as shown in Figure 5f. Though the surface appeared homogeneous for a system size of $240\sqrt{2} \times 240\sqrt{2}$, several locally merged steps were observed at a system size of $40\sqrt{2} \times 40\sqrt{2}$. These locally merged steps cause the kink density of the surface to decrease. Hence, the growth rate in this case was lower than that for the original RSOS model.

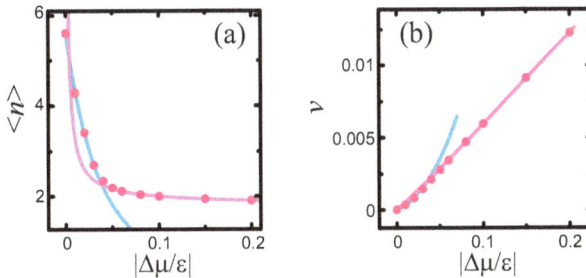

Figure 8. (a) $\Delta\mu$-dependence of $\langle n \rangle$. Light blue line: $\langle n \rangle = 1/(z + 2/N_{step})$ with $z = 0.154 + 4.71|\Delta\mu|/\epsilon + 55.0(|\Delta\mu|/\epsilon)^2$; Pink line: $\langle n \rangle = 1.79 + 0.0288(|\Delta\mu|/\epsilon)^{-0.85}$; (b) $\Delta\mu$-dependence of v. Light blue line: $v = p_1 v_1 = 0.141 p_1 |\Delta\mu|/\epsilon$ with $p_1 = 0.188 + 5.13|\Delta\mu|/\epsilon + 22.5(|\Delta\mu|/\epsilon)^2$. Pink line: $v = 0.0663(|\Delta\mu|/\epsilon)^{1.05}$. $\epsilon_{int}/\epsilon = -1.3$. $k_B T/\epsilon = 0.83$ (step droplet zone I). Circles: system size of $80\sqrt{2} \times 80\sqrt{2}$, averaged over 2×10^8 MCS/site.

6. Discussion

At equilibrium, planar surfaces such as the (001) surface are smooth at temperatures less than the roughening transition temperature T_R. The height–height correlation function $G(r) = \langle (h(r) - \langle h \rangle)^2 \rangle$ of the planar surface is constant for any r on a smooth surface. In contrast, for $T \geq T_R$, planar surfaces are rough, and $G(r)$ is logarithmically divergent with respect to $|r|$ [11,71–74]. For the vicinal surface, $G(r)$ is logarithmically divergent with respect to $|r|$, though the terrace is smooth [71,72]. Hence, in the present case, the height–height correlation function of the surface with a slope of $p_{1,eq}$ is logarithmically divergent, whereas that of the (111) surface—which is the side surface of the macrostep—is constant (non-divergent) at equilibrium.

For $0 < |\Delta\mu|$, the height–height correlation function for a surface with a slope of p_1 is considered to be logarithmically divergent with respect to $|r|$. This is also supported by the exponent β introduced in §5. The growth rate of the rough surface is known to increase linearly as $\Delta\mu$ increases [66,75]. The exponent β being close to 1 is consistent with the results obtained through calculations using the discrete Gaussian model [75].

The kinetically roughened state for $|\Delta\mu|/\epsilon > \Delta\mu_R$ is somewhat different from the state in step droplet zone II [43,44]. In step droplet zone II, the (001) surface is rough because the temperature is higher than its roughening transition temperature. Hence, the correlation length ξ for $G(r)$ on the (001) surface is divergent. Furthermore, $\langle n \rangle$ depends on the surface slope p as $\langle n \rangle = n_0 + C p^2 + \mathcal{O}(p^3)$ in the limit $p \to 0$, where n_0 and C are constants. In contrast, the correlation length ξ for the (001) surface in step droplet zone I is considered to be finite. $\langle n \rangle$ varies with the surface slope p as $\langle n \rangle = n_0 + C p + \mathcal{O}(p^2)$ in the limit $p \to 0$. The locally merged steps remain to some extent (Figure 5f).

The relaxation time required to reach steady state increases as $|\Delta\mu|$ decreases (Figure 6a). The time to relax to steady state is associated with the power law behavior of the surface [55,69,76–78].

The focus of the present study was the dynamics affected by the surface tension. As indicated in the cases of 4H-SiC [1] and Si(113) [21,25–33], elastic step–step repulsion expressed by the formula $\sum_{i\neq j}\sum_{\tilde{y}}g_0/[\tilde{x}_i(\tilde{y}) - \tilde{x}_j(\tilde{y})]^2$ [5–7] is important in real systems near equilibrium. The elastic effects are considered to weaken the step–step attraction in the present study. In the step droplet zone, the elastic step–step repulsion produces two effects: the shift of the zone boundary lines [43,45] and the formation of a regular array of n-merged macrosteps [31,32]. Since the elastic step–step repulsion causes $B_{q,\mathrm{eff}}(T)$ (Equation (4)) and $\gamma_{q,n}(T)$ (Equation (5)) to increase, the zone boundary lines shift to lower temperatures. In the short range, the step–step attraction dominates and sticks steps together, whereas in the long range, the step–step repulsion dominates and separates n^*-merged macrosteps, where n^* is the size of the macrostep with the lowest surface free energy. Thus, the following guidelines were obtained to disperse macrosteps: (1) the temperature should be raised; (2) the elastic step–step repulsion should be increased; and (3) the absolute value of the driving force should be increased.

In real systems, other effects, such as surface diffusion [52–55] and polymorphic effects [19,20], should also be taken into consideration. The combination of these effects and the effect of the discontinuous surface tension will be considered in future work.

7. Conclusions

The effect of the driving force $\Delta\mu$ on the size of a faceted macrostep and the growth rate of the vicinal surface in non-equilibrium steady state were investigated using the Monte Carlo model. Step droplet zone I for the p-RSOS model was the focus of this study.

- As $|\Delta\mu|$ increases, the size of the macrostep $\langle n\rangle$ decreases, whereas the growth rate $|v|$ increases.
- At small $|\Delta\mu|$, the $\Delta\mu$-dependence of $\langle n\rangle$ and v can be explained by the attachment and detachment of elementary steps to and from the macrostep.
- When $|\Delta\mu| \geq \Delta\mu_R$, the macrostep disassembles, and the surface roughens kinetically. $\Delta\mu_R/\epsilon = 0.005$ is the crossover point from the two-surface coexistent state to the rough surface state.
- A scaling function $Y(x)$ was obtained with $x = \ln(|\Delta\mu|/\epsilon)$ and $\langle n\rangle = \langle n\rangle_\infty + \exp[L^{0.38}Y(x)]$.
- For $|\Delta\mu| > \Delta\mu_R$, $\langle n\rangle$ and $|v|$ both show power law behavior.

Acknowledgments: This work was supported by the Japan Society for Promotion of Science (JSPS) KAKENHI Grant Number JP25400413.

Conflicts of Interest: The authors declare no conflict of interest. The founding sponsors had no role in the design of the study; the collection, analysis, or interpretation of data; the writing of the manuscript; or the decision to publish the results.

Abbreviations

The following abbreviations are used in this manuscript:

ECS	Equilibrium crystal shape
TSK	Terrace–step–kink
GMPT	Gruber–Mullins–Pokrovsky–Talapov
RSOS	Restricted solid-on-solid
p-RSOS	Restricted solid-on-solid with a point-contact-type step–step attraction
NN	Nearest-neighbor
NNN	Next-nearest-neighbor
DMRG	Density matrix renormalization group
1D	One-dimensional
PWFRG	Product wave function renormalization group
MCS	Monte Carlo steps
2D	Two-dimensional

Appendix A. Anomalous Surface Tension: Density Matrix Renormalization Group Calculation

The surface tension is the amount of surface free energy per unit normal area. To evaluate the surface free energy of the vicinal surface, the terms related to the Andreev field [13] were added: $\eta = (\eta_x, \eta_y)$. The Hamiltonian for the grand canonical ensemble with respect to the number of steps is [79].

$$\mathcal{H}_{\text{vicinal}} = \mathcal{H}_{\text{p-RSOS}} - \eta_x \sum_{n,m} [h(n+1,m) - h(n,m)] - \eta_y \sum_{n,m} [h(n,m+1) - h(n,m)]. \tag{A1}$$

The Andreev field behaves like a chemical potential with respect to a single step. The Legendre-transformed surface free energy introduced by Bhattacharjee [30] corresponds to the Andreev free energy [11,13].

From a statistical mechanics perspective, the grand partition function $\mathcal{Z}$ is calculated as $\mathcal{Z} = \sum_{\{h(m,n)\}} \exp[-\beta \mathcal{H}_{\text{vicinal}}]$, where $\beta = 1/k_B T$. The summation with respect to $\{h(m,n)\}$ is taken over all possible values of $h(m,n)$. The Andreev free energy $\tilde{f}(\eta)$ [13] is the thermodynamic grand potential and is calculated from the grand partition function $\mathcal{Z}$ as [79]

$$\tilde{f}(\eta) = \tilde{f}(\eta_x, \eta_y) = -\lim_{\mathcal{N} \to \infty} \frac{1}{\mathcal{N}} k_B T \ln \mathcal{Z}, \tag{A2}$$

where $\mathcal{N}$ is the number of lattice points on the square lattice. The Andreev free energy calculated using the DMRG method is shown in Figure A1a as a function of $(\eta_x/\epsilon, \eta_y/\epsilon)$ for $\eta_x = \eta_y$. It should be noted that the profile of the Andreev free energy $\tilde{f}(\eta_x, \eta_y)$ is similar to the ECS $z = z(x,y)$, where $\tilde{f}(\eta_x, \eta_y) = \lambda z(x,y)$, $(\eta_x, \eta_y) = -\lambda(x,y)$, and λ represents the Lagrange multiplier related to the crystal volume.

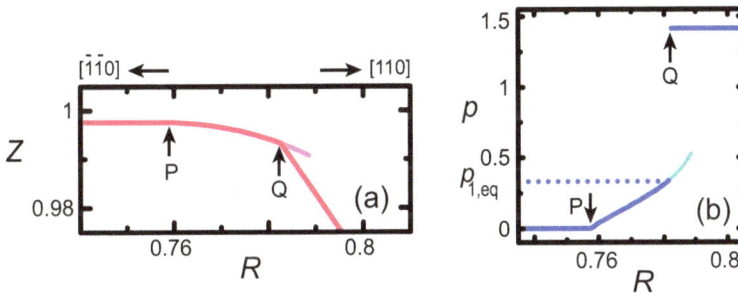

Figure A1. (a) Profile of the Andreev free energy, which is equivalent to the profile of the ECS. This figure was reproduced from Akutsu [45]; (b) $p = |p|$ plotted against R. $p = \tan \theta$, where θ is the tilt angle of the vicinal surface toward the $\langle 111 \rangle$ direction. $k_B T/\epsilon = 0.63$. $\epsilon_{\text{int}}/\epsilon = -0.9$. $Z = \tilde{f}(\eta_x, \eta_y)/\epsilon = (\lambda/\epsilon)z(x,y)$, $R = \pm\sqrt{X^2 + Y^2}$ with $X = Y$ and $(X,Y) = (\eta_x/\epsilon, \eta_y/\epsilon) = -(\lambda x/\epsilon, \lambda y/\epsilon)$, where $z = z(x,y)$ is the ECS and λ is the Lagrange multiplier related to the crystal volume. Pink lines: Andreev free energy calculated using the DMRG method. Blue lines: surface slope p calculated using the DMRG method. The pale pink line near point Q in (a) represents a metastable surface; The pale blue line in (b) represents the surface slope of the metastable surface. The end points of the pale lines show the approximate spinodal points. It is assumed that ϵ_{surf} equals ϵ. Points P and Q indicate the (001) and (111) facet edges, respectively.

The surface gradient p was also calculated using the DMRG method as $p = \langle(h(m+1,n) - h(m,n), h(m,n+1) - h(m,n))\rangle$. The calculated p with $p_x = p_y$ is shown in Figure A1b.

Using the inverse Legendre transform with respect to $\tilde{f}(\eta)$,

$$f(p) = \tilde{f}(\eta) + \eta \cdot p,\qquad(A3)$$

we obtained the surface free energy $f(p)$ per unit xy area. A plot of $f(p)$ is shown in Figure 4b.

References

1. Mitani, T.; Komatsu, N.; Takahashi, T.; Kato, T.; Harada, S.; Ujihara, T.; Matsumoto, Y.; Kurashige, K.; Okumura, H. Effect of aluminum addition on the surface step morphology of 4H–SiC grown from Si–Cr–C solution. *J. Cryst. Growth* **2015**, *423*, 45–49.
2. Cabrera, N.; Coleman, R.V. *The Art and Science of Growing Crystals*; Gilman, J.J., Ed.; John Wiley & Sons: New York, NY, USA; London, UK, 1963.
3. Cabrera, N. The equilibrium of crystal surfaces. *Surf. Sci.* **1964**, *2*, 320–345.
4. Jayaprakash, C.; Rottman, C.; Saam, W.F. Simple model for crystal shapes: Step-step interactions and facet edges. *Phys. Rev. B* **1984**, *30*, 6549–6554.
5. Calogero, F. Solution of a three—Body problem in one dimension. *J. Math. Phys.* **1969**, *10*, 2191–2196.
6. Sutherland, B. Quantum Many—Body Problem in One Dimension: Ground State. *J. Math. Phys.* **1971**, *12*, 246–250.
7. Alerhand, O.L.; Vanderbilt, D.; Meade, R.D.; Joannopoulos, J.D. Spontaneous formation of stress domains on crystal surfaces. *Phys. Rev. Lett.* **1988**, *61*, 1973–1976.
8. Gruber, E.E.; Mullins, W.W. On the theory of anisotropy of crystalline surface tension. *J. Phys. Chem. Solids* **1967**, *28*, 875-887.
9. Pokrovsky, V.L.; Talapov, A.L. Ground state, spectrum, and phase diagram of two-dimensional incommensurate crystals. *Phys. Rev. Lett.* **1979**, *42*, 65–67.
10. Einstein, T.L. Equilibrium Shape of Crystals. *Handbook of Crystal Growth*; Nishinaga, T., Ed.; Elsevier: Amsterdam, The Netherlands; Boston, MA, USA; Heidelberg, Germany; London, UK; Paris, France; Singapore; Sydney, Australia; Tokyo, Japan, 2015; Volume I, p. 216.
11. Akutsu, N.; Yamamoto, T. Rough-Smooth Transition of Step and Surface. *Handbook of Crystal Growth*; Nishinaga, T., Ed.; Elsevier: Amsterdam, The Netherlands; Boston, MA, USA; Heidelberg, Germany; London, UK; Paris, France; Singapore; Sydney, Australia; Tokyo, Japan, 2015; Volume I, p. 265.
12. Landau, L.D.; Lifshitz, E.M. *Statistical Physics*, 2nd ed.; Pergamon: Oxford, UK, 1968.
13. Andreev, A.F. Faceting phase transitions of crystals. *Sov. Phys. JETP* **1981**, *53*, 1063–1069.
14. Von Laue, M. Der Wulffsche Satz für die Gleichgewichtsform von Kristallen. *Z. Kristallogr.* **1944**, *105*, 124–133.
15. Herring, C. Some theorems on the free energies of crystal surfaces. *Phys. Rev.* **1951**, *82*, 87–93.
16. MacKenzie, J.K.; Moore, A.J.W.; Nicholas, J.F. Bonds broken at atomically flat crystal surfaces–I: Face-centred and body-centred cubic crystals. *J. Chem. Phys. Solids* **1962**, *23*, 185–196.
17. Toschev, S. *Crystal Growth: An Introduction*; Hartman, P., Ed.; North-Holland: Amsterdam, The Netherlands; Oxford, UK,1973; p. 328.
18. Rottman, C.; Wortis, M. Equilibrium crystal shapes for lattice models with nearest- and next-nearest-neighbor interactions. *Phys. Rev. B* **1984**, *29*, 328–339.
19. Williams, E.D.; Bartelt, N.C. Thermodynamics of surface morphology. *Science* **1991**, *251*, 393–401.
20. Williams, E.D.; Phaneuf, R.J.; Wei, J.; Bartelt, N.C.; Einstein, T.L. Thermodynamics and statistical mechanics of the faceting of stepped Si (111). *Surf. Sci.* **1993**, *294*, 219–242; Erratum in **1994**, *310*, 451.
21. Yamamoto, T.; Akutsu, Y.; Akutsu, N. Fluctuation of a single step on the vicinal surface-universal and non-universal behaviors. *J. Phys. Soc. Jpn.* **1994**, *63*, 915–925.
22. Akutsu, Y.; Akutsu, N.; Yamamoto, T. Logarithmic step fluctuations in vicinal surface: A Monte Carlo study. *J. Phys. Soc. Jpn.* **1994**, *63*, 2032–2036.
23. Hibino, H.; Ogino, T. Transient Step Bunching on a Vicinal Si(111) Surface. *Phys. Rev. Lett.* **1994**, *72*, 657–660.
24. Ogino, T.; Hibino, H.; Homma, Y. Kinetics and Thermodynamics of Surface Steps on Semiconductors. *Crit. Rev. Solid State Mater. Sci.* **1999**, *24*, 227–263.

25. Song, S.; Mochrie, S.G.J. Tricriticality in the orientational phase diagram of stepped Si (113) surfaces. *Phys. Rev. Lett.* **1994**, *73*, 995–998.

26. Song, S.; Mochrie, S.G.J. Attractive step-step interactions, tricriticality, and faceting in the orientational phase diagram of silicon surfaces between [113] and [114]. *Phys. Rev. B* **1995**, *51*, 10068–10084.

27. Van Dijken, S.; Zandvliet, H.J.W.; Poelsema, B. Anomalous strong repulsive step-step interaction on slightly misoriented Si (113). *Phys. Rev. B* **1997**, *55*, 7864–7867.

28. Jeong, H.C.; Williams, E.D. Steps on surfaces: Experiment and theory. *Surf. Sci. Rep.* **1999**, *34*, 171–294.

29. Lassig, M. Vicinal surfaces and the Calogero-Sutherland model. *Phys. Rev. Lett.* **1996**, *77*, 526–529.

30. Bhattacharjee, S.M. Theory of tricriticality for miscut surfaces. *Phys. Rev. Lett.* **1996**, *76*, 4568–4571.

31. Shenoy, V.B.; Zhang, S.; Saam, W.F. Bunching transitions on vicinal surfaces and quantum n-mers. *Phys. Rev. Lett.* **1998**, *81*, 3475–3478.

32. Shenoy, V.B.; Zhang, S.; Saam, W.F. Step-bunching transitions on vicinal surfaces with attractive step interactions. *Surf. Sci.* **2000**, *467*, 58–84.

33. Einstein, T.L.; Richards, H.L.; Cohen, S.D.; Pierre-Louis, O.; Giesen, M. Terrace-width distributions on vicinal surfaces: Generalized Wigner surmise and extraction of step–step repulsions. *Appl. Surf. Sci.* **2001**, *175–176*, 62–68.

34. Akutsu, N.; Akutsu, Y.; Yamamoto, T. Stiffening transition in vicinal surfaces with adsorption. *Prog. Theory Phys.* **2001**, *105*, 361–366.

35. Akutsu, N.; Akutsu, Y.; Yamamoto, T. Statistical mechanics of the vicinal surfaces with adsorption. *Surf. Sci.* **2001**, *493*, 475–479.

36. Akutsu, N.; Akutsu, Y.; Yamamoto, T. Thermal step bunching and interstep attraction on the vicinal surface with adsorption. *Phys. Rev. B* **2003**, *67*, 125407.

37. Akutsu, N.; Hibino, H.; Yamamoto, T. A Lattice Model for Thermal Decoration and Step Bunching in Vicinal Surface with Sub-Monolayer Adsorbates. *e-J. Surf. Sci. Nanotechnol.* **2009**, *7*, 39–44.

38. Akutsu, N. Thermal step bunching on the restricted solid-on-solid model with point contact inter-step attractions. *Appl. Surf. Sci.* **2009**, *256*, 1205–1209.

39. Akutsu, N. Zipping process on the step bunching in the vicinal surface of the restricted solid-on-solid model with the step attraction of the point contact type. *J. Cryst. Growth* **2011**, *318*, 10–13.

40. Akutsu, N. Non-universal equilibrium crystal shape results from sticky steps. *J. Phys. Condens. Matter* **2011**, *23*, 485004.

41. Akutsu, N. Sticky steps inhibit step motions near equilibrium. *Phys. Rev. E* **2012**, *86*, 061604.

42. Akutsu, N. Pinning of steps near equilibrium without impurities, adsorbates, or dislocations. *J. Cryst. Growth* **2014**, *401*, 72–77.

43. Akutsu, N. Faceting diagram for sticky steps. *AIP Adv.* **2016**, *6*, 035301.

44. Akutsu, N. Effect of the roughening transition on the vicinal surface in the step droplet zone. *J. Cryst. Growth* **2016**, doi:10.1016/j.jcrysgro.2016.10.014.

45. Akutsu, N. Profile of a Faceted Macrostep Caused by Anomalous Surface Tension. *Adv. Condens. Matter Phys.* **2017**, *2021510*.

46. White, S.R. Density matrix formulation for quantum renormalization groups. *Phys. Rev. Lett.* **1992**, *69*, 2863–2866.

47. Nishino, T. Density matrix renormalization group method for 2D classical models. *J. Phys. Soc. Jpn.* **1995**, *64*, 3598–3601.

48. Schollwöck, U. The density-matrix renormalization group. *Rev. Mod. Phys.* **2005**, *77*, 259–315.

49. Nishino, T.; Okunishi, K. Product wave function renormalization group. *J. Phys. Soc. Jpn.* **1995**, *64*, 4084–4087.

50. Hieida, Y.; Okunishi, K.; Akutsu, Y. Magnetization process of a one-dimensional quantum antiferromagnet: The product-wave-function renormalization group approach. *Phys. Lett. A* **1997**, *233*, 464–470.

51. Hieida, Y.; Okunishi, K.; Akutsu, Y. Numerical renormalization approach to two-dimensional quantum antiferromagnets with valence-bond-solid type ground state. *New J. Phys.* **1999**, *1*, 7.

52. Ehrlich, G.; Hudda, F.G. Atomic view of surface self-diffusion: Tungsten on tungsten. *J. Chem. Phys.* **1966**, *44*, 1039–1049.

53. Schwoeble, R.L.; Shipsey, E.J. Step motion on crystal surfaces. *J. Appl. Phys.* **1966**, *37*, 3682–3686.

54. Pimpinelli, A.; Villain, J. *Physics of Crystal Growth*; Cambridge University Press: Cambridge, UK, 1998.

55. Misbah, C.; Pierre-Louis, O.; Saito, Y. Crystal surfaces in and out of equilibrium: A modern view. *Rev. Mod. Phys.* **2010**, *82*, 981–1040.

56. Mermin, N.D.; Wagner, H. Absence of ferromagnetism or antiferromagnetism in one-or two-dimensional isotropic Heisenberg models. *Phys. Rev. Lett.* **1966**, *17*, 1133–1136; Erratum in **1966**, *17*, 1307.

57. Akutsu, Y.; Akutsu, N.; Yamamoto, T. Universal jump of Gaussian curvature at the facet edge of a crystal. *Phys. Rev. Lett.* **1988**, *61*, 424–427.

58. Yamamoto, T.; Akutsu, Y.; Akutsu, N. Universal Behavior of the Equilibrium Crystal Shape near the Facet Edge. I. A Generalized Terrace-Step-Kink Model. *J. Phys. Soc. Jpn.* **1988**, *57*, 453–460.

59. Mikheev, L.V.; Pokrovsky, V.L. Free-fermion solution for overall equilibrium crystal shape. *J. Phys.* **1991**, *1*, 373–382.

60. Akutsu, Y. Exact Landau Free-Energy of Solvable N-State Vertex Model. *J. Phys. Soc. Jpn.* **1989**, *58*, 2219–2222.

61. Okunishi, K.; Hieida, Y.; Akutsu, Y. δ-function Bose-gas picture of S = 1 antiferromagnetic quantum spin chains near critical fields. *Phys. Rev. B* **1999**, *59*, 6806–6812.

62. Markov, I.V. *Crystal Growth for Beginners*, 2nd ed.; World Scientific: Hackensack, NJ, USA; London, UK; Singapore; Hong Kong, China, 2003.

63. Müller-Krumbhaar, H.; Burkhardt, T.W.; Kroll, D.M. A generalized kinetic equation for crystal growth. *J. Cryst. Growth* **1977**, *38*, 13–22.

64. Enomoto, Y.; Kawasaki, K.; Ohta, T.; Ohta, S. Interface dynamics under the anisotropic surface tension. *Phys. Lett.* **1985**, *107*, 319–323.

65. Akutsu, N.; Akutsu, Y. Roughening, faceting and equilibrium shape of two-dimensional anisotropic interface. I. Thermodynamics of interface fluctuations and geometry of equilibrium crystal shape. *J. Phys. Soc. Jpn.* **1987**, *56*, 1443–1453.

66. Burton, W.K.; Cabrera, N.; Frank, F.C. The growth of crystals and the equilibrium structure of their surfaces. *Philos. Trans. R. Soc. Lond. A* **1951**, *243*, 299–358.

67. Krug, J.; Meakin, P. Kinetic roughening of Laplacian fronts. *Phys. Rev. Lett.* **1991**, *66*, 703–706.

68. Uwaha, M.; Saito, Y. Kinetic smoothing and roughening of a step with surface diffusion. *Phys. Rev. Lett.* **1992**, *68*, 224–227.

69. Barabasi, A.L.; Stanley, H.E. *Fractal Concepts in Surface Growth*; Cambridge University Press: Cambridge, UK, 1995.

70. Reichhardt, C.; Reichhardt, C.J.O. Depinning and nonequilibrium dynamic phases of particle assemblies driven over random and ordered substrates: A review. *Rep. Prog. Phys.* **2017**, *80*, 026501.

71. Jayaprakash, C.; Saam, W.F.; Teitel, S. Roughening and facet formation in crystals. *Phys. Rev. Lett.* **1983**, *50*, 2017–2020.

72. Cabrera, N.; Garcia, N. Roughening transition in the interface between superfluid and solid He 4. *Phys. Rev. B* **1982**, *25*, 6057–6059.

73. Saenz, J.; Garcia, N. Classical critical behaviour in crystal surfaces near smooth and sharp edges. *Surf. Sci.* **1985**, *155*, 24–30.

74. Van Beijeren, H.; Nolden, I. *Structure and Dynamics of Surfaces*; Schommers, W., von Blancken-Hagen, P., Eds.; Springer: Berlin/Heidelberg, Germany, 1987; Volume 2, p. 259.

75. Saito, Y. Self-Consistent Calculation of Statics and Dynamics of the Roughening Transition. *Z. Phys. B* **1978**, *32*, 75–82.

76. Pimpinelli, A.; Tonchev, V.; Videcoq, A.; Vladimirova, M. Scaling and Universality of Self-Organized Patterns on Unstable Vicinal Surfaces. *Phys. Rev. Lett.* **2002**, *88*, 206103.

77. Krug, J.; Tonchev, V.; Stoyanov, S; Pimpinelli, A. Scaling properties of step bunches induced by sublimation and related mechanisms. *Phys. Rev. B* **2005**, *71*, 045412.

78. Krasteva, A.; Popova, H.; Akutsu, N.; Tonchev, V. Time scaling relations for step bunches from models with step-step attractions (B1-type models). *AIP Conf. Proc.* **2016**, *1722*, 220015.

79. Akutsu, N.; Akutsu, Y. Thermal evolution of step stiffness on the Si (001) surface: Temperature-rescaled Ising-model approach. *Phys. Rev. B* **1998**, *57*, R4233–R4236.

© 2017 by the author. Licensee MDPI, Basel, Switzerland. This article is an open access article distributed under the terms and conditions of the Creative Commons Attribution (CC BY) license (http://creativecommons.org/licenses/by/4.0/).

crystals

MDPI

Review

Recent Progress in Computational Materials Science for Semiconductor Epitaxial Growth

Tomonori Ito * and Toru Akiyama

Department of Physics Engineering, Mie University, 1577, Kurima-Machiya, Tsu 514-8507, Japan;
akiyama@phen.mie-u.ac.jp
* Correspondence: tom@phen.mie-u.ac.jp; Tel.: +81-59-231-9724

Academic Editor: Hiroki Nada
Received: 19 December 2016; Accepted: 4 February 2017; Published: 9 February 2017

Abstract: Recent progress in computational materials science in the area of semiconductor epitaxial growth is reviewed. Reliable prediction can now be made for a wide range of problems, such as surface reconstructions, adsorption-desorption behavior, and growth processes at realistic growth conditions, using our ab initio-based chemical potential approach incorporating temperature and beam equivalent pressure. Applications are examined by investigating the novel behavior during the hetero-epitaxial growth of InAs on GaAs including strain relaxation and resultant growth mode depending growth orientations such as (111)A and (001). Moreover, nanowire formation is also exemplified for adsorption-desorption behaviors of InP nanowire facets during selective-area growth. An overview of these issues is provided and the latest achievement are presented to illustrate the capability of the theoretical-computational approach by comparing experimental results. These successful applications lead to future prospects for the computational materials design in the fabrication of epitaxially grown semiconductor materials.

Keywords: computer simulation; semiconductor epitaxial growth; strain relaxation; growth mode; InAs/GaAs; nanowire formation; polytypes; InP nanowires; selective-area growth

---

1. Introduction

Computational materials science is now popular among researchers including experimentalists to predict the physical properties of materials. This reflects recent developments in the computer software using electronic theory to explain many experimental results and to predict the physical properties of materials without actually synthesizing them. During the last several decades, semiconductor technology has made it possible to create artificial materials whose physical properties are on the atomic scale such as quantum dots and nanowires fabricated by various epitaxial growth techniques. In particular, crystal growth processes during molecular beam epitaxy (MBE) and metal-organic vapor phase epitaxy (MOVPE) have been paid much attention from a number of theorists to work in the field of semiconductor epitaxial growth, since these epitaxial growth techniques realize highly non-equilibrium conditions where the growth is governed by the diffusive kinetics of the source atoms within the crystalline surfaces in contrast with bulk-growth processes which proceed near thermodynamic equilibrium conditions. Thus, a fundamental understanding of the growth processes in these techniques not only leads to better growth techniques but also gives guiding-principles for the fabrication of the novel materials that are not found in nature.

As a physical process, the growth of thin films is controlled by various growth processes that involve adsorption of atoms or molecules onto a surface, their subsequent diffusion across the surface, dissociation of molecules, and desorption, etc. at the gas-solid interface in the growth front. It is well known that reconstructed structures appear on the growth front (surfaces) of semiconductor materials [1]. Therefore, we have to understand the atomic structures on the surfaces to control the interface mass transfer. To this end, so far a lot of theoretical works have been carried out to investigate the surface structures of semiconductors using ab initio calculation that is a promising tool for clarifying the complicated growth processes because of its ability to calculate electronic structures and total energy [2–4]. Kaxiras et al. [2] studied the lowest-energy geometry for GaAs(111) with different stoichiometries. Qian et al. [3] discussed relationship between the stoichiometry and the surface reconstruction on GaAs(001) by means of chemical potentials. Northrup [4] classified the stable structures on Si(001)H using H chemical potential. However, all these approaches discussed about the static structural stability on the surfaces at 0 K.

Generally, vapor phase epitaxy such as MBE and MOVPE is performed under finite temperatures and gas-pressures. This implies that it is indispensable to consider the ambient conditions to predict the reconstructed structures on the growth surfaces. We [5,6] proposed ab initio-based chemical potential approach that incorporates the free energy of gas phase. Therefore, the theoretical analysis is useful to analyze the influence of temperatures and gas-pressure on the stability of reconstructed surfaces. By the application of this method, epitaxial growth processes can be investigated for various semiconductors including GaAs and GaN [7–9]. Controlling the growth processes, recent developments in semiconductor technology made it possible to create artificial nanostructures such as nanodots, nanowire, and nanosheet fabricated by various thin film growth techniques including MBE and MOVPE. In this article, we focus on the nanostructures to exemplify the feasibility of the chemical potential approach to their actual growth systems such as MBE growth for lattice mismatched InAs and selective-area growth of InP nanowires (NWs). Surface phase diagram calculations as functions of temperature and gas-pressure are performed for InAs(001), InAs(111)A, InP(111)A, $\{1\bar{1}00\}$ and $\{11\bar{2}0\}$ facets of InP NWs with wurtzite (WZ) structure. Furthermore, the growth modes of heteroepitaxial InAs/GaAs system are also investigated using macroscopic free energy theory incorporating surface energy and misfit dislocation formation energy obtained by the ab initio and the empirical interatomic potential calculations. The adsorption-desorption behavior during growth of InP NWs, which are crucial in determining their crystal structure, are also investigated using Monte Carlo (MC) simulations combined with surface phase diagrams calculations. Comparative studies between calculated and experimental results are shown to check the versatility of the chemical potential approach to predictions of phase diagram and novel adatom behavior on various semiconductor surfaces.

2. Computational Methods

We proposed an ab initio-based approach that incorporates the free energy of ideal gas per one particle (chemical potential) μ_{gas} [5,6]. The adsorption-desorption behavior of In atom and As molecule on these surfaces can be determined by comparing μ_{gas} with the adsorption energy E_{ad}. An impinging atom (molecule) can adsorb on the surface if the free energy of the atom (molecule) in the gas phase is larger than its adsorption energy. In contrast, an impinging atom (molecule) desorbs if its gas-phase free energy is smaller than the adsorption energy. The μ_{gas} can be computed using quantum statistical mechanics as functions of temperature and pressure. The adsorption energy can be obtained using ab initio calculations. Exemplifying In atom and As_4 molecule, the chemical potential μ_{gas} of ideal gas is given by the following Equations:

$$\mu_{In} = -k_B T(k_B T/p_{In} \times g \times \zeta_{trans}), \tag{1}$$

$$\mu_{As4} = -k_B T(k_B T/p_{As4} \times g \times \zeta_{trans} \times \zeta_{rot} \times \zeta_{vib}). \tag{2}$$

Here, k_B is the Boltzmann's constant, T the gas temperature, g the degree of degeneracy of electron energy level, p the beam equivalent pressure (BEP) of In atom or As molecule (As_4), and ζ_{trans}, ζ_{rot}, and ζ_{vib} are the partition functions for translational, rotational, and vibrational motions, respectively. In general, both μ_{In} and μ_{As4} decrease with temperature and increases with BEP. The adsorption energy E_{ad} is obtained by

$$E_{ad} = E_{total} - E_{substrate} - E_{atom}, \tag{3}$$

where E_{total} is the total energy of the surface with adatoms, $E_{substrate}$ the total energy of the surface without adatoms, and E_{atom} is the total energy of isolated atoms. We note that the Gibbs free energy of formation vibrational contribution is very small compared with the energy difference between a given structure and the ideal surface. The gas-phase entropy difference is also considerably larger than the surface entropy change, when the temperature or pressure is varied. Therefore, only the entropic effects of the gas phase are considered throughout our theoretical approach. Using the chemical potential and the adsorption energy, the adsorption and desorption behaviors on the surfaces under growth conditions are obtained as functions of T and p, i.e., the structure corresponding to adsorbed surface is favorable when E_{ad} is less than μ_{gas}, whereas desorbed surface is stabilized when μ_{gas} is less than E_{ad}. Based on these results, we obtain adsorption-desorption boundaries for In and As and surface phase diagrams of the InAs WL.

Indeed, this method has been successfully applied to determine the reconstructions on GaAs(001) surfaces [5]. It has been found that the μ_{gas} of Ga atom using Equation (1) decreases with temperature and increases with BEP. By comparing the adsorption energy of Ga atom (-3.3 eV) using Equation (3), the μ_{gas} of Ga atom becomes lower than the adsorption energy when temperatures is higher than 1000 K for Ga-BEP at 1.0×10^{-5} Torr. This suggests that the critical temperature for Ga adsorption is ~1000 K for Ga-BEP at 1.0×10^{-5} Torr. The surface phase diagram for Ga adsorption obtained by the comparison of μ_{gas} with E_{ad} thus suggest that the Ga-droplet appear under low temperature and high Ga-BEP conditions, while GaAs-(4 × 2)β2 surface is stabilized under high temperature and low Ga-BEP conditions. The calculated results are consistent with the observation of Ga-droplet under ~900 K during the MBE growth of GaAs under Ga-rich conditions and Ga desorption above ~970 K after turning off the Ga flux, suggesting that ab initio-based approach that incorporates the free energy of the gas phase is feasible for investigating surface structures, while previous calculations [2–4] are unable to incorporate temperature and pressure.

In this study, the total-energy calculations are performed within the generalized gradient approximation (GGA) [10] within the density functional theory. Norm-conserving pseudopotentials with partial core corrections are used to describe electron-ion interaction [11]. The conjugate-gradient minimization technique is used for both the electronic-structure calculation and the geometry optimization [12]. The geometry is optimized until the remaining forces acting on the atoms are less than 5.0×10^{-3} Ry/Å. The valence wave functions are expanded by the plane-wave basis set with a cutoff energy of 16 Ry. We take the model with slab geometries of six atomic layers for the InAs(111)A and InP(111)A and eight atomic layers for the InAs(001), WZ-InP($1\bar{1}00$) and WZ-InP($11\bar{2}0$) with artificial H atoms [13] and a vacuum region equivalent to nine atomic layer thickness. The k-point sampling corresponding to 36 points in the irreducible part of the 1 × 1 surface Brillouin zone, which provides sufficient accuracy in the total energies, is employed. The computations are carried out using Tokyo Ab initio Program Package [12].

According to our macroscopic theory for growth mode [14], the free energies F for various growth modes such as two-dimensional coherent growth (2D-coherent), 2D growth with misfit dislocation (2D-MD) and (b) 2D growth with stacking-fault tetrahedron (2D-SFT), and Stranski-Krastanov growth (SK-coh) are given by

$$F_{\text{2D-coherent}}(h) = \gamma + M\varepsilon_0^2 h/2, \tag{4}$$

$$F_{\text{2D-MD}}(h) = \gamma + E_d/l_0 - E_d^2/2(M\varepsilon_0^2 l_0^2 h), \tag{5}$$

$$F_{\text{2D-SFT}}(h) = \gamma + M\varepsilon_0^2 h/2 + E_{\text{SFT}}/A_{\text{unit}}, \tag{6}$$

$$F_{\text{SK-coh}}(h) = \gamma(1 + \beta) + M(1 - \alpha)\varepsilon_0^2 h/2, \tag{7}$$

where γ is the surface energy, M the effective elastic constant, ε_0 the intrinsic strain (=0.072) [14], h the layer thickness, E_d the dislocation formation energy, l_0 the average dislocation spacing (=58.76 (Å) [14]), E_{SFT} the SFT formation energy, A_{unit} the area of 14 × 14 planar unit cell used in the system energy calculations for the 2D-MD with l_0 and the 2D-SFT, and β and $-\alpha$ are the effective energy increase in surface energy of the epitaxial layer and the effective decrease in strain energy due to SK-island formation.

In order to obtain the parameter values of M, E_d, and E_{SFT}, we employ a simple formula for estimating system energy E for computational models such as the 2D-coherent, the 2D-MD, and the 2D-SFT as follows.

$$E = E_0 + \Delta E_{SF}, \tag{8}$$

$$E_0 = 1/2 \times \sum_{i,j} V_{ij}, \tag{9}$$

$$V_{ij} = A \exp[-\beta(r_{ij} - R_i)^\gamma][\exp(-\theta r_{ij}) - B_0 \exp(-\lambda r_{ij})G(\eta)/Z_i^\alpha], \tag{10}$$

$$\Delta E_{SF} = K \left[3/2 \times (1 - f_i) \times Z_b^2 / r_{bb} - f_i \times Z_i^2 / r_{ii} \right]. \tag{11}$$

where E_0 is the cohesive energy estimated by Kohr–Das Sarma-type empirical interatomic potential V_{ij} within the second-neighbor interactions [15]. The potential parameters A, β, R_i, γ, θ, B_0, λ, η, and α are determined by reproducing equilibrium interatomic bond lengths, elastic stiffness for zinc blende structure, and relative stability among zinc blende, rocksalt, and CsCl structures [16]. Stacking-fault energy ΔE_{SF} is described as the summation of electrostatic energies consisting of repulsive interaction between covalent bond charges Z_b ($=-2$) and attractive interaction between ionic charges Z_i ($=\pm 3$ for III–V compound semiconductors) depending on ionicity f_i ($=0.298$ for InAs) [17]. The value of coefficient K is determined to be 8.7 (meV·Å) by reproducing energy difference 25.3 (meV/atom) between diamond and hexagonal structures for C with $f_i = 0$ obtained by ab initio calculations. In the system energy calculations, the lattice parameter a of InAs is fixed to be that of GaAs(111) for the 2D-coherent and the 2D-SFT or relaxed value of a for the 2D-MD, and the lattice parameter c and the atomic positions are varied to minimize the system energy in a 14×14 planar unit cell with increase of layer thickness h.

Furthermore, we employ stochastic MC simulations [18] in order clarify the growth mechanism of nanowires. The MC simulations take temperature and pressure into account. In the MC procedure, the probability of adsorption, migration, and desorption ($P_{ad}(x)$, $P_{diff}(x \to x')$, and $P_{de}(x)$, respectively) at adsorption site x are estimated by

$$P_{ad}(x) = \exp\{-\Delta\mu(x)/k_BT\}/[1 + \exp\{-\Delta\mu(x)/k_BT\}], \tag{12}$$

$$P_{diff}(x \to x') = R\exp\{\Delta E(x \to x')/k_BT\}, \tag{13}$$

$$P_{de}(x) = R\exp[-\{E_{de}(x) - \Delta\mu(x)\}/k_BT], \tag{14}$$

where $\Delta\mu(x)$ is the energy difference between adsorption energy $E_{ad}(x)$ and μ_{gas} at adsorption site x, R is diffusion prefactor taken to be $k_BT/\pi\hbar$ [19], $\Delta E(x \to x')$ is the local migration barrier involving the adatom hopping from site x to x' and $E_{de}(x)$ is desorption energy at site x. The probability for surmounting the activation energy is reduced (or enhanced) by a weighting function of $\exp\{\Delta\mu(x)/k_BT\}$, which corresponds to the local-thermal equilibrium desorption probability. In the simulation procedure, we adopt extensive surfaces as growth substrates which are $30,000 \times 30,000$ times as large as the surface unit cells. The atoms and molecules with pressure p are supplied from gas phase to these surfaces at an interval of Δt_{gas} per surface area that is derived from Maxwell's law of velocity distribution given by

$$\Delta t_{gas} = (2mk_BT/p)^{1/2} \tag{15}$$

where m is the mass of the adsorbate. The evaluation of adsorption using Equation (15) are repeated during the simulation. Events for adatoms on the surface are determined by

$$K_{diff}(x \to x_i) = P_{diff}(x \to x_i)/\{\Sigma_j P_{diff}(x \to x_j) + P_{de}(x)\}, \tag{16}$$

$$K_{de}(x \to x_i) = P_{de}(x \to x_i)/\{\Sigma_j P_{diff}(x \to x_j) + P_{de}(x)\}, \tag{17}$$

$$\Delta t = 1/\{\Sigma_j \, P_{\text{diff}}(x \rightarrow x_j) + P_{\text{de}}(x)\}, \tag{18}$$

where $K_{\text{diff}}(x \rightarrow x_i)$ is the relative diffusion probability from site x to another site x_i and $K_{\text{de}}(x)$ is the relative desorption probability at site x. The sum is carried out all over the neighboring site x_j. Events for adatoms are taken place with a time increment Δt expressed in Equation (18).

3. Hetero-Epitaxial Growth of InAs on GaAs

Semiconductor hetero-epitaxial systems grown by molecular beam epitaxy (MBE) are crucial for fabricating low-dimensional semiconductor nanostructures such as three-dimensional (3D) island shaped quantum dots (QDs) in InAs grown on GaAs(001). Despite a constant lattice mismatch, InAs thin films fabricated on the GaAs(111)A exhibit two-dimensional (2D) growth. Many studies have been done to investigate the InAs/GaAs growth using reflection high-energy electron-diffraction (RHEED) and scanning tunneling microscopy (STM). Joyce et al. [20] found that an intermediate wetting layer (WL) is formed on the (001)-oriented substrates prior to the formation of QDs while strain relaxation involving misfit dislocation formation causes a continuous 2D growth mode on the (111)A. For the InAs/GaAs(111), Yamaguchi et al. [21] reported experimental measurements using STM on the mechanisms of strain relaxation in InAs/GaAs(111)A hetero-epitaxy to propose a possible model for the resulting semicoherent interface structure. Ohtake et al. [22] studied strain-relaxation processes in the InAs/GaAs(111)A using rocking-curve analysis of reflection high-energy electron diffraction. Zepeda-Ruiz et al. [23] theoretically analyzed the strain relaxation at the semicoherent interface structure consisting of a network of interacting misfit dislocations. We have also investigated the stacking-fault tetrahedron (SFT) formation in the InAs/GaAs(111)A using empirical interatomic potentials to clarify the contribution of strain relaxation stabilizing it [24]. These previous studies have mainly focused on the strain relaxation at the InAs/GaAs(111)A interface without incorporating surface related phenomena. Taguchi and Kanisawa have studied the surface structures and adsorption-desorption behavior on the InAs(111)A surface using ab initio calculations [25]. Moreover, it should be noted that the stability of surface structures and adsorption behavior have never been clarified on the InAs/GaAs(111)A.

The formation of the WL on the (001) as a function of substrate temperature, reconstruction, and the amount of InAs deposition has been investigated in detail by Belk et al. [26]. Their RHEED results indicate that the surface in InAs/GaAs changes its structure from substrate GaAs(001)-c(4 × 4) to InAs(001)-(2 × 4) via (1 × 3)/(2 × 3) WL with increase of InAs coverage over the temperature range of 350–500 °C. Grabowski et al. [27] observed a complete transition into (2 × 3) at InAs coverage of 0.76 ± 0.07 monolayer (ML) with many missing dimers at 450 °C in STM. Eisele et al. [28] found that the (4 × 3) surface unit cells are arranged either in-line or brick-lined at about 2/3 ML of InAs deposited on the GaAs(001)-c(4 × 4) surface. Moreover, Konishi and Tsukamoto [29] found that the domains of (1 × 3)/(2 × 3) and (2 × 4) are distributed in an ordered pattern closely related to the density of QD precursors just after nucleation on the basis of in-situ STM observations. Although many studies have been also done for surface structures on the InAs/GaAs(001) from theoretical viewpoints, there have been only a few theoretical studies for the (2 × 3) surfaces. Kratzer et al. [30] showed that the (2 × 3) reconstruction can be regarded as the main building unit of the InAs WL. In contrast with InAs/GaAs(111), there have been very few studies for the strain relaxation on the InAs WL. We have successfully investigated various semiconductor surface structures and the elemental growth processes on them using ab initio-based approach incorporating MBE growth conditions such as temperature T and beam equivalent pressure p [8,9,31–34]. In this chapter, we systematically discuss the novel behavior in hetero-epitaxial growth of InAs on GaAs(111)A and GaAs(001) surfaces including adsorption/desorption behavior and strain relaxation during MBE growth.

3.1. Surface Reconstructions on InAs(111)A WL

Figure 1 shows top view of InAs (111)A-(2 × 2) with In vacancy (V_{In}), As-adatom, and As-trimer (T_{As}) WL surface structures considered in this study. Figure 2 shows the calculated surface phase diagrams for (a) InAs(111)A WL surface (InAs(111)-WL) and (b) fully relaxed (FR) surface without interface structure of InAs/GaAs (InAs(111)-FR) as functions of temperature T and BEP of As_4 molecules p_{As4}. Here, we assume that the BEP of In atom p_{In} in gas phase is $p_{In} \sim 10^{-7}$ Torr. The (2×2) with T_{As} is stabilized only at low temperatures, while the (2×2) with V_{In} appears at high temperatures. The stable temperature ranges for the (2×2) with V_{In} (beyond 550–600 K) on the InAs(111)-WL are consistent with MBE growth temperature range of 723–773 K (black area) at which the (2×2) with V_{In} is observed experimentally [20,35,36]. It should be noted that the InAs(111)-WL with As-adatom surface newly appears between stable regions of the (2×2) with V_{In} and the (2×2) with T_{As}, where the As adatom stably resides in the interstitial site bonding with three substrate In atoms. The large decrease in the charge density is found at surface As atoms on the In-vacancy surface [31]. This is because the electronegativity of In on the surface is smaller than that of Ga at the interface [37]. This results in breaking the electron counting model (ECM) [38,39] to destabilize the In-vacancy surface. The As-adatom surface makes up this deficiency in charge density, where the charge density around the As adatom is transferred to the substrate In atom to strengthen the interatomic bond between the In and As atoms. Therefore, the As-adatom surface appears as a stable phase between V_{In} and T_{As} on the InAs(111)-WL, while the InAs(111)-FR without interface Ga does not favor the As-adatom surface.

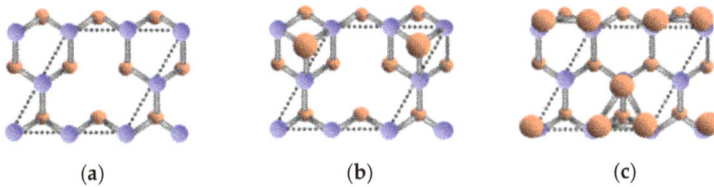

Figure 1. Schematics of top view of InAs(111)A-(2×2) with (**a**) In-vacancy; (**b**) As-adatom; and (**c**) As-trimer surfaces considered in this study. Blue and orange atoms denote In and As, respectively.

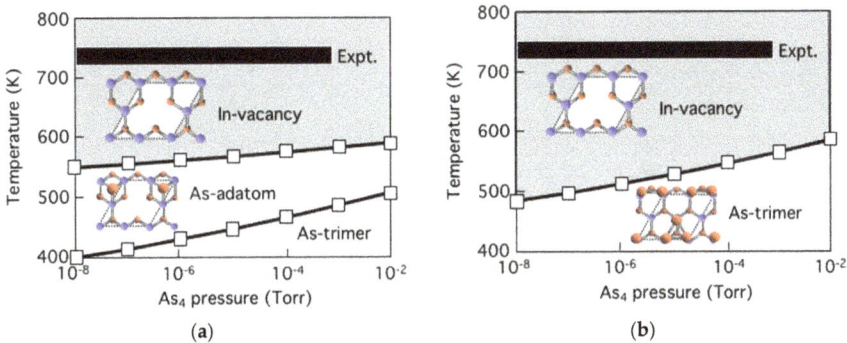

Figure 2. Calculated surface phase diagrams for (**a**) InAs(111)A WL surface and (**b**) fully relaxed surface without interface structure of InAs/GaAs as functions of temperature and BEP of As_4. Black area denotes the temperature range of the conventional MBE growth.

3.2. Growth Process on InAs(111)A WL

Figure 3 shows the calculated adsorption-desorption boundaries for In atom without (dotted line) and with simultaneous As adsorption (solid line) on the (2×2) with V_{In} of (a) the InAs(111)-WL and (b) the InAs(111)-FR as functions of T and p_{In}. These figures reveal that In adsorption on the In-vacancy surface does not occur without simultaneous As adsorption under conventional growth temperatures in the range of 723–773 K (black area). This is because In adsorption energies on the In vacancy surface dramatically decrease after As adsorption such as −1.76 eV without As to −4.17 eV with As on the InAs(111)-FR and −1.46 eV to −3.27 eV for the InAs(111)-WL. These findings thus imply that In adsorption is promoted only when As atoms are adsorbed on the InAs(111)A-(2 × 2) surface and the InAs growth proceeds with the adsorption of As atoms. This reveals that the self-surfactant effect is crucial for the growth processes similarly to the GaAs growth [40,41]. Moreover, it is found that relative stability among the surface lattice sites and adsorption energies for In adatom do not strongly depend on the layer thickness [31]. Therefore, it is expected that the strain in the WL does not significantly affect the adsorption-desorption transition curve and the kinetic behavior of In adatom. Considering the fact that the misfit dislocation is formed at 4–7 ML of InAs [23,24], these results suggest that strain accumulated in the WL is not significant for the hetero-epitaxial growth on the InAs(111)A to keep 2D growth at the initial growth stage.

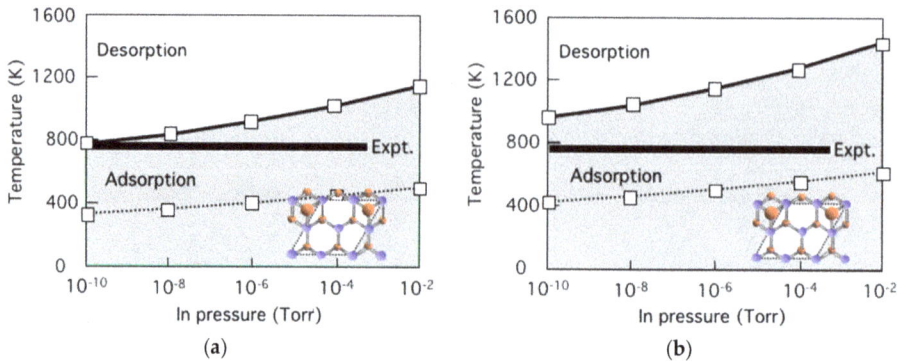

Figure 3. Calculated adsorption-desorption boundaries for In atom without (dotted line) and with simultaneous As adsorption (solid line) on the (2 × 2) with In-vacancy of (**a**) the InAs(111)-WL and (**b**) the InAs(111)-FR as functions of temperature and BEP of In. Black area denotes the temperature range of the conventional MBE growth.

3.3. Strain Relaxation on InAs(111)A WL

In order to clarify the strain relaxation in the InAs/GaAs(111)A system, we employ the computational models including 2D growth with MD (2D-MD), 2D growth with SFT (2D-SFT), and the 2D coherent growth (2D-coherent). Figure 4 shows the calculated energy difference ΔE_{MD} between the 2D-coherent and the 2D-MD as a function of layer thickness h [42]. Here, the MD core with five- and seven-member rings (5/7 core) are inserted at the interface between InAs and GaAs(111), as shown in Figure 4a. The 5/7 core is often found in transmission electron microscopy (TEM) observations [43], and is recognized to be the stable core structure in the MD formation energy calculations for compound semiconductors [44–46]. This reveals that the ΔE_{MD} changes its sign from positive to negative at $7 \leq h \leq 8$ ML, where the strain in InAs thin layers is relaxed to stabilize the 2D-MD. The critical layer thickness $7 \leq h \leq 8$ ML for the MD generation agrees well with its value of h about 7 ML estimated from People-Bean's formula to minimize energy in the entire crystal at thermodynamic equilibrium [47]. Using the calculated results shown in Figure 4a, the parameter values used in Equations (4) and (5) are determined to be $M = 2.63 \times 10^{10}$ (N/m^2) and $E_d = 0.675$ (eV/Å).

The calculated energy difference ΔE_{SFT} between the 2D-coherent and the 2D-SFT as a function of layer thickness is shown in Figure 5 [42] along with the schematic of the SFT consisting of the face with stacking-fault and the ridge corresponding to stair-rod dislocation along the (110) direction, similarly to the SFT in Si [48]. The ΔE_{SFT} with negative at $h \geq 4$ ML suggests that the SFT formation acts as a strain relaxation mechanism near the surface as well as the MD formation at the interface in InAs/GaAs(111)A system. The ΔE_{SFT} results from the competition between energy profit in the face region and energy deficits in the face and the ridge regions. The face region dramatically decreases the system energy due to strain relaxation, inducing upward displacements of atoms in the SFT that overwhelm the energy deficit due to the stacking-fault formation. In the ridge region, however, the stair-rod dislocation increases system energy due to its energetically unfavorable dimers along the ridge line, shown in Figure 5b. The calculated results shown in Figure 5b approximately determine the parameter value of $E_{SFT} = 0.014h - 0.0011h^2$ (eV) as a function of layer thickness h.

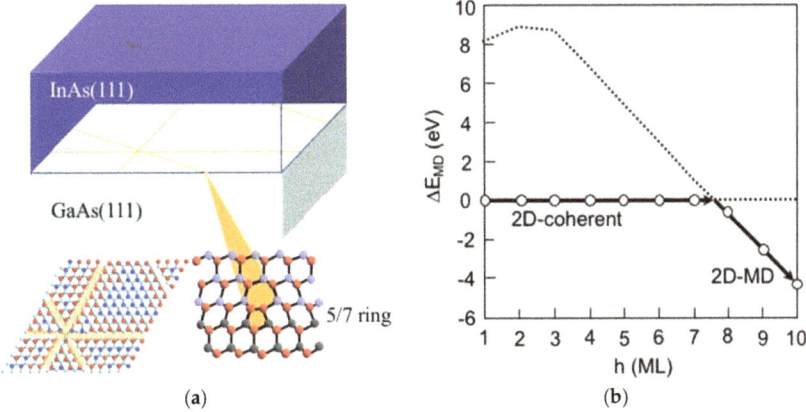

(a)

(b)

Figure 4. (**a**) Schematic of the 2D growth with dislocation (5/7-ring core structure) network formation at the interface; (**b**) Calculated energy difference between the 2D coherent growth (2D-coherent) and the growth with MD (2D-MD) as a function of layer thickness h.

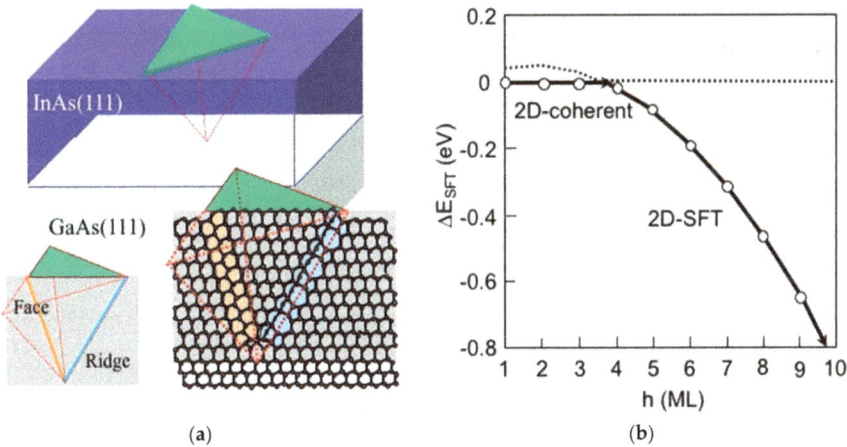

(a)

(b)

Figure 5. (**a**) Schematic of the 2D growth with formation of stacking-fault tetrahedron (SFT) consisting of ridge and face; (**b**) Calculated energy difference between the 2D coherent growth (2D-coherent) and the growth with SFT (2D-SFT) as a function of layer thickness h.

3.4. Hetero-Epitaxial Growth of InAs on GaAs(111)A

Figure 6 depicts the calculated free energy differences ΔF between the 2D-coherent and various growth modes such as the 2D-MD and the 2D-SFT as a function of layer thickness h. It should be noted that the ΔF for the 2D-MD becomes negative at h about 6 ML different from $7 \leq h \leq 8$ ML shown in Figure 4b. This means that the first isolated dislocation formation occurs at $h \sim 6$ ML, and the dislocation spacing l gradually decreases according to $l = l_0/(1 - h_c{}^{MD}/h)$ approaching the average dislocation spacing l_0 (=58.76 Å) with increase of h, where $h_c{}^{MD}$ is estimated by $E_d/(M\varepsilon_0{}^2 l_0)$ [21]. Figure 6 implies that the InAs growth on the GaAs(111)A proceeds along the lower energy path from the 2D-coherent ($h \leq 4$ ML) to the 2D-MD ($h \geq 7$ML) via the 2D-SFT (4 ML $\leq h \leq 7$ ML). This is consistent with STM observations, where faulted triangle domains containing stacking-faults appear with the MD network at 5 ML on the InAs/GaAs(111) [21]. Furthermore, a similar process was found in molecular dynamics simulations for (111)-oriented heteroepitaxial Al films forming a local disorder zone like the SFT near surface layers followed by the MD nucleation [49]. Experimental results, however, indicate that strain is gradually relieved beyond 1–2 ML, contradicting our calculated results with 4 ML [21,22]. This discrepancy can be interpreted by considering the fact that our computational model is not optimized for 2D-SFT. The 14 × 14 unit cell used in this study is not set for the SFT, but for the MD with geometrically optimized dislocation spacing. If the SFT spacing is optimized, the ΔE_{SFT} becomes lower to give smaller layer thickness for the SFT formation. Moreover, employing the MD consisting of faulted and unfaulted domains observed by STM [21], further strain relaxation may occur at smaller layer thickness [23]. Consequently, our calculated results suggest that the first strain relaxation occurs near the surface due to the SFT formation before the MD formation at the interface. Assuming $\gamma = 42$ (meV/Å^2) for the InAs(111)A, $\beta = 0.084$, and $\alpha = 0.748$ extracted from the previously reported results for the InAs/GaAs(001) [50], the calculated free energy difference ΔF between the 2D-coherent and the SK-coherent is also shown in Figure 6. Although the ΔF for the SK-coherent becomes negative at 5 ML $\leq h \leq 6$ ML that is energetically competitive with the 2D-MD, the SK-coherent does not appear due to the 2D-SFT preceding the SK-coherent.

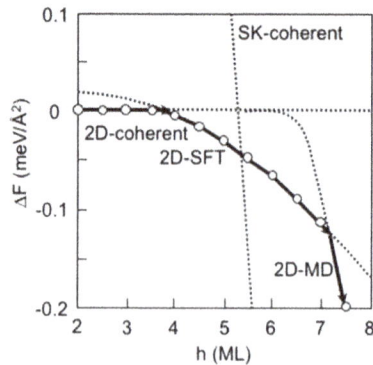

Figure 6. Calculated free energy differences ΔF for various growth modes for the InAs/GaAs(111)A as a function of layer thickness h.

3.5. Surface Reconstructions on InAs(001) WL

The RHEED and the STM observations for InAs on GaAs(001) have clarified that the surface changes its structure from initial GaAs(001)-c(4 × 4)α surface to final InAs(001)-(2 × 4)α2 via InAs(001)-(1 × 3)/(2 × 3) with increase of InAs coverage [26]. On the basis of these results, the surface structural change is schematically shown in Figure 7 obtained by counting the number of In and As atoms deposited on the initial GaAs(001)-c(4 × 4)α with Ga-As surface dimers. The intermediate InAs(001)-(2 × 3) surface with In$_{0.375}$Ga$_{0.625}$As interface reasonably appears after 0.625 ML InAs deposited on the GaAs(001)-c(4 × 4)α. This is consistent with experimental findings where 2/3 ML InAs deposition creates the (2 × 3) surface with In$_{1/3}$Ga$_{2/3}$As [26,51–53]. Further deposition with 0.708 ML InAs realizes the (2 × 4) surface that is also consistent with the RHEED observations where the (2 × 4) appears supplying ~1.3–1.4 ML InAs on the GaAs(001)-c(4 × 4)α [26]. Here it should be noted that the desorption of 0.375 ML As is indispensable to realize the (2 × 4) surface during the structural change from the (2 × 3) to the (2 × 4). Figure 8a shows the calculated surface phase diagram for the InAs(001)-(2 × 3) and (n × 3) WL surfaces (n = 4, 6, and 8) as functions of T and p_{As4}. Here the (2 × 3) surface is fully covered by As-dimers on the surface while one As-dimer is missing every n As-dimers on the (n × 3) surfaces to approach satisfying the ECM at n = 8. This implies that the (2 × 3) surface is unstable at growth conditions, since desorption energy of As-dimer on the (2 × 3) surface is very small such as 1.12 eV to easily desorb from the WL surface to change its structure from the (2 × 3) to the (8 × 3) even at low temperatures. Figure 8b shows the calculated surface phase diagram for the InAs(001)-(2 × 4) WL surfaces as functions of T and p_{As4}. The phase boundary between the (001)-(2 × 4)α2 and the (001)-(2 × 4)β2 is ranging from 470 to 600 K. The stable temperature range of these structures is consistent with experimental conditions during MBE growth [52,54]. This reveals that (2 × 4)α2 is stable at the conventional growth conditions such as $T\sim$700–750 K and $p_{As4}\sim10^{-7}$–10^{-6} Torr.

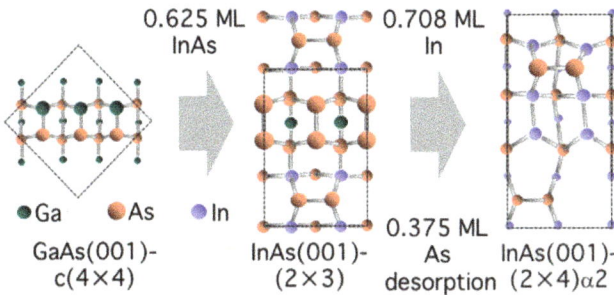

Figure 7. Schematic of structural change during MBE growth from initial GaAs(001)-c(4 × 4) to InAs(001)-(2 × 4)α2 via InAs(001)-(2 × 3). Green, blue, and orange atoms denote Ga, In, and As, respectively.

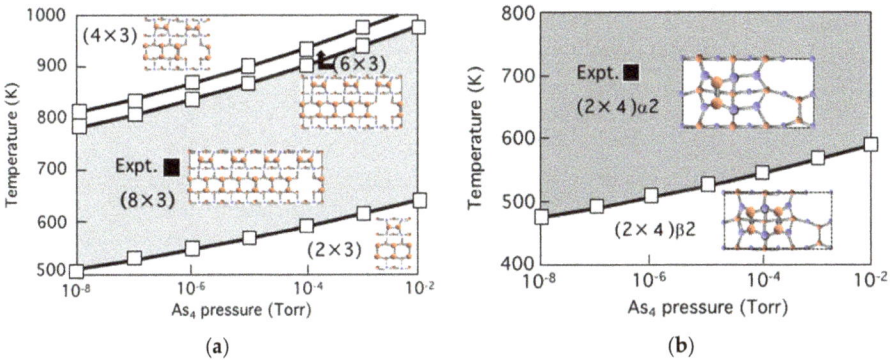

Figure 8. Calculated surface phase diagrams for InAs(001) WL surfaces such as (**a**) (2 × 3) and (**b**) (2 × 4) as functions of temperature and BEP of As$_4$. Black square denotes the t conventional MBE growth conditions.

3.6. Growth Process on InAs(001) WL

Calculated adsorption-desorption boundaries of In on the InAs(001)-(n × 3) WL surfaces is shown in Figure 9a as functions of T and p_{In}. This indicates that In adsorption does not occur on the (2 × 3) but is allowed on the (4 × 3), the (6 × 3), and the (8 × 3) at growth conditions. This is because the adsorption energy for In atom is very large such as −1.68 eV on the (2 × 3) in contrast with −3.87 eV on the (4 × 3), −3.66 eV on the (6 × 3), and −3.33 eV on the (8 × 3). This is because the stable adsorption site of In on the (2 × 3) is around the center position between upper As-dimer and lower As-dimer quite different from the missing As-dimer site on the (n × 3) as shown in Figure 9a. Figure 9b shows the calculated adsorption-desorption boundaries for As dimer on the stable In-adsorbed InAs(001)-(n × 3) WL surfaces as functions of T and p_{As4}. This implies that In adsorption does not induce As desorption, indispensable for InAs growth to change the surface structure from the (2 × 3) to the (2 × 4)α2, on the (4 × 3) and (6 × 3). Although As dimer desorption can be found on the (8 × 3) shown in Figure 9b schematically, it should be noted that newly appeared missing dimer is quickly occupied by In adatom to form the surface structure equivalent to the In-adsorbed (4 × 3) also schematically shown in Figure 9b. Moreover, it is found that further In adsorption also no longer occurs to prevent InAs growth on these (n × 3) WL surfaces.

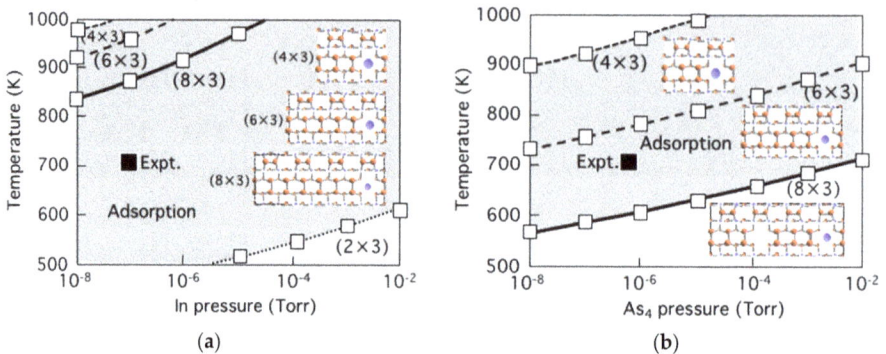

Figure 9. Calculated adsorption-desorption boundaries for (**a**) In atom and (**b**) As-dimer on the InAs(001)-(n × 3) WL surfaces as functions of temperature and BEP of In and As₄, respectively.

According to the results shown in Figure 9, another growth processes should be considered to allow In adsorption with simultaneous As desorption. On the (4 × 3), for example, it is found that In-As dimers stably reside on the (4 × 3) instead of As dimers at the conventional growth conditions [34]. Desorption of As dimer after forming In-As dimer in the missing dimer region is due to the fact that desorption energy of As dimer (1.44 eV) dramatically decreases from 2.84 eV on the initial (4 × 3) and 2.38 eV on the In-adsorbed (4 × 3). This is because adsorption of In-As dimer with bond length ~1.5 Å larger than As dimer largely deforms As dimer to destabilize it on the (4 × 3). Furthermore, In-As dimer adsorption replacing As dimer increases the number of In-As interatomic bond energetically more favorable than As-As interatomic bond to stabilize the (4 × 3). Figure 10 summarizes the sequence between for As dimer desorption and In-As dimer formation reflecting the change in the number of electrons in the surface dangling bonds (EC) on the (n × 3). It should be noted that In adsorption on the (8 × 3) induces As-dimer desorption to enhance further In adsorption to form In adsorbed (4 × 3) with subsequent As adsorption to form In-As dimer on the (4 × 3). Therefore, considering this In-As dimer formation process, the (8 × 3) is equivalent to the (4 × 3). Finally, the (4 × 3) consisting of three In-As dimers and one missing dimer, and the (6 × 3) with three In-As and two As dimers and one missing dimer appear on the InAs(001) WL surface. If we assume the existence ratio of the (4 × 3) and the (6 × 3) units satisfying 0.375 ML desorption of As to realize the (2 × 4)α2 as shown in Figure 7, the calculated $f_{(4×3)} = 0.70$ and $f_{(6×3)} = 0.30$ can be favorably compared with $f_{(4×3)} = 0.83$ and $f_{(6×3)} = 0.17$ estimated by directly counting the number of the units from Figure 2 in Ref. 27. Therefore, the (4 × 3) with three In-As dimers is most likely to appear among the (n × 3) WL surfaces during MBE growth. In order to realize the (2 × 4) α2 on the (4 × 3), further 0.5 ML In adsorption is necessary. It should be noted, however, In atoms are no longer incorporated on the (4 × 3) unless the strain is relaxed similar to the (2 × 4) α2 [33]. These results suggest that strain relaxation might occur at InAs coverage less than about 0.9 ML, contradicting well-known SK growth mode, to continue InAs growth.

Surfaces	(4×3)		(6×3)		(8×3)	
EC without In		-2		-1		0
EC with In		+1		+2		+3
As-dimer desorption	900 - 1060 K		720 - 880 K		570 - 700 K	
In-As dimer formation					(4×3)	
Resultant surface		3 In-As		3 In-As 2 As-As		6 In-As

Figure 10. Summary of the growth processes on the (*n* × 3) surfaces obtained in this study.

3.7. Strain Relaxation on InAs(001) WL

Figure 11 shows the calculated energy *E* for the 2D-coherent, the 2D-MD with 5/7-ring core, and 2D-MD with 8-ring core as a function of layer thickness *h* along with computational model used in this study. Here, the MD cores are inserted at the interface between InAs and GaAs(001) similarly to the InAs/GaAs(111)A. This reveals that the energy difference ΔE_{MD} between the energies for the 2D-coherent and the 2D-MD with 5/7-core changes its sign from positive to negative at *h*~1.3 ML, where the strain in InAs thin layers is relaxed to stabilize the 2D-MD with 5/7-core [55]. Using the calculated results shown in Figure 11b, the parameter values used in the free energy formula are determined to be $M = 5.72 \times 10^{11}$ (N/m²) and $E_d = 1.51$ (eV/Å). These values easily lead the layer thickness for the first MD generation h_c^{2D-MD} using the relationship of $h_c^{2D-MD} = E_d/(M\varepsilon_0^2 l_0)$ [14]. The calculated results suggest that the first MD formation occurs at 0.5 ML and the MD spacing l^{2D-MD} gradually decreases with increase of the layer thickness according to $l^{2D-MD} = l_0/(1 - h_c^{2D-MD}/h)$ to effectively relax the strain accumulated in the InAs layers. Therefore, the MD formation is one of possible mechanisms of eliminating lattice strain to proceed the growth on the InAs(001)-(*n* × 3) WL surfaces.

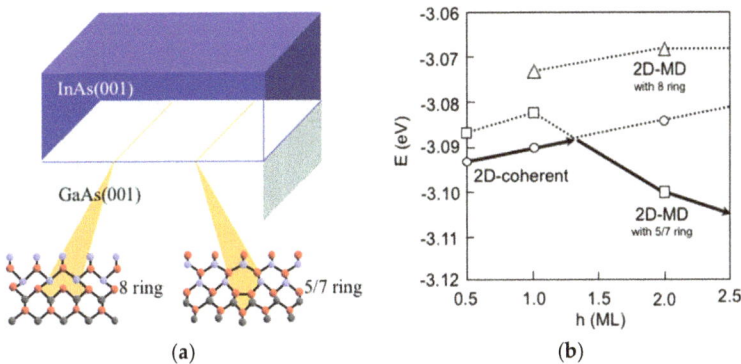

Figure 11. (a) Schematic of the 2D growth with dislocation (8-ring core and 5/7-ring core) formation at the interface; (b) Calculated energy for the 2D coherent growth (2D-coherent) and the 2D growth with MD (2D-MD with 8-ring and 2D-MD with 5/7-ring) as a function of layer thickness *h*.

3.8. Hetero-Epitaxial Growth of InAs on GaAs(001)

On the basis of physical quantities $\gamma = 51$ (meV/Å^2) for the InAs(001), $\beta = 0.076$, and $\alpha = 0.188$ extracted from the previously reported results for the InAs/GaAs(001) [50], Figure 12 shows the calculated free energy differences ΔF between the 2D-coherent and various growth modes such as the 2D-MD and the SK-coherent as a function of layer thickness h. It is found that the ΔF for the 2D-MD becomes negative at 0.46 ML, while negative ΔF appears at 0.75 ML for the SK-coherent. This suggests that the initial 2D coherent growth mode changes into the 2D growth mode with MD formation. Therefore, strain relaxation occurs at early growth stage before the SK island formation. This is not consistent with well-known QD formation mechanism due to the SK island formation but is consistent with the fact that strain accumulated in InAs(001)-($n \times 3$) WL should be relaxed at InAs coverage less than about 0.9 ML to continue InAs growth as pointed out in Section 3.6. Although further careful investigations incorporating surface reconstructions as a submonolayer phenomenon are necessary for confirming the strain relaxation for the InAs/GaAs(001), the calculated results for the InAs/GaAs systems suggest that the strain relaxation due to the MD formation to produce the 2D growth is competitive with the SK-island formation inducing the 3D growth during MBE growth.

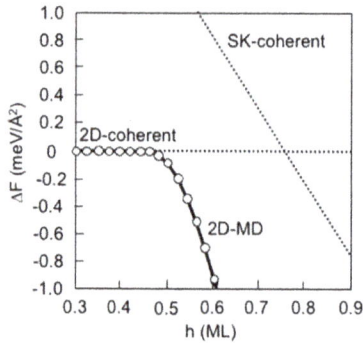

Figure 12. Calculated free energy differences ΔF for various growth modes for the InAs/GaAs(001) as a function of layer thickness h.

3.9. Quantum Dot Formation of InAs

Using Equations (5) and (7), the boundary between the 2D-MD and the SK-coherent is simply described as $\beta/\alpha = 1/(2\gamma)(E_d/l_0)$. Using the MD formation energy $E_d = 0.675$ (eV/Å) for InAs/GaAs(111)A and 1.51 (eV/Å) for InAs/GaAs(001) obtained by our empirical potential calculations, Figure 13 shows the growth mode boundaries for the InAs/GaAs system depending growth orientations as functions of β/α and surface energy γ. This reveals that the stable region of the 2D-MD in the InAs/GaAs(111)A is larger than that in the InAs/GaAs(001) because of its smaller MD formation energy. Furthermore, the larger the β/α and γ, the more favourable the 2D-MD. This is because the energy increase in surface energy $\gamma\beta$ tends to overwhelm the decrease in strain energy $-M\alpha\varepsilon_0^2 h/2$ due to SK-island formation. According to these facts, the surface energy γ in addition to E_d is a crucial factor for determining the growth mode, since the change in γ is related not only to γ itself but also to β/α through $\gamma\beta$. Figure 13 also includes the boundary denoted by dotted line for the InAs/GaAs(110) with $E_d = 0.96$ (eV/Å) obtained by ab initio calculations [14]. The InAs/GaAs(110) system has been extensively examined using RHEED, TEM, and STM for a wide range of film thickness where the growth follows a layer-by-layer mode irrespective of thickness and strain relaxation occurs solely by the formation of MD [20,56]. However, self-assembled InAs QD has been realized on GaAs(110) using MBE with buffer layer insertion [57] and MOVPE with optimizing growth temperature [58], where the strain reduction is considered to be a crucial origin of the QD formation. This novel QD formation can be interpreted by considering the change in γ. It is known that the strain reduction lowers the surface energy by 10–20 (meV/Å^2) [50]. Assuming $\beta/\alpha = 0.2$ and $\gamma = 50$ (meV/Å^2) (denoted by closed diamond in Figure 13), the 2D-MD is favourable without strain relaxation while decrease in γ due to strain relaxation with the buffer-layer insertion easily changes favourable growth mode from the 2D-MD to the SK-coherent found in the (110). Although the dependence of β/α and γ on strain relaxation should be rigorously investigated using ab initio calculations, Figure 13 gives an insight on the growth modes among different oriented InAs/GaAs systems such that (001), (110), and (111)A favours the 3D growth, the 2D and the 3D growths depending on the strain relaxation, and the 2D growth, respectively. Consequently, the QD formation on the InAs/GaAs can be qualitatively interpreted by considering γ and E_d.

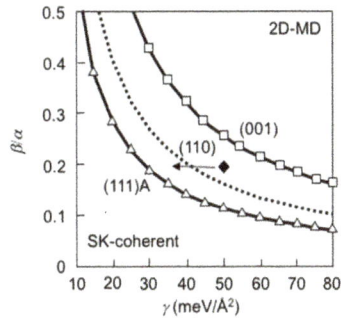

Figure 13. Calculated growth mode boundaries as functions of β/α and γ for the InAs/GaAs systems with different orientations. Open square and open triangle denote the results for the (001)- and the (111)A-oriented InAs/GaAs, respectively. The growth mode boundary (dotted line) and the data (closed diamond) for the InAs/GaAs(110) are also shown in this figure.

4. Growth Processes of InP NWs

Semiconductor NWs have great potential to understand the fundamental roles of reduced dimensionality and size in optical, electrical, and mechanical properties and their wide range of potential applications in nanoscale devices. So far, semiconductor NWs have been fabricated by various growth methods such as MOVPE [59,60], MBE [61–63], and laser assisted catalytic growth [64–66]. In particular, InP NWs are intriguing targets in fiber optic communications, high-speed electronic applications, and photovoltaic devices, such as p–n junction light-emitting diodes [67], field effect transistors [67], one-dimensional polarization-sensitive photodetectors [65], solar cells [68], and so on. It has been reported that semiconductor NWs grown in the [111] direction have some structural characteristics different from those of bulk crystals: semiconductor NWs often incorporate the WZ segments as stacking faults in the zinc blende (ZB) structure [69–83], which is called a rotational twin structure, as shown in Figure 14. It has also been suggested that rotational twins significantly affect the electronic properties of the NWs: discontinuity of the electron wave function occurs at the stacking fault, leading to a reduced mobility of carriers [84]. Therefore, the structural stability of InP NWs has been intensively discussed because of the small energy difference between the ZB and WZ structures in bulk form [85].

Despite these experimental findings, twinning formation processes in InP NWs have not been clarified from theoretical viewpoint. The relative stability between the ZB and WZ structures in InP NWs have been clarified using empirical interatomic potential calculations: the WZ structure can be stabilized when the contribution of NW side facets to the cohesive energy is prominent, and the WZ structure is more stable than the ZB structure for NWs with small diameters [86,87]. Glas et al. have theoretically suggested that the growth processes affect the formation of the WZ structure in

NWs [88,89]. However, elementary processes such as adsorption and desorption behaviors of In and P atoms at the growth forefront have never been studied at present. In particular, it is still unclear how the atoms are adsorbed at the lattice sites of the WZ structure at the top layer of NWs under growth conditions. Another important characteristics of InP NWs is that the plane direction of substrates depends on experimental methods. InP NWs fabricated by selective-area MOVPE (SA-MOVPE) are grown on the InP(111)A substrates [90,91]. On the other hand, InP(111)B substrates are used for InP NWs synthesized by vapor–liquid–solid (VLS) growth [70,85,92]. Mohan et al. have reported that InP NWs with the pure WZ structure are synthesized by the SA-MOVPE [90]. Kitauchi et al. have revealed that the crystal structure of InP NWs grown by the SA-MOVPE depends on growth conditions such as temperature and partial pressure of the supplied gas [91]. InP NWs grown for high temperature and low P pressure take the pure WZ structure, whereas those with rotational twins are fabricated for low temperature and high P pressure. Furthermore, these results suggest the formation of the WZ crystal phase even in (111)A planar layers. These experimental results imply that the growth process of InP layers on InP(111)A surface in itself affects the crystal structure of InP NWs. In spite of these experimental findings, temperature and pressure dependence of structural stability of InP layers on InP(111)A surface still remains unclear. In the SA-MOVPE, it can be considered that not only side facets but also the growth mechanism of InP layers on the surface is quite important for determining their crystal structure. However, little is known about the elemental growth processes of InP layers on InP(111) surface. The knowledge for adsorption and desorption behavior of In and P atoms on InP(111) surfaces is lacking.

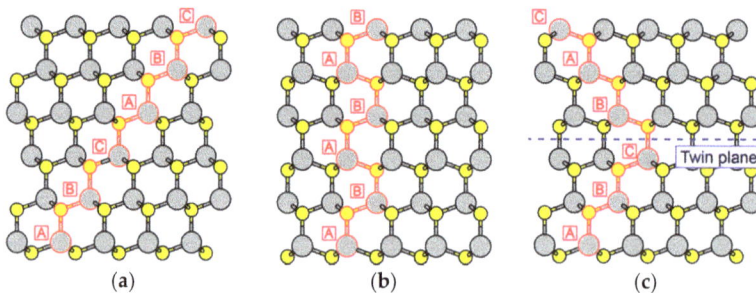

Figure 14. Schematics of crystal structures of III-V compound semiconductors. Large and small circles represent cations and anions, respectively. Stacking sequences of (a) zinc blende (ZB) and (b) wurtzite (WZ) structures are $\cdots ABCABC \cdots$ and $\cdots ABAB \cdots$, respectively, while (c) rotational twins are formed in semiconductor nanowires (NWs). Dashed line denote a twin plane.

Furthermore, Kitauchi et al. have also revealed that InP NWs are composed of $\{1\bar{1}00\}$ side facets (see Figure 15a) for high temperature (660 °C) and low V/III ratio, where the lateral growth does not occur. On the other hand, the growth in the lateral direction takes place for low temperatures (600 °C) and high V/III ratio, and InP NWs with lateral growth consist of $\{11\bar{2}0\}$ side facets (see Figure 15b). It has also been reported that InP NWs consisting of $\{11\bar{2}0\}$ side facets without lateral growth are fabricated for intermediate temperature (625 °C) and V/III ratio. A similar structural characteristic has been found in InP NWs grown by both the SA-MOVPE and the VLS growth. InP NWs grown by VLS growth consist of $\{1\bar{1}00\}$ ($\{11\bar{2}\}$ in cubic crystal) or {111} side facets without the lateral growth [76,79,84]. In addition, it has been reported that GaAs NWs fabricated by VLS growth are composed of $\{11\bar{2}0\}$ ($\{1\bar{1}0\}$ in cubic crystal) side facets with the lateral growth [78]. However, temperature and pressure dependence of the shapes in semiconductor NWs still remains unclear.

In this chapter, focusing on the top layer of growing NWs, we discuss the adsorption and desorption behaviors of In and P atoms during SA-MOVPE, taking account of the effects of NW side facets. By using our ab initio based approach incorporating growth conditions such as temperature

T and pressure p, [8,9] we determine whether the adatoms are adsorbed at the lattice site of the ZB or the WZ structure on the top layer. We also discuss the growth processes of InP epitaxial layers on InP(111)A P-stabilized surface. In particular, adsorption–desorption behavior on InP(111)A surface is calculated by our ab initio-based approach. We demonstrate temperature and pressure dependence of structural stability of InP layers, leading to the formation of rotational twins at low temperature and high V/III ratio conditions. Furthermore, focusing on the side facets of InP NWs grown by the SA-MOVPE, we investigate the initial growth processes on both $\{1\bar{1}00\}$ and $\{11\bar{2}0\}$ side facets surfaces. Assuming the WZ structure is assumed because the WZ structure is feasible compared with the ZB structure in InP NWs. In particular, adsorption–desorption behavior on these facets is calculated by our first-principles-based approach which incorporates the growth conditions such as temperature and pressure. We compare the growth rates of $(1\bar{1}00)$ and $(11\bar{2}0)$ surfaces using Monte Carlo (MC) simulations and clarify temperature and pressure dependence of the shapes in InP NWs.

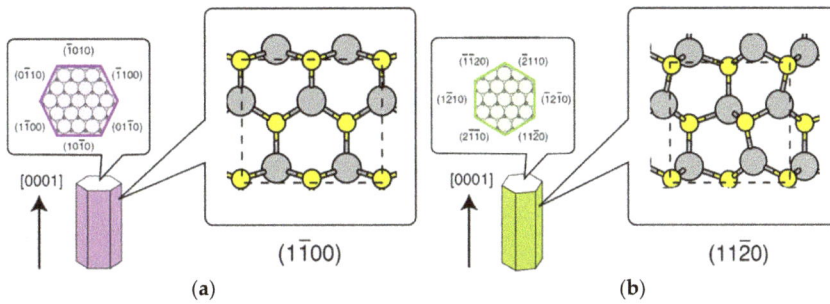

Figure 15. Schematics of InP NWs consisting of (a) $\{1\bar{1}00\}$ and (b) $\{11\bar{2}0\}$ side facets. Cross-sectional views and side views of NW facets are also shown. Dashed red lines in the top views represent unit cells. Large gray and small yellow circles represent In and P atoms, respectively.

4.1. Nanowire Growth on the Basis of Classical Nucleaton Theory

In this section, we briefly explain the conventional nanowire growth model using classical nucleation theory, which have been successfully applied for NWs synthesis with WZ structure by the VLS growth [79,88,89,93]. During the VLS growth, we assume a two-dimensional nucleus with semicircles shape at the interface between top layer of NW and liquid catalysis, and NWs are formed according to the formation of two dimensional nuclei. The Gibbs free energy of a nucleus with ZB structure $\Delta G(r)$ is given by

$$\Delta G(r) = -\pi r^2 \Delta \mu_{\text{liquid-solid}}/2 + \pi r^2 \sigma_{\text{int}}/2 + r\Gamma_{\text{step}}, \tag{19}$$

where r is radius of nucleus with semicircle shape, $\Delta\mu_{\text{liquid-solid}}$ is chemical potential difference (per unit area) between liquid and solid phases, σ_{int} is the interface energy per area between nucleus and nanowire top layer, and Γ_{step} is the contribution of the step energy caused by a two-dimensional nucleus. According to Equation (19), the critical radius for nucleation r^* satisfy the condition, $\partial G(r)/\partial r = 0$. Therefore, r^* and activation energy for nucleation ΔG^* are written as

$$r^* = \Gamma_{\text{step}}/\pi(\Delta\mu_{\text{liquid-solid}} - \sigma_{\text{int}}), \tag{20}$$

$$\Delta G^* = \Gamma_{\text{step}}^2/(2\pi\Delta\mu_{\text{liquid-solid}}). \tag{21}$$

On the other hand, the activation enegy for nulceation with WZ structure, ΔG^*_{WZ}, is also written as

$$\Delta G^*_{\text{WZ}} = \Gamma^{WZ}_{\text{step}}^2/\{2\pi\Delta\mu_{\text{liquid-solid}} - \sigma^{WZ}_{\text{int}}\}\}, \tag{22}$$

where $\Gamma^{WZ}{}_{step}$ is the contribution of the step energy caused by two dimensional growth with WZ structure, and $\sigma^{WZ}{}_{int}$ is the interface energy for WZ structure. The WZ structure is formed satisfying $\Delta G^* < \Delta G^*{}_{WZ}$, and the boundary between ZB and WZ structure in NWs are obtained using the step energy difference $\Delta\Gamma_{step} = \Gamma_{step} - \Gamma^{WZ}{}_{step}$, as

$$(1 - \Delta\Gamma_{step}/\Gamma_{step})^{-2} = (1 - \sigma^{WZ}{}_{int}/\Delta\mu_{liquid\text{-}solid})^{-1} \qquad (23)$$

Using this relationship, we find that NWs with WZ structure is stabilized when $\Delta\Gamma_{step}$ and/or $\Delta\mu_{liquid\text{-}solid}$ is large. Figure 16 shows the estimated phase diagram for crystal structure of InP NWs as functions of $\Delta\mu_{liquid\text{-}solid}$ and $\Delta\Gamma_{step}/\Gamma_{step}$ using the energy difference between ZB and WZ structures in bulk InP (6.8 meV/atom) [86] for $\sigma^{WZ}{}_{int}$. Actually, this figure manifests that the WZ structure is stabilized if $\Delta\mu_{liquid-solid}$ is large and $\Delta\Gamma_{step}/\Gamma_{step}$ takes positive value. Similar results have been obtained if we take more sophisticated nucleation model into account to evaluate the free energies [88,89,91]. These results suggest that the stability of the step and driving force for nucleation during the VLS growth are crucial for the formation of NWs with WZ structure.

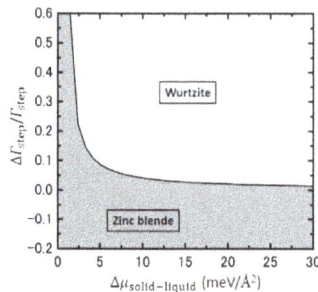

Figure 16. Phase diagram for crystal structure of InP NWs as functions of chemical potential difference between liquid and solid phases $\Delta\mu_{liquid\text{-}solid}$ and normalized step energy difference between ZB and WZ structures $\Delta\Gamma_{step}/\Gamma_{step}$, using the energy difference between ZB and WZ structures in bulk InP (6.8 meV/atom) [87] for interface energy per area between nucleus and nanowire top layer for WZ structure $\sigma^{WZ}{}_{int}$. White and grey areas denote the stable regions for WZ and ZB structures, respectively.

4.2. Effect of Side Facets on Adsorption–Desorption Behaviors at Top Layers in InP NWs

In order to clarify the effects of NW side facets during the SA-MOVPE, we have to consider adsorption-desorption behavior on two kinds of growth regions. One is a central area on the top layer of InP NWs, where we adopt the (2×2) surface with P trimer as a calculation model. It is expected that the the (2×2) surface with P trimer appears because it satisfies the ECM [38,39], and previous theoretical calculations also support the stabilization of the P trimer on the InP(111)A surface under P-rich conditions [94]. Since the InP NWs are grown under high P pressures [91], this surface could be a possible surface reconstruction during the growth of NWs. In addition, investigating adsorption–desorption behavior on the (2×2) is quite important since the adsorption site of In atoms determines the crystal structure. Another growth area we have to consider is a side area, where the calculation model of the side area consists of both (111)A and $(1\bar{1}0)$ surfaces. Here, we assume the (2×2) surface with P trimer on the top surface of the side area. For the central area of NWs, the calculated adsorption energy of P atom is a positive value (2.26 eV). This indicates that the adsorption of a single P atom does not occur even at 0 K. Thus, it is reasonable to consider the adsorption of a P atom on an In-preadsorbed surface, which corresponds to the self-surfactant effect [40,41]. We also find that the adsorption of a P atom with one and two In atoms is also inhibited under the typical growth conditions. The adsorption occurs only below 200 °C Figure 17 shows the calculated adsorption of a P atom with three In atoms (3 In-P), which has the highest adsorption probability [90].

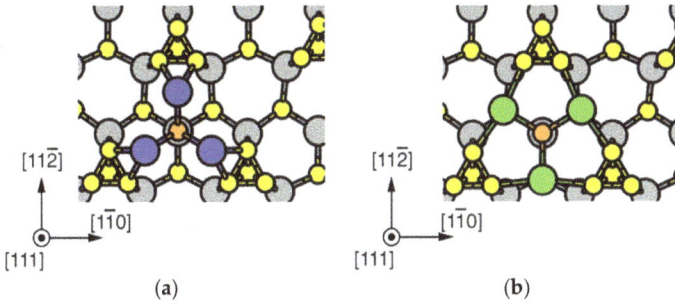

Figure 17. Top views of (**a**) 3In-P occupying the ZB lattice sites (3In-ZB-P) and (**b**) 3In-P occupying the WZ lattice sites (3In-WZ-P) on the InP(111)A surface. Large and small circles represent In and P atoms, respectively. Blue and green circles denote the In adatoms that occupy the lattice sites of the ZB and WZ structures, respectively. Orange circles represent the P adatoms. Reprinted with permission from [95]. Copyright (2011) by the Japan Society of Applied Physics.

This kind of adsorption process is similar to those on the GaAs(111)A surface, in which the surface with three Ga and one As atom appears in the growth process on the GaAs(111)A-(2 × 2) surface [96]. The calculated adsorption energy difference between the three In atoms occupying the ZB lattice sites (3In-ZB-P shown in Figure 17a) and those occupying the WZ lattice sites (3In-WZ-P shown in Figure 17b) is very small (0.07 eV). Figure 18 displays the calculated phase diagrams of the 3In-ZB-P and 3In-WZ-P adsorbed surfaces, respectively, as functions of temperature and P_2 pressure. The pressure of In is taken to be 3.3×10^{-3} Torr, which corresponds to a typical growth condition [90,91]. These surface phase diagrams for the adsorption P atoms exhibits little difference in the adsorption probability between 3In-ZB-P and 3In-WZ-P. The small energy difference is due to similar bonding configurations, where both 3In-ZB-P and 3In-WZ-P form nine In-P bonds regardless of the adsorption site. These results imply that the WZ structure as well as the ZB structure can be formed when nucleation occurs away from the NW side facets.

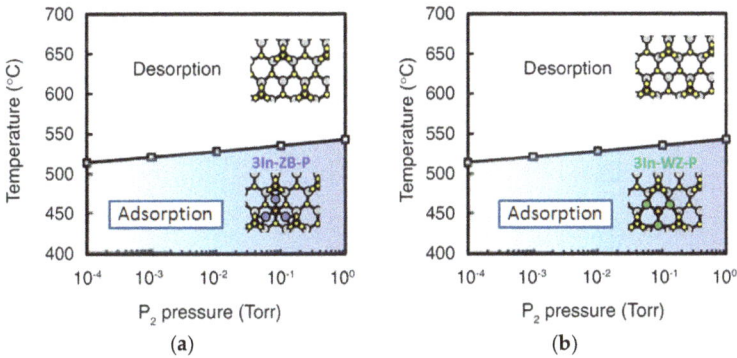

Figure 18. Calculated phase diagrams of (**a**) 3In-ZB-P and (**b**) 3In-WZ-P adsorbed surfaces as functions of temperature and P_2 pressure. In pressure is taken to be 3.3×10^{-3} Torr. Colored regions correspond to the atom-adsorbed regions. Reprinted with permission from [95]. Copyright (2011) by the Japan Society of Applied Physics.

The calculations for the adsorption–desorption behavior in the side area demonstrate that the calculated adsorption energy is positive though the adsorption of a P atom at its lattice site is necessary.

Therefore, we can expect the self-surfactant effect [40,41] in the same way as in the central area. The calculations for the adsorption of a P atom with In atoms demonstrate that the adsorption probability of two P atoms with four In adatoms has the highest value. Figure 19 shows the geometries of two P adatoms with four In adatoms occupying the ZB (4In-ZB-2P) and WZ (4In-WZ-2P) lattice sites near the side facet, respectively. The adsorption energy of 4In-ZB-2P is found to be higher than that of 4In-WZ-2P by 0.46 eV. There is a two-coordinated In adatom in 4In-ZB-2P, shown in Figure 20a, whereas all the In adatoms in 4In-WZ-2P are three-coordinated. The three-coordinated In adatom at the side facet in 4In-WZ-2P originates from the formation of an In-P bond with the P atom in the second layer, as shown in Figure 20b. Therefore, the formation of the three-coordinated In adatom results in the stabilization of the surface with 4In-WZ-2P. Figure 21 shows the calculated phase diagrams for the adsorption of the surface with 4In-ZB-2P and 4In- WZ-2P, respectively. These surface phase diagrams suggest that the stable temperature range of the surface with 4In-WZ-2P is higher than that with 4In-ZB-2P by 25 K. It is thus expected that the WZ structure is preferentially formed when the nucleation occurs near the NW side facets.

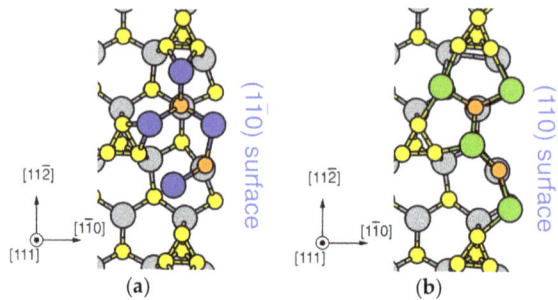

Figure 19. Top views of (**a**) 4In-2P occupying the ZB lattice sites (4In-ZB-2P) and (**b**) 4In-2P occupying the WZ lattice sites (4In-WZ-2P) near the side facet. There is a two-coordinated In adatom indicated by the red circle in 4In-ZB-2P, whereas all the In adatoms in 4In-WZ-2P are three-coordinated. The three-coordinated In adatom at the side facet in 4In-WZ-2P is due to the formation of In-P bonds with the P atom in the second layer. Reprinted with permission from [95]. Copyright (2011) by the Japan Society of Applied Physics.

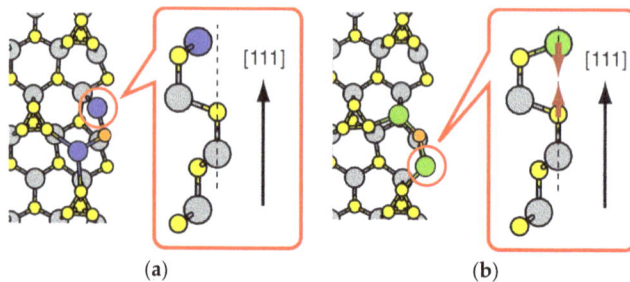

Figure 20. Top views of (**a**) 4In-ZB-2P and (**b**) 4In-WZ-2P adsorbed surfaces in the side area. There is a two-coordinated In adatom indicated by the red circle in 4In-ZB-2P, whereas all the In adatoms in 4In-WZ-2P are three-coordinated. The three-coordinated In adatom at the side facet in 4In-WZ-2P is due to the formation of In-P bonds with the P atom in the second layer. Reprinted with permission from [96]. Copyright (2011) by the Japan Society of Applied Physics.

To determine the predominant area for the nucleation, we estimate the difference in adsorption probability between the central and side areas. By comparing the phase diagram shown in Figure 18 with that in Figure 21b, it is concluded that the difference in the adsorption probability between the central and side areas to be negligible, so that adsorption occurs in both the central and side areas. Since the ratio of the central area is large for NWs with large diameters, the crystal structure in NWs with large diameters is mainly dominated by the nucleation at the central area. The small energy difference between the ZB and WZ structures at the central area could results in the formation of rotational twins for NWs with large diameters. In contrast, most of the top layer consists of side facets for NWs with small diameters. The crystal structure in NWs with small diameters is thus dominated by the nucleation on the side facets, resulting in the preference of NWs with WZ structure under the growth conditions. The preference of the WZ structure over the ZB structure in InP NWs with small diameters is qualitatively consistent with the experimental results [90,91].

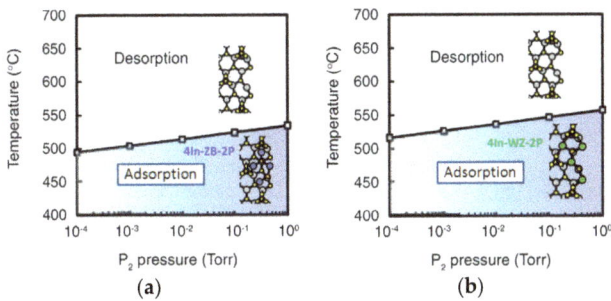

Figure 21. Calculated phase diagrams of (a) 4In-ZB-2P and (b) 4In-WZ-2P adsorbed surfaces as functions of temperature and P_2 pressure. In pressure is taken to be 3.3×10^{-3} Torr. Colored regions correspond to the atom-adsorbed regions. Reprinted with permission from [95]. Copyright (2011) by the Japan Society of Applied Physics.

4.3. Nanowire Growth Processes under Epitaxial Growth Condirions

Figure 22 shows the lattice sites of In atoms for the WZ and ZB structures on the (2×2) surface with P trimer [91]. Focusing on these adsorption sites, we discuss adsorption–desorption behavior of In and P atoms under growth conditions. On the basis of experimental results, we here consider two types of growth conditions: high temperature and low V/III ratio (660 °C and V/III ratio is 18, condition (A) hereafter) and low temperature and high V/III ratio (600 °C and V/III ratio is 55, condition (B) hereafter) [91]. Figure 23 shows contour plots of adsorption energies for In and P atoms, respectively [97]. The values of contour plots are measured from the values of chemical potentials μ_{gas} in Equations (1) and (2) under the growth condition (B) which are -2.5 and -1.2 eV for In and P, respectively. There are four adsorption sites for an In atom labeled 1, 2, 3, and 4 in Figure 23a. Three of them (sites 1, 2, and 3) are symmetrically equivalent and correspond to the adsorption sites for the ZB structure, and the other (site 4) corresponds to the adsorption site for the WZ structure. The adsorption energy difference among the adsorption sites for the WZ and ZB structure is negligible (0.06 eV). The positive values of adsorption energy (0.60 and 0.66 eV) indicate that the adsorption hardly occurs under typical growth conditions [90,91]. The calculated adsorption probability estimated using Equation (12) of an In atom under the growth condition (B) is found to be less than 0.03%. However, even if the adsorption probability is 0.03%, the adsorption occurs. This is because of the stabilization of the P trimer. The P trimer on (2×2) surface hardly desorbs under typical growth conditions. Indeed, the MC simulations for adsorption, migration and desorption of an In atom on the P trimer surface demonstrate that the adsorption of an In atom occurs even under the growth conditions (A) and (B), and surface lifetime and diffusion length of an In adatom are relatively large. The relatively large values in lifetime and diffusion length is due to low migration barriers on the (2×2) surface with P trimer less than 0.13 eV.

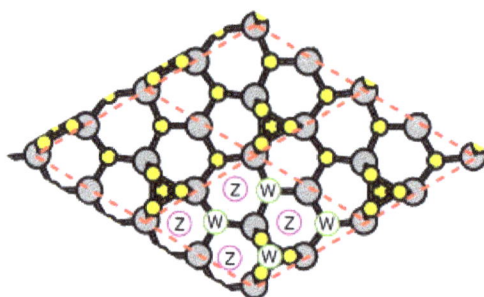

Figure 22. Top view of the (2 × 2) surface with P trimer surface. Red dashed lines correspond to the (2 × 2) unit cells. Large and small circles represent In and P atoms, respectively. Symbols of W and Z stand for In lattice sites in the WZ and ZB structures, respectively. Reprinted with permission from [97]. Copyright (2013) by Elsevier.

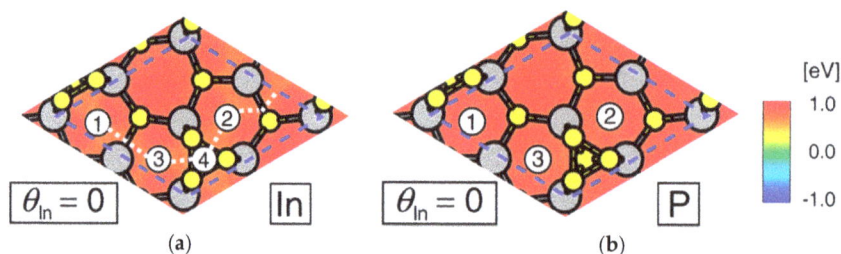

Figure 23. Contour plots of adsorption energies for (**a**) In and (**b**) P atoms on the P trimer surface. These contour plots are indicated by colors with respect to μ_{gas} under the growth condition (B) which is displayed in green region. Atoms tend to adsorb (desorb) in blue (red) colored regions under the growth condition (B). Encircled numbers represent adsorption sites. The coverage of In (θ_{In}) and a kind of atom for contour plots are denoted at the lower left and right, respectively. Dotted white lines in (**a**) denotes a possible migration path of In adatom. Reprinted with permission from [98]. Copyright (2013) by Elsevier.

The migration path of In adatom corresponds to the orange-colored region in Figure 23a. Moreover, this also supports that the number of atoms consisting of the critical nuclei on InP(111)A surface is estimated to be small (10 at most) [98]. The small number of atoms implies that nucleation on the surface can be formed under typical growth conditions. Figure 23b shows that there are three symmetrically equivalent adsorption sites (labeled by encircled numbers such as 1, 2, and 3) for a P atom. The calculated adsorption energies at these sites in Figure 23b are 0.91 eV. The adsorption probability of a P atom under growth condition (B) (5×10^{-4}%) is much lower than that of an In atom. From these results, the adsorption of an In atom is obviously feasible compared with that of a P atom on the (2 × 2) surface with P trimer. It is thus concluded that the crystal growth on the (2 × 2) surface starts with the adsorption of an In atom. Further investigations for the adsorption of In and P atoms on the In-adsorbed surfaces clarify that there are two types of plausible growth processes depending on the growth conditions.

Figure 24 shows the growth process which is feasible under the growth conditions (A) and (B), leading to the formation of an InP monolayer belonging to the WZ structure. First, the adsorption of an In atom occurs at the site 4 on the surface with In coverage of zero ($\theta_{In} = 0$). There are three symmetrically equivalent and stable adsorption sites for an In atom on the surface with $\theta_{In} = 1/4$, which correspond to the WZ structure. The calculated adsorption energy is −0.54 eV. The surface with $\theta_{In} = 2/4$ represents the surface after adsorption of an In atom at the site 1 on the surface with $\theta_{In} = 1/4$. After the formation of the surface with $\theta_{In} = 2/4$, the adsorption of an In atom with calculated adsorption energy of -0.06 eV subsequently occurs at one of the two symmetrically equivalent. The surface after the adsorption of a P atom is the (2 × 2) surface with In vacancy surface which is well known as a stable surface satisfying the ECM [38,39]. It should be noted that this In vacancy surface has an InP monolayer belonging to the WZ structure. In order to verify whether this process is likely under the growth conditions, we have calculated phase diagrams as functions of temperature and pressures. Figure 25a illustrates the adsorption–desorption behavior of an In atom on the surface with $\theta_{In} = 1/4$ shown in Figure 24. This calculated phase diagram manifests that the adsorption of an In atom easily occurs and stabilizes the surface under both the growth conditions (A) and (B). The adsorption of an In atom at $\theta_{In} = 1/4$ originates from the fact that the In-adsorbed surface corresponding to $\theta_{In} = 2/4$ satisfies the ECM [38,39]. As shown in Figure 25b, the adsorption of an In atom occurs on the surface with $\theta_{In} = 2/4$ under the growth condition (B) (red triangle), while the growth condition (A) is located near the boundary between adsorption and desorption (red circle). However, the adsorption possibility of an In atom under the growth condition (A) is relatively high (15%). The adsorption of an In atom certainly occurs even under the growth condition (A). In addition, once the adsorption of an In atom occurs, a P atom adsorbs easily and stabilizes the surface as shown in Figure 25c. The growth process shown in Figure 24 indeed occurs under both the growth conditions (A) and (B), suggesting that an InP monolayer belonging to the WZ structure is formed.

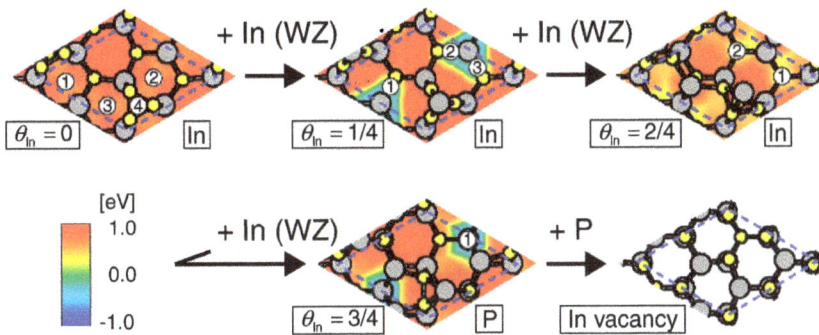

Figure 24. Growth process which is plausible under the growth conditions (A) and (B). The adsorption of a P atom occurs on the surface with $\theta_{In} = 3/4$. This growth process results in formation of an InP monolayer belonging to the WZ structure. Reprinted with permission from [97]. Copyright (2013) by Elsevier.

Figure 26 shows the growth process which is feasible only under the growth condition (B), leading to the formation of an InP monolayer belonging to either the WZ or ZB structure. The adsorption of an In atom occurs at the adsorption site for the ZB structure (the site 3 in Figure 26) on the P trimer surface ($\theta_{In} = 0$). There are two symmetrically equivalent adsorption sites for a P atom on the surface with $\theta_{In} = 1/4$, which are located at near P lattice sites, whose adsorption energy is -0.02 eV. The P-adsorbed surface with $\theta_{In} = 1/4$ represents the surface after adsorption of a P atom at the site 2. The adsorption of an In atom subsequently occurs after the formation of the P-adsorbed surface. Although there are three adsorption sites for an In atom on the P-adsorbed surface, these three sites are no longer equivalent, so that calculated adsorption energies at the sites 1, 2, and 3 are different with each other (-0.07, -0.05, and 0.37 eV, respectively). The positive value at the site 3 indicates that the adsorption of an In atom hardly occurs. Consequently, this growth process splits into two sub-processes with almost the same probabilities after the adsorption of a P atom. One of the two sub-processes starts with the adsorption of an In atom at the site 1 which corresponds to the adsorption site for the WZ structure. The adsorption of an In atom (the calculated adsorption energy is -2.00 eV) at site 1 easily occurs on the surface with $\theta_{In} = 2/4$ in this sub-process, leading to the formation of the In vacancy surface which has an InP monolayer belonging to the WZ structure. The other sub-process begins with the adsorption of an In atom at the site 2, which corresponds to the adsorption site for the ZB structure, on the P-adsorbed surface with $\theta_{In} = 1/4$. The calculated adsorption energy of an In atom at site 1 on the surface with $\theta_{In} = 2/4$ is -1.95 eV, resulting in the formation of the In vacancy surface which has an InP monolayer belonging to the ZB structure.

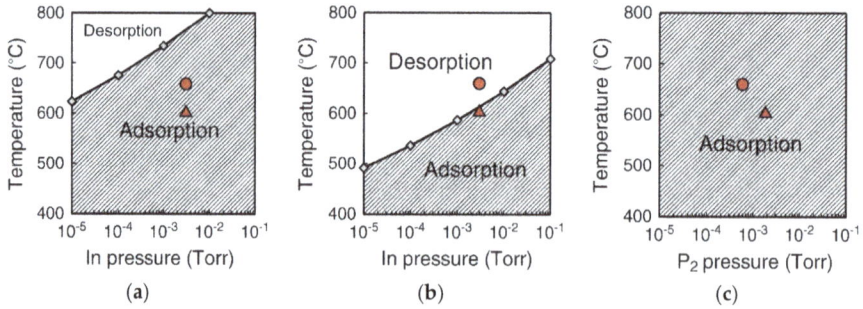

Figure 25. Calculated phase diagrams as functions of temperature and pressure in the growth process shown in Figure 21. Panels (**a**) and (**b**) show adsorption–desorption behavior of an In atom on the surface with $\theta_{In} = 1/4$ and $2/4$, respectively. Panel (**c**) illustrates adsorption–desorption behavior of a P atom on the surface with $\theta_{In} = 3/4$. Regions with hatched lines correspond to atom-adsorbed regions. Red circles and triangles represent the growth conditions (A) and (B), respectively. Reprinted with permission from [97]. Copyright (2013) by Elsevier.

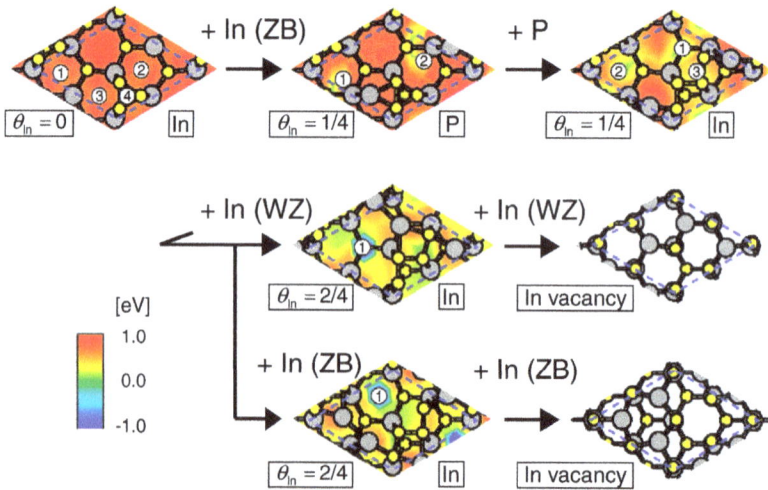

Figure 26. Growth process which is feasible only under the growth condition (B). The adsorption of a P atom occurs on the surface with $\theta_{In} = 1/4$. This growth process splits into two sub-processes after the adsorption of a P atom, and results in the formation of an InP monolayer belonging to the WZ or ZB structure. Reprinted with permission from [97]. Copyright (2013) by Elsevier.

Figure 27 shows calculated phase diagram for the adsorption of a P atom on the surface with $\theta_{In} = 1/4$ in the growth process shown in Figure 26. The phase diagram indicates that a P atom can adsorb on the surface with $\theta_{In} = 1/4$ only under the growth condition (B). Therefore, this growth process can occur only under the growth condition (B). The adsorption of a P atom on the surface with $\theta_{In} = 1/4$ under the growth condition (B) is predominant for the growth process shown in Figure 26, which is quite different from that in the growth process shown in Figure 24. This is because the P-adsorbed surface satisfies the ECM [38,39], and is stabilized only for low temperatures and high P pressures. Although the adsorption probability of a P atom under the growth condition (A) might not be zero, the growth process shown in Figure 24 must be dominant under the growth condition (A) as shown in Figure 25a). The growth process shown in Figure 26 occurs only under the growth condition (B), leading to the formation of InP layers with both WZ and ZB structures. Both WZ and ZB structures can be grown with the same probability under growth condition (B).

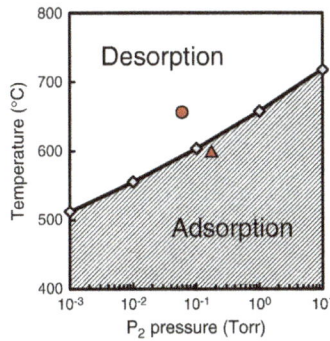

Figure 27. Calculated phase diagrams functions of temperature and pressure for adsorption–desorption behavior of a P atom in the growth process shown in Figure 26. Red circle and triangle represent the growth conditions (A) and (B), respectively. Reprinted with permission from [98]. Copyright (2013) by Elsevier.

Figure 28 shows the relative adsorption probabilities of a P atom, $P_{re}(P)$, on the surface with $\theta_{In} = 1/4$ as functions of temperature and pressure estimated by the MC simulations. Here, we note that $P_{re}(P) + P_{re}(In) = 100\%$, where $P_{re}(In)$ is relative adsorption probabilities of an In atom. Because of high V/III ratio, the adsorption of a P atom on the surface at $\theta_{In} = 1/4$ is dominant under the growth condition (B) compared with that of an In atom, leading to the growth process shown in Figure 26. From these results, it can be concluded that the growth process shown in Figure 24 is dominant for high temperatures and low P pressures, while the growth process shown in Figure 26 is dominant for low temperatures and high P pressures. Hence, it can also be concluded that InP layers grown for high temperatures and low P pressures (low V/III ratio) take the WZ structure, while those grown for low temperatures and high P pressures (high V/III ratio) include both the WZ and ZB layers with rotational twins. These results are consistent with the experimental result [91]. It is thus concluded that adsorption of P atoms depending on temperature and pressure is an important factor determining crystal structures of InP layers on InP(111)A surface. If the nucleation occurs on the InP(111)A top layer of InP NWs far from side facets, the growth processes depending on the growth conditions could strongly affect the structural stability of InP NWs. We believe that this scenario is applicable to clarify the crystal structure of InP NWs on InP(111)A substrates depending on growth conditions [91].

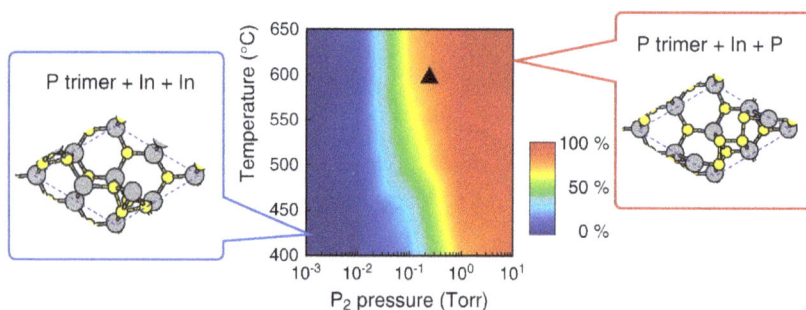

Figure 28. Relative adsorption probabilities of a P atom, $P_{re}(P)$, on the surface with $\theta_{In} = 1/4$ as functions of temperature and pressure. It should be noted that $P_{re}(P) + P_{re}(In) = 100\%$. In pressure is fixed at 3.3×10^{-3} Torr which corresponds to both the growth conditions (A) and (B). Triangle represents the growth condition (B). Reprinted with permission from [97]. Copyright (2013) by Elsevier.

4.4. Effects of Growth Condition on InP NW Shape

Figure 29 shows the calculated adsorption sites and migration barriers for an In atom on $(1\bar{1}00)$ and $(11\bar{2}0)$ surfaces for clarifying the effects of growth condition on the shape of InP NWs. There are two symmetrically equivalent adsorption sites on both $(1\bar{1}00)$ and $(11\bar{2}0)$ surfaces. The calculated adsorption energy difference between $(1\bar{1}00)$ and $(11\bar{2}0)$ surfaces is almost zero and the value of the adsorption energy is taken to be -1.90 eV. The adsorption probability calculated by Equation (12) for the low temperature and high V/III ratio is found to be less than 0.03%. This low probability indicates that it is difficult for an In atom to adsorb on the surface. However, the MC simulations support that the adsorption of an In atom occurs even under those growth conditions.

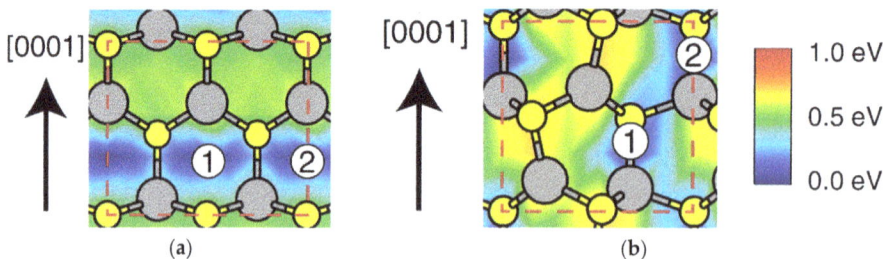

Figure 29. Adsorption sites and migration barriers for an In atom on (a) $(1\bar{1}00)$ and (b) $(11\bar{2}0)$ surfaces. Encircled numbers and contour plots represent adsorption sites and migration barriers, respectively. Large gray and small yellow circles represent In and P atoms, respectively. Reprinted with permission from [99]. Copyright (2013) by Elsevier.

Figure 30 shows temperature dependence of surface lifetime τ and diffusion length L_{diff}, respectively, for an In adatom on $(1\bar{1}00)$ and $(11\bar{2}0)$ surfaces obtained by the MC simulations. Here, surface lifetime τ is defined by the time during the migration, and diffusion length L_{diff} is the distance between the adsorption site and the desorption site. Furthermore, In pressure is fixed at 3.3×10^{-3} Torr, which corresponds to the experimental growth conditions [90,91]. It is found that once the adsorption of an In atom occurs, the In adatom can migrate a certain distance (101–103 nm) with certain life time of 10^{-8}–10^{-6} s under the experimental growth conditions. The values of τ and L_{diff} are long enough to grow at relatively low temperatures. Furthermore, both τ and L_{diff} on $(1\bar{1}00)$ surface are larger than those on $(11\bar{2}0)$ surface, although there is little orientation dependence in the adsorption energy.

This is because the migration barriers on $(1\bar{1}00)$ surface (0.27 eV) is smaller than that on $(11\bar{2}0)$ surface (0.32 eV) as shown in Figure 29. For a P atom, the calculated adsorption sites are the same as those for an In atom, and calculated adsorption energies on $(1\bar{1}00)$ and $(11\bar{2}0)$ surfaces are 0.36 and 0.43 eV, respectively. These positive values indicate that the adsorption of a P atom does not occur even at 0 K. This is because the P-adsorbed surfaces have more excess electrons, which does not satisfy the ECM [38,39], compared to the In-adsorbed surfaces. Hence, the growth on both $(1\bar{1}00)$ and $(11\bar{2}0)$ surfaces starts from the adsorption of an In atom. In order to proceed the lateral growth on the surfaces, either In adatoms on the surface encountering each other caused by migration or additional adsorption on the In-adsorbed surfaces is necessary before vaporization of adsorbed In adatoms.

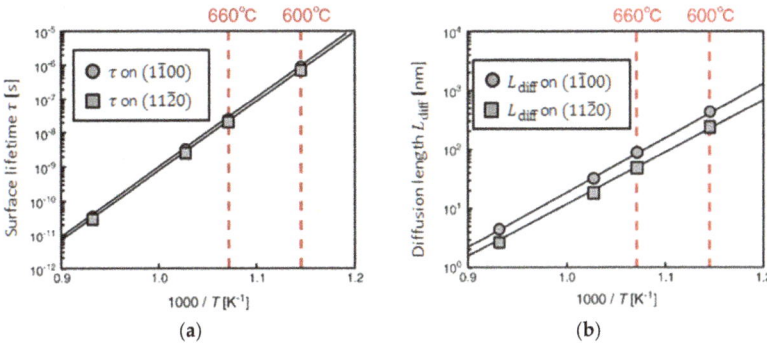

Figure 30. Temperature dependence of (**a**) surface lifetime τ and (**b**) diffusion length L_{diff} for an In adatom on $(1\bar{1}00)$ and $(11\bar{2}0)$ surfaces. In pressure is fixed at 3.3×10^{-3} Torr, which corresponds to the experimental growth conditions [90,91]. Reprinted with permission from [99]. Copyright (2013) by Elsevier.

Figure 31 shows the calculated adsorption sites and migration barriers for In and P atoms on the In-adsorbed $(1\bar{1}00)$ surface. There are two symmetrically equivalent adsorption sites for In and P atoms on the In-adsorbed $(1\bar{1}00)$ surface. The calculated adsorption energies at these sites for In and P atoms are -2.75 and -1.30 eV, respectively. Using these adsorption energies, the phase diagrams for adsorption–desorption behavior of In and P atoms on the In-adsorbed $(1\bar{1}00)$ surface are obtained as shown in Figure 32. The surface phase diagram shown in Figure 32a demonstrates that an In atom can adsorb on the surface under the experimental growth conditions which are symbolized by red circles, squares, and triangles in Figure 32. In contrast, a P atom can adsorb only at the low temperature and high V/III ratio denoted by the red triangle in Figure 32b.

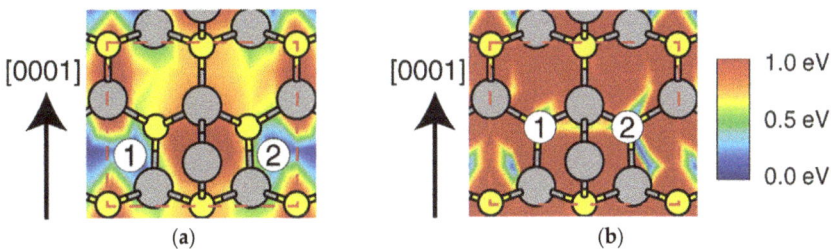

Figure 31. Calculated adsorption sites and migration barriers on the In-adsorbed $(1\bar{1}00)$ surface for (**a**) In and (**b**) P atoms. Encircled numbers and contour plots represent adsorption sites and migration barriers, respectively. Large gray and small yellow circles represent In and P atoms, respectively. Reprinted with permission from [99]. Copyright (2013) by Elsevier.

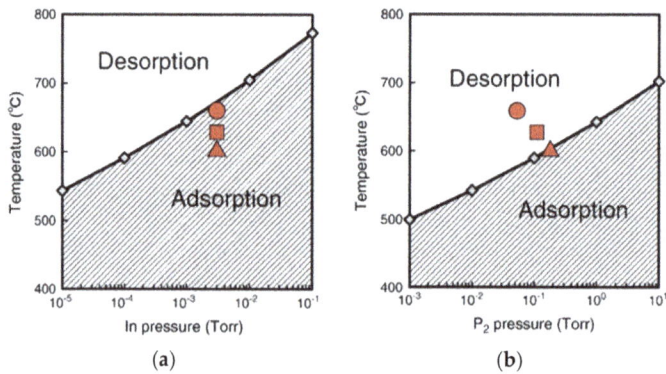

Figure 32. Calculated phase diagrams as functions of temperature and pressure on the In-adsorbed (1$\bar{1}$00) surface for adsorption of (**a**) In and (**b**) P atoms. Regions with hatched lines correspond to atom-adsorbed regions. Red circles, squares, and triangles represent the growth conditions for the high temperature and low V/III ratio, the intermediate temperature and V/III ratio, and the low temperature and high V/III ratio, respectively. In pressure is fixed at 3.3×10^{-3} Torr which corresponds to the experimental growth conditions [90,91]. Reprinted with permission from [99]. Copyright (2013) by Elsevier.

Figure 33 shows the calculated adsorption sites and migration barriers for In and P atoms on the In-adsorbed (11$\bar{2}$0) surface. There is only one adsorption site for each atom on the In-adsorbed (11$\bar{2}$0) surface. The calculated adsorption energies for In and P atoms are −2.43 and −1.27 eV, respectively, which are slightly higher than those on the In-adsorbed (1$\bar{1}$00) surface. Figure 34 shows the calculated phase diagrams for adsorption–desorption behavior of In and P atoms on the In-adsorbed (11$\bar{2}$0) surface. The calculated phase diagrams indicate lower adsorption probabilities on the In-adsorbed (11$\bar{2}$0) surface compared with those on the In adsorbed (1$\bar{1}$00) surface. Under specific experimental condition (red circles in Figure 34) at 600 °C with V/III ratio of 55, however, the calculated adsorption probabilities for In and P atoms are 27% and 40%, respectively. These adsorption probabilities imply that this adsorption could occur at the low temperatures with high V/III ratio.

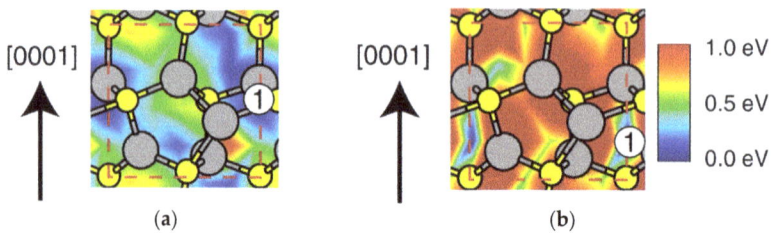

Figure 33. Calculated adsorption sites and migration barriers on the In-adsorbed (11$\bar{2}$0) surface for (**a**) In and (**b**) P atoms. Encircled numbers and contour plots represent adsorption sites and migration barriers, respectively. Large gray and small yellow circles represent In and P atoms, respectively. Reprinted with permission from [100]. Copyright (2013) by Elsevier.

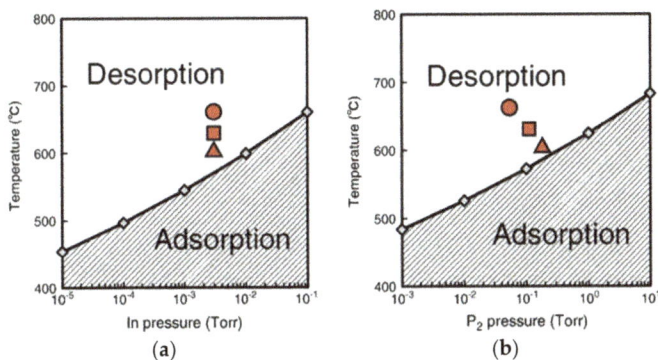

Figure 34. Calculated phase diagrams as functions of temperature and pressure on the In-adsorbed $(11\bar{2}0)$ surface for adsorption of (**a**) In and (**b**) P atoms. Regions with hatched lines correspond to atom-adsorbed regions. Red circles, squares, and triangles represent the growth conditions for the high temperature and low V/III ratio, the intermediate temperature and V/III ratio, and the low temperature and high V/III ratio, respectively. In pressure is fixed at 3.3×10^{-3} Torr which corresponds to the experimental growth conditions [90,91]. Reprinted with permission from [99]. Copyright (2013) by Elsevier.

Table 1 shows relative adsorption probabilities of a P atom on the In-adsorbed $(1\bar{1}00)$ and $(11\bar{2}0)$ surfaces, $P_{re}^{(1-100)}$(P) and $P_{re}^{(11-20)}$(P), respectively, which are percentages calculated by $n_P/(n_{In} + n_P)$. Here, n_{In} and n_P are the number of the formation of In and P adatoms using MC simulations, respectively, on the In-adsorbed surfaces during the MC sampling ($n_{In} + n_P = 1000$). Both $P_{re}^{(1-100)}$ (P) and $P_{re}^{(11-20)}$(P) increase with temperature up to 550 °C, because In adatoms hardly encounter each other as seen in the decrease in τ and L_{diff} in Figure 30. On the In-adsorbed $(1\bar{1}00)$ surface, $P_{re}^{(1-100)}$ (P) shifts to decrease around 550–600 °C. This is because a P atom hardly adsorbs beyond 600 °C. On the In-adsorbed $(11\bar{2}0)$ surface, a P atom as well as an In atom hardly adsorbs beyond 600 °C. These results demonstrate that the probability of P/In atoms is higher, indicating that the adsorption of a P atom is dominant on the In-adsorbed surfaces. Although adsorption probability of an In atom on the In-adsorbed $(1\bar{1}00)$ surface is higher than that of a P atom as shown in Figure 32, P atoms frequently attach on the surface owing to high V/III ratio, as seen in the estimated Δt_{gas} using Equation (15) shown in Table 2. It is found that in most cases, the adsorption of a P atom occurs on the In-adsorbed surfaces. From the calculated results by the MC simulations, we can deduce the initial growth process, such as a sequence of adsorption, migration, and desorption of In atoms, which iteratively occur on $(1\bar{1}00)$ and $(11\bar{2}0)$ surfaces, and the adsorption of P atom supplied from gas phase with In adatoms. Therefore, there are two reasons for the growth rate difference between $(1\bar{1}00)$ and $(11\bar{2}0)$ surfaces. One is the longer τ and L_{diff} of In adatoms on $(1\bar{1}00)$ surface as shown in Figure 30 because In adatoms are necessary for the adsorption of a P atom. The other is higher adsorption probability of a P atom on the In-adsorbed $\{1\bar{1}00\}$ surface compared with that on $(11\bar{2}0)$ surface, as shown in Figures 29b and 31b. The latter is more important than the former, since adsorption behavior of P atoms is sensitive around the critical temperature and pressure. It is expected that at 600 °C P atoms tend to adsorb on the In-adsorbed $(1\bar{1}00)$ surface whereas P atoms tend to desorb on the In-adsorbed $(11\bar{2}0)$ surface. The difficulty in P adsorption on the In-adsorbed $(11\bar{2}0)$ surface beyond 600 °C causes the growth rate difference. If we consider successful epitaxial growth of semiconductor NWs, the growth in the axial direction should be faster than that in the lateral directions. The difference in growth rate in InP.

Table 1. Relative adsorption probabilities of a P atom on the In-adsorbed ($1\bar{1}00$) and ($11\bar{2}0$) surfaces, $P_{re}^{(1-100)}$(P) and $P_{re}^{(11-20)}$(P), respectively, obtained by MC simulations. It should be noted that $P_{re}^{(1-100)}$(In) + $P_{re}^{(1-100)}$(P) = $P_{re}^{(11-20)}$(In) + $P_{re}^{(11-20)}$(P) = 100%, where $P_{re}^{(1-100)}$ (In) and $P_{re}^{(11-20)}$(In) are relative adsorption probabilities of an In atom on the In-adsorbed surfaces. Here, we assume that pressure is 0.1 Torr.

Temperature (°C)	$P_{re}^{(1-100)}$ (P) (%)	$P_{re}^{(11-20)}$ (P) (%)
500	67.3	72.3
550	72.1	77.1
650	69.1	76.9

Table 2. Values of Δt_{gas} for In and P obtained by Equation (15). Here, we assume that In and P pressures are 3.3×10^{-3} and 0.1 Torr, respectively.

Temperature (°C)	Δt_{gas} for In (s)	Δt_{gas} for P (s)
500	4.53×10^{-13}	1.11×10^{-14}
550	4.67×10^{-13}	1.15×10^{-14}
650	4.81×10^{-13}	1.18×10^{-14}

NWs can be clarified by comparing with the results on InP(111)A surface discussed in Section 4.3 [98]. Although the calculated adsorption energy difference of an In atom on bare (111)A, ($1\bar{1}00$), and ($11\bar{2}0$) surfaces is insignificant, τ and L_{diff} of an In adatom on the (111)A surface are longer than those on other surfaces. The remarkable values in τ and L_{diff} on the (111)A surface originates from lower migration barriers on the (111)A, as shown in Table 3. Furthermore, the calculated adsorption energies of an In and a P atom on an In-adsorbed (111)A surface are lower than those on the In-adsorbed ($1\bar{1}00$) and ($11\bar{2}0$) surfaces.

Table 3. Values of τ and L_{diff} obtained by MC simulations, and migration barriers of In adatom on bare (111)A, ($1\bar{1}00$), and ($11\bar{2}0$) surfaces using Figures 23 and 29. Here, we assume that temperature and In pressure are 600 °C and 3.3×10^{-3} Torr, respectively. The lowest values for migration barriers are taken on each surface.

Surface Orientation	τ ($\times 10^{-7}$ s)	L_{diff} ($\times 10^2$ nm)	Migration Barrier (eV)
(111)A	22.4	16.3	0.13
($1\bar{1}00$)	8.57	4.36	0.27
($11\bar{2}0$)	6.97	2.38	0.32

The adsorption energies of an In and a P atom on the In-adsorbed (111)A surface are −3.04 and −1.32 eV, respectively, while those of a single In (P) atom on the In-adsorbed ($1\bar{1}00$) and ($11\bar{2}0$) surfaces are −2.75 (−1.30) and −2.45 (−1.27) eV, respectively. Therefore, it is clear that the growth in the axial direction is faster than that in the lateral direction. The MC calculations indeed demonstrate that the difference in growth rate between ($1\bar{1}00$) and ($11\bar{2}0$) surfaces can be realized depending on the growth conditions. It is considered that facets with higher growth rate disappear and those with lower growth rate remain [100]. Thus, the lower growth rate of ($11\bar{2}0$) surface results in the formation of InP NWs consisting of {$11\bar{2}0$} no side facets for lower temperatures. Using surface phase diagrams shown in Figures 32b and 34b and MC calculation results, we can deduce NW growth modes which are related to the shape of InP NWs.

Figure 35 shows the phase diagram for growth modes of InP NWs as functions of temperature and P pressure, derived from the comparisons of the calculated results among (111)A, ($1\bar{1}00$), and ($11\bar{2}0$) surfaces. Region I in Figure 35 corresponds to the growth of InP NWs consisting of {$1\bar{1}00$} side facets without lateral growth. In this region, due to its growth conditions with high temperatures and low P pressures, P atoms hardly adsorb either on ($1\bar{1}00$) or ($11\bar{2}0$) surfaces, and the {$1\bar{1}00$} side facets emerge due to their low surface energy compared with ($11\bar{2}0$) surface [101,102]. Region II in Figure 32 corresponds to the growth of InP NWs consisting of {$11\bar{2}0$} side facets without lateral growth. In this region, P atoms adsorb on {$1\bar{1}00$} surface, resulting in annihilation of {$1\bar{1}00$} side facets and emergence of the {$11\bar{2}0$} side facets. It should be noted that the lateral growth of {$11\bar{2}0$} side facets does not occur since P atoms hardly adsorb on the surface in this region. Finally, in region III of Figure 32, P atoms tend to adsorb on both ($1\bar{1}00$) and ($11\bar{2}0$) surfaces due to low temperatures and high P pressures. This leads to the lateral growth and the formation of NWs with large diameters. In this region, InP NWs consisting of {$11\bar{2}0$} side faces are preferentially grown due to the low growth rate on ($11\bar{2}0$) surface compared with that on ($1\bar{1}00$) surface. The experimentally reported growth modes of InP NWs under the conditions at the red circle, square, and triangle in Figure 35. Trends of the calculated phase diagram qualitatively agrees with the experimental results [91], although the criteria for adsorption of a P atom (boundary lines in Figure 35) could be underestimated. We note that the possibility of In adsorption on the In-adsorbed ($1\bar{1}00$) surface, which facilitates the growth of ($1\bar{1}00$) surface for high temperatures cannot be ruled out (see Figure 32a). Even if it is true, the boundary between regions I and II in Figure 35 might shift toward higher temperatures, resulting in expansion of region II. However, our conclusions still agree with the experiments qualitatively. It is thus expected that the shape of InP NWs grown by the SA-MOVPE can be predicted on the basis of this phase diagram.

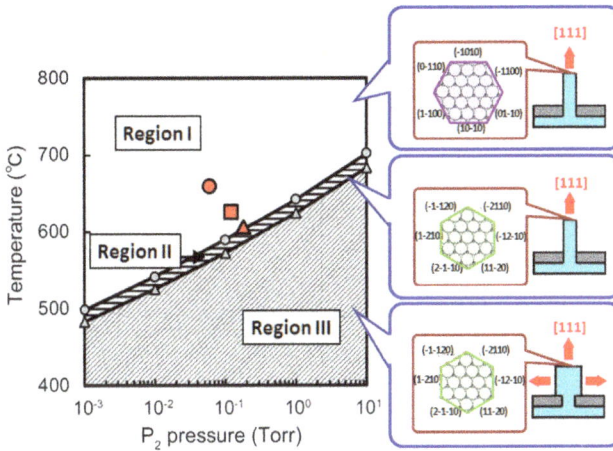

Figure 35. Phase diagram for growth modes of InP NWs as functions of temperature and pressure. Red circle, square, and triangle represent the experimental growth conditions for the high temperature and low V/III ratio, the intermediate temperature and V/III ratio, and the low temperature and high V/III ratio, respectively [90,91].

5. Concluding Remarks

In this article, we have exemplified the applicability of the chemical potential approach with the aid of the macroscopic theory and MC simulations for fundamental understanding of the growth processes of nanostructures for InAs and InP. The power of this approach is not only on understanding the epitaxial growth processes but also on producing fundamental data such as surface phase diagrams as functions of growth conditions. Surface phase diagrams and elemental growth processes have been extensively investigated using similar approach for various semiconductors except InAs and InP such as GaAs [18,41,103–108] and III-Nirides [7–9,109–128]. Figure 36 summarizes the calculated surface phase diagrams with different orientations for AlN and GaN under H-rich conditions during

MOVPE growth. It is found that the phase diagrams change whole situation depending on materials and orientations. This suggests that MOVPE growth stably proceeds without depending on growth conditions on polar (000$\bar{1}$), nonpolar (1$\bar{1}$00), and nonpolar (11$\bar{2}$0) for AlN, and polar (000$\bar{1}$), semipolar (1$\bar{1}$01), and semipolar (11$\bar{2}$2) for GaN, since these phase diagrams are simple with small number of phases. On the other hand, AlN semipolar surfaces give complicated phase diagrams strongly depending on growth conditions in contrast with those of GaN. Therefore, careful control of growth conditions is necessary to keep stable AlN MOVPE growth on the semipolar surfaces. This is just an example for the availability of the chemical potential approach and surface phase diagrams as fundamental data for the epitaxial growth.

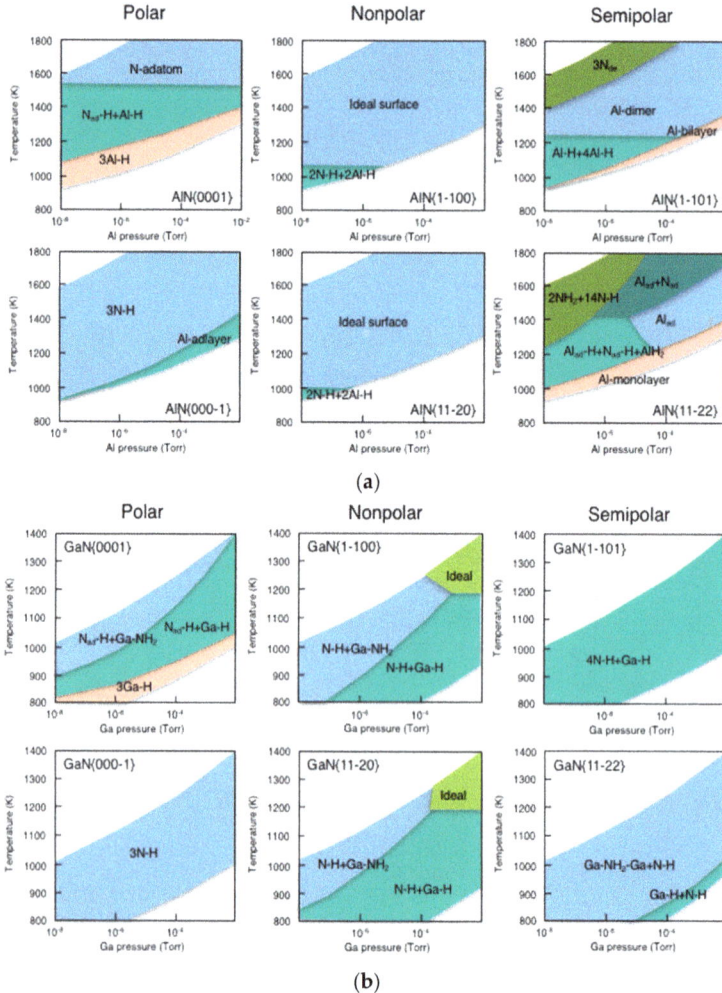

Figure 36. Surface phase diagrams for H-adsorbed (**a**) AlN and (**b**) GaN with different orientations as functions of temperature and pressure.

The applicability of our approach has also been exemplified for various NWs including SiC, GaP, GaAs, InAs, ZnS, and ZnSe [98,101,102,129–131]. Furthermore, owing to ab initio calculations our studies are easily able to be extended to computational modeling and predictions of the electronic and optical properties of NWs. Such kinds of studies have been carried out for GaN and ZnO NWs, as seen in Figure 37 [131]. The calculated results demonstrate that the band gap energy of NWs becomes large compared with the calculated bulk energy gap due to quantum confinement effects. Moreover, the calculated imaginary part of their dielectric functions exhibit strong anisotropy and there are several side peaks near the absorption edge caused by valence electronic states around the highest-occupied band involved in the large dipole matrix elements. These calculated results provide a firm theoretical framework to predict microscopic properties of various semiconductor NWs. Consequently, the chemical potential approach enables us to interpret and predict epitaxial growth for semiconductors from a quantum mechanical viewpoint, such as hetero-epitaxial growth and nanowire growth in understanding their surface reconstructions through surface phase diagrams and growth processes including adsorption-desorption and migration behavior of atoms and molecules on the surfaces. Success will lead to more realistic simulations of epitaxial growth to give guiding principles for their fabrications as functions of growth conditions such as temperature and gas pressure. Recent progress in fabrication techniques will make it possible to create various new materials, including nanosheet in addition to nanodots and nanowires by controlling the atomic arrangements during epitaxial growth. Under these circumstances, the chemical potential approach will be an essential technique in the future of materials design to interpret and predict dynamic changes of the material properties because of its availability of predicting atomic arrangements during epitaxial growth.

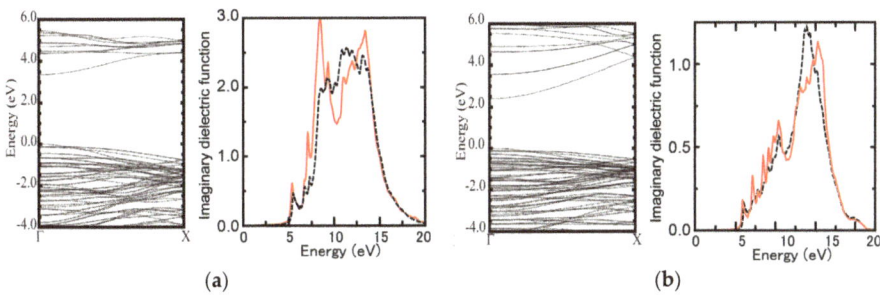

(a) (b)

Figure 37. Calculated band structure and imaginary part of dielectric function of (**a**) GaN and (**b**) ZnO NWs with diameter of 1.3 nm. Energies are measured from the highest occupied state. Solid and dashed lines in dielectric function represent the imaginary part of dielectric function polarized along the nanowire axis and that in the orthogonal plane.

Acknowledgments: This work was supported in part by the JSPS KAKENHI Grant No. 21560032, No. 24560025, No. JP16K04962, No. JP16H06418, and CREST-JST (No. 16815710). One of the authors (Toru Akiyama) would like to thank the Toyota Physical and Chemical Research Institute for financial support. The computations were performed using Research Center for Computational Science (National Institutes of Natural Sciences) and Research Institute for Information Technology (Kyushu University).

Author Contributions: All the authors participated in conceiving and designing the review articles; Tomonori Ito and Toru Akiyama designed the calculations for the hetero-epitaxial growth and nanowire growth, respectively. Toru Akiyama conducted the calculations.

Conflicts of Interest: The authors declare no conflict of interest.

References

1. Däweritz, L.; Hey, R. Reconstruction and defect structure of vicinal GaAs(001) and $Al_xGa_{1-x}As$ surfaces during MBE growth. *Surf. Sci.* **1990**, *236*, 15–22. [CrossRef]
2. Kaxiras, E.; Bar-Yam, Y.; Joannopoulos, J.D.; Pandey, K.C. Ab initio theory of polar semiconductor surfaces. I. Methodology and the (2 × 2) reconstructions of GaAs(111). *Phys. Rev. B* **1987**, *35*, 9625–9635. [CrossRef]
3. Qian, G.X.; Martin, R.M.; Chadi, D.J. Stoichiometry and surface reconstruction: An ab initio study of GaAs(001) surfaces. *Phys. Rev. Lett.* **1988**, *60*, 1962–1965. [CrossRef] [PubMed]

4. Northrup, J.E. Structure of Si(100)H: Dependence on the H chemical potential. *Phys. Rev. B* **1991**, *44*, 1419–1422. [CrossRef]

5. Kangawa, Y.; Ito, T.; Taguchi, A.; Shiraishi, K.; Ohachi, T. A new theoretical approach to adsorption-desorption behavior of Ga on GaAs surfaces. *Surf. Sci.* **2001**, *493*, 178–181. [CrossRef]

6. Kangawa, Y.; Ito, T.; Hiraoka, Y.S.; Taguchi, A.; Shiraishi, K.; Ohachi, T. Theoretical approach to influence of As$_2$ pressure on GaAs growth kinetics. *Surf. Sci.* **2002**, *507*, 285–289. [CrossRef]

7. Ito, T.; Akiyama, T.; Nakamura, K. Ab initio-based approach to reconstruction, adsorption and incorporation on GaN surfaces. *Semicond. Sci. Technol.* **2012**, *27*, 024010. [CrossRef]

8. Kangawa, Y.; Akiyama, T.; Ito, T.; Shiraishi, K.; Nakayama, T. Surface stability and growth kinetics of compound semiconductors: An ab initio-based approach. *Materials* **2013**, *6*, 3309–3360. [CrossRef]

9. Ito, T.; Kangawa, Y. Ab Initio-Based Approach to Crystal Growth: Chemical Potential Analysis. In *Handbook of Crystal Growth*, 2nd ed.; Nishinaga, T., Rudolph, P., Kuech, T.F., Eds.; Elsevier: Amsterdam, The Netherlands, 2015; Volume IA, pp. 477–520.

10. Perdew, J.P.; Burke, K.; Ernzerhof, M. Generalized gradient approximation made simple. *Phys. Rev. Lett.* **1997**, *77*, 3865–3868. [CrossRef] [PubMed]

11. Louie, S.G.; Froyen, S.; Cohen, M.L. Nonlinear ionic pseudopotentials in spin-density-functional calculations. *Phys. Rev. B* **1982**, *26*, 1738–1742. [CrossRef]

12. Yamauchi, J.; Tsukada, M.; Watanabe, S.; Sugino, O. First-principles study on energetics of c-BN(001) reconstructed surfaces. *Phys. Rev. B* **1996**, *54*, 5586–5603. [CrossRef]

13. Shiraishi, K. A new slab model approach for electronic structure calculation of polar semiconductor surface. *J. Phys. Soc. Jpn.* **1990**, *59*, 3455–3458. [CrossRef]

14. Shiraishi, K.; Oyama, N.; Okajima, K.; Miyagishima, N.; Takeda, K.; Yamaguchi, H.; Ito, T.; Ohno, T. First principles and macroscopic theories of semiconductor epitaxial growth. *J. Cryst. Growth* **2002**, *237–239*, 206–211. [CrossRef]

15. Khor, K.E.; das Sarma, S. Proposed universal interatomic potential for elemental tetrahedrally bonded semiconductors. *Phys. Rev. B* **1988**, *38*, 3318–3822. [CrossRef]

16. Ito, T.; Khor, K.E.; das Sarma, S. Systematic approach to developing empirical potentials for compound semiconductors. *Phys. Rev. B* **1990**, *41*, 3893–3896. [CrossRef]

17. Ito, T. Simple criterion for wurtzite-zinc-blende polytypism in semiconductors. *Jpn. J. Appl. Phys.* **1998**, *37*, L1217–L1220. [CrossRef]

18. Kangawa, Y.; Ito, T.; Taguchi, A.; Shiraishi, K.; Irisawa, T.; Ohachi, T. Monte Carlo simulation for temperature dependence of Ga diffusion length on GaAs(001). *Appl. Surf. Sci.* **2002**, *190*, 517–520. [CrossRef]

19. Itoh, M.; Bell, G.R.; Avery, A.R.; Jones, T.S.; Joyce, B.A.; Vvedensky, D.D. Island nucleation and growth on reconstructed GaAs(001) surfaces. *Phys. Rev. Lett.* **1998**, *81*, 633–636. [CrossRef]

20. Joyce, B.A.; Sudijono, J.L.; Belk, J.G.; Yamaguchi, H.; Zhang, X.M.; Dobbs, H.T.; Zangwill, A.; Vvedensky, D.D.; Jones, T.S. A scanning tunneling microscopy-reflection high energy electron diffraction-rate equation study of the molecular beam epitaxial growth of InAs on GaAs(001), (110) and (111)A-quantum dots and two-dimensional modes. *Jpn. J. Appl. Phys.* **1997**, *36*, 4111–4117. [CrossRef]

21. Yamaguchi, H.; Belk, J.G.; Zhang, X.M.; Sudijono, J.L.; Fay, M.R.; Jones, T.S.; Pashley, D.W.; Joyce, B.A. Atomic-scale imaging of strain relaxation via misfit dislocation in highly mismatched semiconductor heteroepitaxy: InAs/GaAs(111)A. *Phys. Rev. B* **1997**, *55*, 1337–1340. [CrossRef]

22. Ohtake, A.; Ozeki, M. Strain relaxation in InAs/GaAs(111)A heteroepitaxy. *Phys. Rev. Lett.* **2000**, *84*, 4665–4668. [CrossRef] [PubMed]

23. Zepeda-Ruiz, L.A.; Maroudas, D.; Weinberg, W.H. Semicoherent interface formation and structure in InAs/GaAs(111)A heteroepitaxy. *Surf. Sci.* **1998**, *418*, L68–L72. [CrossRef]

24. Joe, H.; Akiyama, T.; Nakamura, K.; Kanisawa, K.; Ito, T. An empirical potential approach to the structural stability of InAs stacking-fault tetrahedron in InAs/GaAs(111). *J. Cryst. Growth* **2007**, *301–302*, 837–840. [CrossRef]

25. Taguchi, A.; Kanisawa, K. Stable reconstruction and adsorbates of InAs(111)A surface. *Appl. Surf. Sci.* **2006**, *252*, 5263–5266. [CrossRef]

26. Belk, J.G.; McConville, C.F.; Sudijono, J.L.; Jones, T.S.; Joyce, B.A. Surface alloying at InAs/GaAs interfaces grown on (001) surfaces by molecular beam epitaxy. *Surf. Sci.* **1997**, *387*, 213–226. [CrossRef]

27. Grabowski, J.; Prohl, C.; Höpfner, B.; Dähne, M.; Elsele, H. Evolution of the InAs wetting layer on GaAs(001)-c(4 × 4) on the atomic scale. *Appl. Phys. Lett.* **2009**, *95*, 233118. [CrossRef]

28. Eisele, H.; Höpfner, B.; Prohl, C.; Grabowski, J.; Dähne, M. Atomic structure of the (4 × 3) reconstructed InGaAs monolayer on GaAs(001). *Surf. Sci.* **2010**, *604*, 283–289. [CrossRef]

29. Konishi, T.; Tsukamoto, S. Statistical analysis of surface reconstruction domains on InAs wetting layer preceding quantum dot formation. *Nanoscale Res. Lett.* **2010**, *5*, 1901–1904. [CrossRef] [PubMed]

30. Kratzer, P.; Penev, E.; Scheffler, M. Understanding the growth mechanisms of GaAs and InGaAs thin films by employing first-principles calculations. *Appl. Surf. Sci.* **2003**, *216*, 436–446. [CrossRef]

31. Ito, T.; Ishimure, N.; Akiyama, T.; Nakamura, K. Ab initio-based approach to adsorption-desorption behavior on the InAs(111)A heteroepitaxially grown on GaAs substrate. *J. Cryst. Growth* **2011**, *318*, 72–75. [CrossRef]

32. Ito, T.; Ogasawara, K.; Sugitani, T.; Akiyama, T.; Nakamura, K. Ab initio-based approach to elemental growth process on the InAs wetting layer grown on GaAs substrate. *J. Cryst. Growth* **2013**, *362*, 2–5. [CrossRef]

33. Ito, T.; Hirai, K.; Akiyama, T.; Nakamura, K. Ab initio-based approach to novel behavior of InAs wetting layer grown on GaAs(001). *J. Cryst. Growth* **2013**, *378*, 13–16. [CrossRef]

34. Ito, T.; Akiyama, T.; Nakamura, K. Ab initio-based approach to structural change in InAs(001)-(2 × 3) wetting layer surfaces during MBE growth. *e-J. Surf. Sci. Nanotechnol.* **2015**, *13*, 190–194. [CrossRef]

35. Yamaguchi, H.; Homma, Y.; kanisawa, K.; Hirayama, H. Drastic improvement in surface flatness properties by using GaAs(111)A substrates in molecular beam epitaxy. *Jpn. J. Appl. Phys.* **1999**, *38*, 635–644. [CrossRef]

36. Yamaguchi, H.; Sudijiono, J.L.; Joyce, B.A.; Jones, T.S.; Gatzke, C.; Stradling, R.A. Thickness-dependent electron accumulation in InAs thin films on GaAs(111)A: A scanning tunneling-spectroscopy study. *Phys. Rev. B* **1998**, *58*, 4219–4222. [CrossRef]

37. Phillips, J.C. Ionicity of the chemical bond in crystals. *Rev. Mod. Phys.* **1970**, *42*, 317–356. [CrossRef]

38. Farrell, H.H.; Harbison, J.P.; Peterson, L.D. Molecular-beam epitaxy growth mechanisms on GaAs(100) surfaces. *J. Vac. Sci. Technol. B* **1987**, *5*, 1482–1489. [CrossRef]

39. Pashley, M.D. Electron counting model and its application to island structures on molecular-beam epitaxy grown on GaAs(001) and ZnSe(001). *Phys. Rev. B* **1989**, *40*, 10481–10487. [CrossRef]

40. Shiraishi, K.; Ito, T. Ga-adatom-induced As rearrangement during GaAs epitaxial growth: Self-surfactant effect. *Phys. Rev. B* **1998**, *57*, 6301–6304. [CrossRef]

41. Tatematsu, H.; Sano, K.; Akiyama, T.; Nakamura, K.; Ito, T. Ab initio-based approach to initial growth processes on GaAs(111)B-(2 × 2) surfaces: Self-surfactant effect of Ga adatoms revisited. *Phys. Rev. B* **2008**, *77*, 233306-1-4. [CrossRef]

42. Ito, T.; Akiyama, T.; Nakamura, K. Theoretical investigations for strain relaxation and growth mode of InAs thin layers on GaAs(111)A. *Condens. Matter* **2016**, *1*, 4. [CrossRef]

43. Potin, V.; Ruterana, P.; Nouet, G.; Pond, R.C.; Morkoç, H. Mosaic growth of GaN on (0001) sapphire: A high-resolution electron microscopy and crystallographic study of threading dislocations from low-angle to high-angle grain boundaries. *Phys. Rev. B* **2000**, *61*, 5587–5589. [CrossRef]

44. Ichimura, M. Stillinger-Weber potentials for III-V compound semiconductors and their application to the critical thickness calculation for InAs/GaAs. *Phys. Status Solidi A* **1996**, *153*, 431–437. [CrossRef]

45. Béré, A.; Serra, A. Atomic structure of dislocation cores in GaN. *Phys. Rev. B* **2002**, *65*, 205323. [CrossRef]

46. Kawamoto, K.; Suda, T.; Akiyama, T.; Nakamura, K.; Ito, T. An empirical potential approach to dislocation formation and structural stability in GaN$_x$As$_{1-x}$. *Appl. Surf. Sci.* **2005**, *244*, 182–185. [CrossRef]

47. People, B.; Bean, J.C. Calculation of critical layer thickness versus lattice mismatch for GexSi1-x/Si strained-layer heterostructures. *Appl. Phys. Lett.* **1985**, *47*, 322–324. [CrossRef]

48. Kobayashi, R.; Nakayama, T. First-principles study on crystal and electronic structures of stacking-fault tetrahedron in epitaxial Si films. *J. Cryst. Growth* **2004**, *278*, 500–504. [CrossRef]

49. Zhou, N.; Zhou, L. A fusion-crystalization mechanism for nucleation of misfit dislocations in FCC epitaxial films. *J. Cryst. Growth* **2006**, *289*, 681–685. [CrossRef]

50. Wang, L.G.; Kratzer, P.; Moll, N.; Scheffler, M. Size, shape, and stability of InAs quantum dots on the GaAs(001) substrate. *Phys. Rev. B* **2000**, *62*, 1897–1904. [CrossRef]

51. Kita, T.; Wada, O.; Nakayama, T.; Murayama, M. Optical reflectance study of the wetting layers in (In, Ga)As self-assembled quantum dot growth on GaAs(001). *Phys. Rev. B* **2002**, *66*, 195312. [CrossRef]

52. Patella, F.; Nufris, S.; Arciprete, F.; Fanfoni, M.; Placidi, E.; Segarlata, A.; Balzarotti, A. Tracing the two- to three-dimensional transition in the InAs/GaAs(001) heteroepitaxial growth. *Phys. Rev. B* **2003**, *67*, 205308. [CrossRef]

53. Xu, C.; Temko, Y.; Suzuki, T.; Jacobi, K. InAs wetting layer evolution on GaAs(001). *Surf. Sci.* **2005**, *580*, 30–38. [CrossRef]

54. Yamaguchi, H.; Horikoshi, Y. Unified model for structure transition and electrical properties of InAs(001) surfaces studied by scanning tunneling microscopy. *Jpn. J. Appl. Phys.* **1994**, *33*, L1423–L1426. [CrossRef]

55. Kaida, R.; Akiyama, T.; Nakamura, K.; Ito, T. Theoretical study for misfit dislocation formation at InAs/GaAs(001) interface. *J. Cryst. Growth* **2016**. [CrossRef]

56. Joyce, B.A.; Vvedensky, D.D. Self-organized growth on GaAs surfaces. *Mater. Sci. Eng. R* **2004**, *46*, 127–176. [CrossRef]

57. Blumin, M.; Ruda, H.E.; Savelyev, I.G.; Shik, A.; Wang, H. Self-assembled InAs quantum dots and wires grown on a cleaved-edge GaAs(110) surface. *J. Appl. Phys.* **2006**, *99*, 093518. [CrossRef]

58. Aierken, A.; Hakkarainen, T.; Sopanen, M.; Riikonen, J.; Sormunen, J.; Mattila, M.; Lipsanen, H. Self-assembled InAs island formation on GaAs(110) by metalorganic vapor phase epitaxy. *Appl. Surf. Sci.* **2008**, *254*, 2072–2076. [CrossRef]

59. Yazawa, M.; Koguchi, M.; Muto, A.; Ozawa, M.; Hiruma, K. Effect of one monolayer of surface gold atoms on the epitaxial growth of InAs nanowhiskers. *Appl. Phys. Lett.* **1992**, *61*, 2051–2053. [CrossRef]

60. Hiruma, K.; Yazawa, H.; Katuyama, T.; Ogawa, K.; Koguchi, M.; Kakibayashi, H. Growth and optical properties of nanometer-scale GaAs and InAs whiskers. *J. Appl. Phys.* **1995**, *77*, 447–462. [CrossRef]

61. Li, H.; Wu, J.; Wang, Z. High-density InAs nanowires realized in situ on (100) InP. *Appl. Phys. Lett.* **1999**, *75*, 1173–1175. [CrossRef]

62. Wu, Z.H.; Mei, X.Y.; Kim, D.; Blumin, M.; Ruda, H.E. Growth of Au-catalyzed ordered GaAs nanowire arrays by molecular-beam epitaxy. *Appl. Phys. Lett.* **2002**, *81*, 5177–5179. [CrossRef]

63. Wu, Z.H.; Sun, M.; Mei, X.Y.; Ruda, H.E. Total reflection amorphous carbon mirrors for vacuum ultraviolet free electron lasers. *Appl. Phys. Lett.* **2004**, *84*, 657–659. [CrossRef]

64. Morales, A.M.; Lieber, C.M. A Laser ablation method for the synthesis of crystalline semiconductor nanowires. *Science* **1998**, *279*, 208–211. [CrossRef] [PubMed]

65. Wang, J.; Gudiksen, M.S.; Duan, X.; Cui, Y.; Lieber, C.M. Highly polarized photoluminescence and photodetection from single indium phosphide nanowires. *Science* **2001**, *293*, 1455–1457. [CrossRef] [PubMed]

66. Gudiksen, M.S.; Lauhon, L.J.; Wang, J.; Smith, D.C.; Lieber, C.M. Growth of nanowire superlattice structures for nanoscale photonics and electronics. *Nature* **2002**, *415*, 617–620. [CrossRef] [PubMed]

67. Duan, X.; Huang, Y.; Cui, Y.; Wang, J.; Lieber, C.M. Indium phosphide nanowires as building blocks for nanoscale electronic and optoelectronic devices. *Nature* **2001**, *409*, 66–69. [CrossRef] [PubMed]

68. Goto, H.; Nosaki, K.; Tomioka, K.; Hara, S.; Hiruma, K.; Motohisa, J.; Fukui, T. Growth of core–shell InP nanowires for photovoltaic a lication by selective-area metal organic vapor phase epitaxy. *Appl. Phys. Express* **2009**, *2*, 035004. [CrossRef]

69. Ohlsson, B.J.; Björk, M.T.; Magnusson, M.H.; de ert, K.; Samuelson, L.; Wallenberg, L.R. Size-, shape-, and position-controlled GaAs nano-whiskers. *Appl. Phys. Lett.* **2001**, *79*, 3335–3337. [CrossRef]

70. Bhunia, S.; Kawamura, T.; Watanabe, Y.; Fujikawa, S.; Tokushima, K. Metalorganic vapor-phase epitaxial growth and characterization of vertical InP nanowires. *Appl. Phys. Lett.* **2003**, *83*, 3371–3373. [CrossRef]

71. Motohisa, J.; Noborisaka, J.; Takeda, J.; Inari, M.; Fukui, T. Catalyst-free selective-area MOVPE of semiconductor nanowires on (111)B oriented substrates. *J. Cryst. Growth* **2004**, *272*, 180–185. [CrossRef]

72. Li, Q.; Gong, X.; Wang, C.; Wang, J.; Ip, K.; Hark, S. Size-dependent periodically twinned ZnSe nanowires. *Adv. Mater.* **2004**, *16*, 1436–1440. [CrossRef]

73. Johansson, J.; Karlsson, L.S.; Svensson, C.P.T.; Martensson, T.; Wacaser, B.A.; Deppert, K.; Samuelson, L.; Seifert, W. Structural properties of <111>B-oriented III–V nanowires. *Nat. Mater.* **2006**, *5*, 574–580. [CrossRef]

74. Tomioka, K.; Motohisa, J.; Hara, S.; Fukui, T. Crystallographic Structure of InAs nanowires studied by transmission electron microscopy. *Jpn. J. Appl. Phys.* **2007**, *46*, L1102–L1104. [CrossRef]

75. Tomioka, K.; Mohan, P.; Noborisaka, J.; Hara, S.; Motohisa, J.; Fukui, T. Growth of highly uniform InAs nanowire arrays by selective-area MOVPE. *J. Cryst. Growth* **2007**, *298*, 644–647. [CrossRef]

76. Cornet, D.M.; Mazzetti, V.G.M.; la Pierre, R.R. Onset of stacking faults in InP nanowires grown by gas source molecular beam epitaxy. *Appl. Phys. Lett.* **2007**, *90*, 013116. [CrossRef]

77. Spirkoska, D.; Arbiol, J.; Gustafsson, A.; Conesa-Boj, S.; Glas, F.; Zardo, I.; Heigoldt, M.; Gass, M.H.; Bleloch, A.L.; Estrade, S. Ga-assisted catalyst-free growth mechanism of GaAs nanowires by molecular beam epitaxy. *Phys. Rev. B* **2008**, *77*, 155326.

78. Plante, M.C.; la Pierre, R.R. Au-assisted growth of GaAs nanowires by gas source molecular beam epitaxy: Tapering, sidewall faceting and crystal structure. *J. Cryst. Growth* **2008**, *310*, 356–363. [CrossRef]

79. Algra, R.E.; Verheijen, M.A.; Borgström, M.T.; Feiner, L.; Immink, G.; van Enckevort, W.J.P.; Vlieg, E.; Bakkers, E.P.A.M. Twinning superlattices in indium phosphide nanowires. *Nature* **2008**, *456*, 369–372. [CrossRef] [PubMed]

80. Paiman, S.; Gao, Q.; Tan, H.H.; Jagadish, C.; Pemasiri, K.; Montazeri, M.; Jackson, H.E.; Smith, L.M.; Yarrison-Rice, J.M.; Zhang, X.; et al. The effect of V/III ratio and catalyst particle size on the crystal structure and optical properties of InP nanowires. *Nanotechnology* **2009**, *20*, 225606. [CrossRef] [PubMed]

81. Caroff, P.; Dick, K.A.; Johansson, J.; Messing, M.E.; Deppert, K.; Samuelson, L. Controlled polytypic and twin-plane superlattices in III–V nanowires. *Nat. Nanotechnol.* **2009**, *4*, 50–55. [CrossRef] [PubMed]

82. Cantoro, M.; Brammertz, G.; Richard, O.; Bender, H.; Clemente, F.; Leys, M.; Degroote, S.; Caymax, M.; Heyns, M.; de Gendt, S. Controlled III/V nanowire growth by selective-area vapor-phase epitaxy. *J. Electrochem. Soc.* **2009**, *156*, H860–H868. [CrossRef]

83. Vogel, A.T.; de Boor, J.; Wittemann, J.V.; Mensah, S.L.; Werner, P.; Schmidt, V. Fabrication of high-quality InSb nanowire arrays by chemical beam epitaxy. *Cryst. Growth Des.* **2011**, *11*, 1896–1900. [CrossRef]

84. Bao, J.; Bell, D.C.; Capasso, F.; Wagner, J.B.; Måttensson, T.; Trägårdh, J.; Samuelson, L. Optical properties of rotationally twinned InP nanowire heterostructures. *Nano Lett.* **2008**, *8*, 836–841. [CrossRef] [PubMed]

85. Yeh, C.-Y.; Lu, Z.W.; Froyen, S.; Zunger, A. Predictions and systematizations of the zinc-blende-wurtzite structural energies in binary octet compounds. *Phys. Rev. B* **1992**, *45*, R12130–R12133. [CrossRef]

86. Akiyama, T.; Sano, K.; Nakamura, K.; Ito, T. An empirical potential a roach to wurtzite-zinc-blende polytypism in group III–V semiconductor nanowires. *Jpn. J. Appl. Phys.* **2006**, *45*, L275–L278. [CrossRef]

87. Akiyama, T.; Nakamura, K.; Ito, T. Structural stability and electronic structures of InP nanowires: Role of surface dangling bonds on nanowire facets. *Phys. Rev. B* **2006**, *73*, 235308. [CrossRef]

88. Glas, F.; Harmand, J.-C.; Patriarche, G. Why does wurtzite Form in nanowires of III–V zinc blende Semiconductors? *Phys. Rev. Lett.* **2007**, *99*, 146101. [CrossRef] [PubMed]

89. Glas, F. A simple calculation of energy changes upon stacking fault formation or local crystalline phase transition in semiconductors. *J. Appl. Phys.* **2008**, *104*, 093520. [CrossRef]

90. Mohan, P.; Motohisa, J.; Fukui, T. Controlled growth of highly uniform, axial/radial direction-defined, individually addressable InP nanowire arrays. *Nanotechnology* **2005**, *16*, 2903–2907. [CrossRef]

91. Kitauchi, Y.; Kobayashi, Y.; Tomioka, K.; Hara, S.; Hiruma, K.; Fukui, T.; Motohisa, J. Structural transition in indium phosphide nanowires. *Nano Lett.* **2010**, *10*, 1699–1703. [CrossRef] [PubMed]

92. Wagner, R.S.; Elis, W.C. Vapor-liquid-solid mechanism of single crystal growth. *Appl. Phys. Lett.* **1964**, *4*, 89–90. [CrossRef]

93. Johansson, J.; Karlsson, L.S.; Dick, K.A.; Bolinsson, J.; Wacaser, B.A.; Deppert, K.; Samuelson, L.; Seifert, W. Effects of supersaturation on the crystal structure of gold seeded III–V nanowires. *Cryst. Growth Des.* **2009**, *9*, 766–773. [CrossRef]

94. Akiyama, T.; Kondo, T.; Tatematsu, H.; Nakamura, K.; Ito, T. Ab initio approach to reconstructions of the InP(111)A surface: Role of hydrogen atoms passivating surface dangling bonds. *Phys. Rev. B* **2008**, *78*, 205318. [CrossRef]

95. Yamashita, T.; Akiyama, T.; Nakamura, K.; Ito, T. Theoretical investigation of effect of side facets on adsorption–desorption behaviors of In and P atoms at top layers in InP nanowires. *Jpn. J. Appl. Phys.* **2011**, *50*, 55001. [CrossRef]

96. Taguchi, A.; Shiraishi, K.; Ito, T.; Kangawa, Y. Theoretical investigations of adatom adsorptions on the As-stabilized GaAs(111)A surface. *Surf. Sci.* **2001**, *493*, 173–177. [CrossRef]

97. Yamashita, T.; Akiyama, T.; Nakamura, K.; Ito, T. Theoretical investigation on temperature and pressure dependence of structural stability of InP thin layers grown on InP(111)A surface. *Surf. Sci.* **2013**, *610*, 16–21. [CrossRef]

98. Yamashita, T.; Sano, K.; Akiyama, T.; Nakamura, K.; Ito, T. Theoretical investigations on the formation of wurtzite segments in group III–V semiconductor nanowires. *Appl. Surf. Sci.* **2008**, *254*, 7668–7671. [CrossRef]

99. Yamashita, T.; Akiyama, T.; Nakamura, K.; Ito, T. Growth of side facets in InP nanowires: First-principles-based approach. *Surf. Sci.* **2013**, *609*, 207–214. [CrossRef]

100. Kangawa, Y.; Akiyama, T.; Ito, T.; Shiraishi, K.; Kakimoto, K. Theoretical approach to structural stability of c-GaN: How to grow cubic GaN. *J. Cryst. Growth* **2009**, *311*, 3106–3109. [CrossRef]

101. Yamashita, T.; Akiyama, T.; Nakamura, K.; Ito, T. Effects of facet orientation on relative stability between zinc blende and wurtzite structures in group III–V nanowires. *Jpn. J. Appl. Phys.* **2010**, *49*, 55003. [CrossRef]

102. Yamashita, T.; Akiyama, T.; Nakamura, K.; Ito, T. Theoretical investigation on the structural stability of GaAs nanowires with two different types of facets. *Phys. E* **2010**, *42*, 2727–2730. [CrossRef]

103. Ito, T.; Ishizaki, H.; Akiyama, T.; Nakamura, K.; Shiraishi, K.; Taguchi, A. An ab initio-based approach to phase diagram calculations for GaAs(001) surfaces. *e-J. Surf. Sci. Nanotechnol.* **2005**, *3*, 488–491. [CrossRef]

104. Ito, T.; Nakamura, T.; Akiyama, T.; Nakamura, K. An ab initio-based approach to phase diagram calculations for GaAs(001)-(2 × 4)γ surfaces. *Appl. Surf. Sci.* **2008**, *254*, 7663–7667. [CrossRef]

105. Akiyama, T.; Tatematsu, H.; Nakamura, K.; Ito, T. A first-principles surface-phase diagram study for Si-adsorption processes on GaAs(111)A surface under low As-pressure condition. *Surf. Sci.* **2010**, *604*, 171–174. [CrossRef]

106. Murase, I.; Akiyama, T.; Nakamura, K.; Ito, T. Ab initio-based approach to initial incorporation of Bi on GaAs(001)-c(4 × 4)α surfaces. *J. Cryst. Growth* **2013**, *378*, 21–24. [CrossRef]

107. Sugitani, T.; Akiyama, T.; Nakamura, K.; Ito, T. Ab initio-based approach to incorporation of nitrogen on GaAs(001) surfaces. *J. Cryst. Growth* **2013**, *378*, 29–33. [CrossRef]

108. Ito, T.; Sugitani, T.; Akiyama, T.; Nakamura, K. Ab initio-based approach to N-pair formation on GaAs(001)-(2 × 4) surface. *e-J. Surf. Sci. Nanotechnol.* **2014**, *12*, 6–10. [CrossRef]

109. Ito, T.; Akiyama, T.; Nakamura, K. An ab initio-based approach to phase diagram calculations for GaN(0001) surfaces. *Appl. Surf. Sci.* **2008**, *254*, 7659–7662. [CrossRef]

110. Ito, T.; Akiyama, T.; Nakamura, K. An ab initio-based approach to the stability of GaN(0001) under Ga-rich condition. *J. Cryst. Growth* **2009**, *311*, 3093–3096. [CrossRef]

111. Akiyama, T.; Ammi, D.; Nakamura, K.; Ito, T. Surface reconstructions on GaN and InN semipolar (10-1-1) surfaces. *Jpn. J. Appl. Phys.* **2009**, *48*, 100201-1-3. [CrossRef]

112. Akiyama, T.; Ammi, D.; Nakamura, K.; Ito, T. Stability of magnesium-incorporated semipolar GaN(10-1-1) surfaces. *Jpn. J. Appl. Phys.* **2009**, *48*, 110202-1-3. [CrossRef]

113. Yamashita, T.; Akiyama, T.; Nakamura, K.; Ito, T. Surface reconstructions on GaN and InN semipolar (11-22) surfaces. *Jpn. J. Appl. Phys.* **2009**, *48*, 120201-1-3. [CrossRef]

114. Akiyama, T.; Yamashita, T.; Nakamura, K.; Ito, T. Ab initio-based study for adatom kinetics on semipolar GaN(11-22) surfaces. *Jpn. J. Appl. Phys.* **2009**, *48*, 120218-1-3. [CrossRef]

115. Yamashita, T.; Akiyama, T.; Nakamura, K.; Ito, T. Surface reconstructions on GaN and InN semipolar (20-21) surfaces. *Jpn. J. Appl. Phys.* **2010**, *49*, 0180014-1-2. [CrossRef]

116. Ito, T.; Ito, T.; Akiyama, T.; Nakamura, K. Ab initio-based Monte Carlo simulation study for the structural stability of AlN grown on 4H-SiC(11–20). *e-J. Surf. Sci. Nanotechnol.* **2010**, *8*, 52–56. [CrossRef]

117. Akiyama, T.; Yamashita, T.; Nakamura, K.; Ito, T. Stability and indium incorporation on In$_{0.25}$Ga$_{0.75}$N surfaces under growth conditions: First-principles calculations. *Jpn. J. Appl. Phys.* **2010**, *49*, 030212-1-3. [CrossRef]

118. Akiyama, T.; Ammi, D.; Nakamura, K.; Ito, T. Surface reconstruction and magnesium incorporation on semipolar GaN(1-101) surfaces. *Phys. Rev. B* **2010**, *81*, 245317-1-6. [CrossRef]

119. Akiyama, T.; Nakamura, K.; Ito, T. Stability of hydrogen on nonpolar and semipolar nitride surfaces: Role of surface orientation. *J. Cryst. Growth* **2011**, *318*, 79–83. [CrossRef]

120. Akiyama, T.; Nakamura, K.; Ito, T. Stability of carbon incorporated semipolar GaN(1-101). *Jpn. J. Appl. Phys.* **2011**, *50*, 080216-1-3.

121. Akiyama, T.; Obara, D.; Nakamura, K.; Ito, T. Reconstructions on AlN polar surfaces under hydrogen rich conditions. *Jpn. J. Appl. Phys.* **2012**, *51*, 018001-1-2. [CrossRef]

122. Akiyama, T.; Saito, Y.; Nakamura, K.; Ito, T. Stability of nitrogen incorporated Al$_2$O$_3$ surfaces: Formation of AlN layers by oxygen desorption. *Surf. Sci.* **2012**, *606*, 221–225. [CrossRef]

123. Akiyama, T.; Obara, D.; Nakamura, K.; Ito, T. Reconstructions on AlN nonpolar surfaces in the presence of hydrogen. *Jpn. J. Appl. Phys.* **2012**, *51*, 048002-1-2. [CrossRef]

124. Akiyama, T.; Nakamura, K.; Ito, T. Ab initio-based study for adatom kinetics on AlN 0001) surfaces during metal-organic vapor-phase epitaxy growth. *Appl. Phys. Lett.* **2012**, *100*, 251601-1-3. [CrossRef]
125. Akiyama, T.; Saito, Y.; Nakamura, K.; Ito, T. Nitridation of Al$_2$O$_3$ surfaces: Chemical and structural change triggered by oxygen desorption. *Phys. Rev. Lett.* **2013**, *110*, 026101-1-5. [CrossRef] [PubMed]
126. Takemoto, Y.; Akiyama, T.; Nakamura, K.; Ito, T. Systematic theoretical investigations on surface reconstructions and adatom kinetics on AlN semipolar surfaces. *e-J. Surf. Sci. Nanotechnol.* **2015**, *13*, 239–243. [CrossRef]
127. Takemoto, Y.; Akiyama, T.; Nakamura, K.; Ito, T. Ab initio-based study for surface reconstructions and adsorption behavior on semipolar AlN(11-22) surfaces during metal-organic vapor-phase epitaxial growth. *Jpn. J. Appl. Phys.* **2015**, *54*, 085502-1-5. [CrossRef]
128. Akiyama, T.; Takemoto, Y.; Nakamura, K.; Ito, T. Theoretical investigations for initial growth processes on semipolar AlN(11-22) surfaces under metal-organic vapor-phase epitaxiy growth conditions. *Jpn. J. Appl. Phys.* **2016**, *55*, 05FM06-1-4. [CrossRef]
129. Ito, T.; Sano, K.; Akiyama, T.; Nakamura, K. A simple approach to polytypes of SiC and its application to nanowires. *Thin Solid Films* **2006**, *508*, 243–246. [CrossRef]
130. Akiyama, T.; Sano, K.; Nakamura, K.; Ito, T. An Empirical Interatomic Potential Approach to Structural Stability of ZnS and ZnSe Nanowires. *Jpn. J. Appl. Phys.* **2007**, *46*, 1783–1787. [CrossRef]
131. Akiyama, T.; Freeman, A.J.; Nakamura, K.; Ito, T. Electronic structures and optical properties of GaN and ZnO nanowires from first principles. *J. Phys. Conf. Ser.* **2008**, *100*, 052056. [CrossRef]

© 2017 by the authors. Licensee MDPI, Basel, Switzerland. This article is an open access article distributed under the terms and conditions of the Creative Commons Attribution (CC BY) license (http://creativecommons.org/licenses/by/4.0/).

Article

Kinetics and Morphology of Flow Induced Polymer Crystallization in 3D Shear Flow Investigated by Monte Carlo Simulation

Chunlei Ruan

School of Mathematics and Statistics, Henan University of Science and Technology, Luoyang 471023, China; ruanchunlei622@mail.nwpu.edu.cn

Academic Editor: Hiroki Nada
Received: 2 January 2017; Accepted: 8 February 2017; Published: 11 February 2017

Abstract: To explore the kinetics and morphology of flow induced crystallization of polymers, a nucleation-growth evolution model for spherulites and shish-kebabs is built based on Schneider rate model and Eder model. The model considers that the spherulites are thermally induced, growing like spheres, while the shish-kebabs are flow induced, growing like cylinders, with the first normal stress difference of crystallizing system being the driving force for the nucleation of shish-kebabs. A two-phase suspension model is introduced to describe the crystallizing system, which Finitely Extensible Non-linear Elastic-Peterlin (FENE-P) model and rigid dumbbell model are used to describe amorphous phase and semi-crystalline phase, respectively. Morphological Monte Carlo method is presented to simulate the polymer crystallization in 3D simple shear flow. Roles of shear rate, shear time and shear strain on the crystallization kinetics, morphology, and rheology are analyzed. Numerical results show that crystallization kinetics, morphology and rheology in shear flow are qualitatively in agreement with the theoretical, experimental and other numerical works which verifies the validity and effectiveness of our model and algorithm. To our knowledge, this is the first time that a model and an algorithm revealing the details of crystal morphology have been applied to the flow induced crystallization of polymers.

Keywords: polymer crystallization; flow induced crystallization; Morphological Monte Carlo simulation; shish-kebabs

1. Introduction

Polymer crystallization is an important factor affecting the microstructure and determining the mechanical properties of the products [1,2]. Usually, polymers are processed with techniques such as extrusion and injection molding. During the manufacturing processing, polymers experience complex flow and thermal condition with the internal chains changing and folding to form different types of crystalline structures. Hence, studies related to the crystalline structures forming and the kinetics of crystallization under different flow and thermal condition are important.

Polymer crystallization in the flow field is also called flow induced crystallization (FIC) [2]. The experimental studies of FIC show that crystallization occuring in the flow field not only accelerates the crystallization rate, but also leads to different types of crystalline structures when compared with the quiescent crystallization [2], namely, both spherulite and shish-kebab structures, a typical oriented crystalline structure under strain, where the extended molecular chains form the shish and remaining molecular chains fold to form the lamellar structure which looks like kebabs, are found in FIC while only spherulite structure is found in quiescent crystallization. Based on the experimental results, many researchers proposed different analytical models for FIC which are mostly based on the Nakamura equation and the Avrami–Kolmogorov equation [3]. For example, Doufas et al. [4],

Tanner [5], and Ziabicki [6] applied a multiplying factor function of stress, shear rate, and orientation, respectively, to modify the crystallization kinetic constant in the original Nakamura model to take into account the effect of flow on crystallization. Eder [1], Kosher and Fulchiron [7], and Zheng and Kennedy [8] molded the effect of flow on crystallization by considering nucleation and modified the original Avrami–Kolmogorov model. The modified Nakamura model does predict well in FIC, however, it has the disadvantage that it cannot reveal the details of crystal morphology. The Avrami–Kolmogorov model, which is based on the morphology evolution, has the disadvantage of lower accuracy at the later stage of polymer crystallization. Eder [1] proposed a mathematical model based upon the crystal morphology to consider the effect of flow on crystallization. Through considering spherulites as the growing spheres and shish-kebabs as the growing cylinders, they obtained a series of differential equations using the Schneider rate equations [9]. Zuidema et al. [10] modified the shear rate in the Eder model by recoverable strain as the driving force for flow induced nucleation. Their work has taken a huge step in revealing the microstructures of the polymer products. However, they did not give the method to capture the details of the nucleation-growth-impingement of crystals. Therefore, their work requires using the crystallization kinetics equation. Boutaous et al. [11] used the Schneider rate equation to describe the growth of thermally and flow induced nuclei and explored the contribution of thermal and flow effects on the global crystallization kinetics under different shear flow. They applied Avrami model to describe the kinetics and took the crystal structure induced by flow as spherulite.

In order to avoid using crystallization kinetics model, morphological simulation is needed. In the morphological simulation, relative crystallinity is transferred to the volume fraction of crystals [12]. Thus far, there have been many studies on the morphological simulation of polymer crystallization. Examples include: Raabe [12–14], Lin et al. [15], and Spina et al. [16,17] presented a cellular automaton method to simulate the kinetics and topology of spherulite growth for polymer crystallization; Liu et al. [18,19] used a level set method to capture the growth and impingement of spherulites during the polymer cooling stage; Micheletti and Burge [20] and Ruan et al. [21,22] applied a pixel coloring method to model and simulate the crystallization of polymer and short fiber reinforced polymer; and Ketdee and Anantawaraskul [23] and Ruan et al. [24] presented the Monte Carlo simulation in study of crystallization kinetics and morphology development in polymer crystallization. However, we shall mention that these works were mainly concentrated on spherulite structure. Our work [24] was an exception. In our previous work [24], we applied a Monte Carlo method to capture the evolution of both spherulites and shish-kebabs and calculate the crystallization kinetics in polymer crystallization. The work was carried out with parametric study where the effects of nucleation density and growth rate of spherulites, nucleation density and length growth rate of shish-kebabs on the crystallization were examined. The work was in an ideal case, parameters of both spherulites and shish-kebabs were keeping constant to allow the simulation. This was not the case in the real manufacturing process.

In this paper, we focus our attention on the more realistic shear flow which exists universally in the manufacturing process and experiments. Based on the Schneider rate model and Eder model, the morphology evolution model of both spherulites and shish-kebabs is deduced. By using this model and the Monte Carlo method, polymer crystallization in 3D simple shear flow is simulated. Effects of shear rate, shear time and shear strain on the crystallization kinetics, crystal morphology, and rheology of the system are discussed.

2. Mathematical Model and Numerical Method

2.1. Morphology Evolution Model for Spherulites and Shish-Kebabs

In the flow field, polymers experience the complex thermal and flow condition, and different crystalline structures like spherulites and shish-kebabs are presented. Both types of crystals contribute to the crystallization kinetics. Like many other work, here we assume the spherulites are thermally induced and the shish-kebabs are flow induced.

For the spherulite structure, Schneider et al. [9] considered the spherulites as the growing spheres and deduced a series of differential equations. These equations, also known as the Schneider rate equations, are listed as follows [9]:

$$
\begin{aligned}
\dot{\phi}_3 &= 8\pi a & (\phi_3 = 8\pi N_s) \\
\dot{\phi}_2 &= G_s \phi_3 & (\phi_2 = 4\pi R_{tot}) \\
\dot{\phi}_1 &= G_s \phi_2 & (\phi_1 = S_{tot}) \\
\dot{\phi}_0 &= G_s \phi_1 & (\phi_0 = V_{tot})
\end{aligned}
\tag{1}
$$

where N_s, R_{tot}, S_{tot}, V_{tot} are the total number, total radius, total surface area and the total volume of spherulites, respectively; a is the nucleation rate; and G_s is the growth rate of spherulites.

For the shish-kebab structure, Eder [1] considered the shish-kebabs as the growing cylinders and obtained a series of differential equations. These equations, known as Eder model, can be described as follows [1]

$$
\begin{aligned}
\dot{\psi}_3 + \tfrac{\psi_3}{\tau_n} &= 8\pi R_1 & (\psi_3 = 8\pi N_{s-k}) \\
\dot{\psi}_2 + \tfrac{\psi_2}{\tau_l} &= \psi_3 R_2 & (\psi_2 = 4\pi L_{tot}) \\
\dot{\psi}_1 &= G_{s-k,r} \psi_2 & (\psi_1 = \tilde{S}_{tot}) \\
\dot{\psi}_0 &= G_{s-k,r} \psi_1 & (\psi_0 = \tilde{V}_{tot})
\end{aligned}
\tag{2}
$$

where N_{s-k}, L_{tot}, $\tilde{S}_{tot}$, $\tilde{V}_{tot}$ are the total number, total length, total surface area and the total volume of shish-kebabs, respectively; τ_n is the temperature dependent relaxation time for the nuclei formation; $R_1 = \dot{\gamma}^2 g_n / \dot{\gamma}_n^2$ is a driving force for nucleation of shish-kebabs, with $\dot{\gamma}$ the shear rate and $g_n / \dot{\gamma}_n^2$ the fitted parameters; τ_l is the temperature and shish-length-dependent relaxation time for the shish during axial growth; $R_2 = \dot{\gamma}^2 g_l / \dot{\gamma}_l^2$ is a driving force of length growth of shish-kebabs, with $g_l / \dot{\gamma}_l^2$ the fitted parameters; and $G_{s-k,r}$ is the radius growth rate of shish-kebabs.

The equivalent differential equations of spherulites can be deduced from Equation (1):

$$
\begin{aligned}
\dot{N}_s &= N_s \\
\dot{R}_{tot} &= 2N_s G_s \\
\dot{S}_{tot} &= 4\pi G_s R_{tot} \\
\dot{V}_{tot} &= G_s S_{tot}
\end{aligned}
\tag{3}
$$

From Equation (3), we know that two parameters can define the crystallization of spherulites, namely, the nucleation density of spherulites N_s and the growth rate of spherulites G_s. Different kinds of nucleation models for spherulites were proposed by researchers, which are mostly based on data fitting.

Here, we adopt the model proposed by Koscher and Fulchiron [7] and use the following equation to describe the nucleation density of spherulites

$$
N_s(T) = \exp\left(\tilde{a}\Delta T + \tilde{b}\right)
\tag{4}
$$

In Equation (4), nucleation density is a function of supercooling temperature ΔT which is defined as $\Delta T = T_m^0 - T$ with T_m^0 the equilibrium melting temperature, and $\tilde{a}$ and $\tilde{b}$ are the empirical parameters. Equation (4) clearly shows that the nucleation of spherulites is induced by thermal condition.

As reported by researchers [3,25], growth rate of spherulites does not seem to be strongly influenced by flow. Here, Hoffman–Lauriten expression [26] is used to describe it, namely,

$$
G_s(T) = G_0 \exp\left[-\frac{U^*}{R_g(T - T_\infty)}\right] \exp\left(-\frac{K_g}{T\Delta T}\right)
\tag{5}
$$

where G_0 and K_g are constants, U^* is the energy parameter similar to an apparent activation energy of motion, R_g is the gas constant and $T_\infty = T_g - 30\,^\circ\text{C}$ is considered as the temperature at which no further molecular displacement is possible.

The equivalent differential equations of shish-kebabs can be deduced from Equation (2):

$$\begin{aligned}
\dot{N}_{s-k} + \frac{N_{s-k}}{\tau_n} &= R_1 \\
\dot{L}_{tot} + \frac{L_{tot}}{\tau_l} &= 2N_{s-k}R_2 \\
\dot{\tilde{S}}_{s-ktot} &= 4\pi G_{s-k,r}L_{tot} \\
\dot{\tilde{V}}_{tot} &= G_{s-k,r}\tilde{S}_{tot}
\end{aligned} \tag{6}$$

Under the assumption that $\tau_l = \infty$ [10], we obtain the following expressions

$$\begin{aligned}
N_{s-k} &= N_{s-k} \\
\dot{L}_{tot} &= 2N_{s-k}R_2 = 2N_{s-k}G_{s-k,l} \\
\dot{\tilde{S}}_{s-ktot} &= 4\pi G_{s-k,r}L_{tot} \\
\dot{\tilde{V}}_{tot} &= G_{s-k,r}\tilde{S}_{tot}
\end{aligned} \tag{7}$$

From Equation (7), we know that three parameters can define the crystallization of shish-kebabs, namely, the nucleation density of shish-kebabs N_{s-k}, the length growth rate of shish-kebabs $G_{s-k,l}$ and the radius growth rate of shish-kebabs $G_{s-k,r}$. According to Eder [1], length growth rate of shish-kebabs $G_{s-k,l}$ can be written as

$$G_{s-k,l} = R_2 = \dot{\gamma}^2 g_l / \dot{\gamma}_l^2 \tag{8}$$

The radius growth rate of shish-kebabs $G_{s-k,r}$ is often assumed to be equal to the growth rate of spherulites G_s [10], namely

$$G_{s-k,r} = G_s \tag{9}$$

Due to the fact that the driving force for the nucleation density of shish-kebabs N_{s-k} is not well understood, several approaches are found in literatures. We have explained these in the Introduction Section. Here, we adopt the model proposed by Koscher and Fulchiron [7], which is

$$\dot{N}_{s-k} = CN_1 \tag{10}$$

where C is a constant, N_1 is the first normal stress difference of the system. Equation (10) shows that the nucleation of shish-kebabs is induced by flow condition.

2.2. Amorphous Phase and Semi-Crystalline Phase Model

The first normal stress difference appears in Equation (10); hence, it is necessary to give the mathematical model of the crystallizing system. Here, we adopt the idea of Zheng and Kennedy [8] and use a two-phase suspension model to deal with the crystallizing system. According to Zheng and Kennedy [8], the crystallizing system can be treated as a suspension of semi-crystalline phase growing and spreading in a matrix of amorphous material. The amorphous phase can be described as the FENE-P dumbbell model and the semi-crystalline phase can be described as the rigid dumbbell model.

In the amorphous phase, the matrix can be treated as the elastic dumbbell model, which is two beads connected by a spring. This dumbbell model obeys the well-known Fokker–Planck equation. There are three kinds of numerical methods to solve the Fokker–Planck equation: deterministic method, stochastic method and macroscopic method [27]. In the macroscopic method, through the moment operation in Fokker–Planck equation, the relating constitutive equation is obtained. However, this constitutive equation never closed and needs the closure approximation. The familiar closure approximations are Finitely Extensible Non-linear Elastic-Peterlin (FENE-P), Finitely Extensible Non-linear Elastic-Chilcott-Rallison (FENE-CR), Finitely Extensible Non-linear

Elastic-Lielens (FENE-L), Finitely Extensible Non-linear Elastic-Lielens-Simplified (FENE-LS), etc. [27]. Here, the FENE-P model is used, which is given by [8,27]

$$\lambda_a(T)\overset{\nabla}{C} + [\frac{1}{1 - trC/b}C - I] = 0 \tag{11}$$

where C is the conformation tensor, $\lambda_a(T)$ is the relaxation time of the fluid, I is the unit tensor, $tr(\bullet)$ is the trace of the tensor, and $\overset{\nabla}{C} = DC/Dt - (\nabla u)^T \cdot C - C \cdot (\nabla u)$ is the upper-convected derivative of C. The relaxation time of the fluid $\lambda_a(T)$ is a function of temperature and can be calculated by the shift factor $a_T(T)$ as follows [8]

$$\lambda_a(T) = a_T(T)\lambda_{a,0} = \exp[\frac{E_a}{R_g}(\frac{1}{T} - \frac{1}{T_0})]\lambda_{a,0} \tag{12}$$

where $\lambda_{a,0}$ is the relaxation time at the reference temperature T_0. E_a/R_g is the constant and can be determined by experiment. The stress contributed by amorphous phase can be written as [8,23]

$$\tau_a = nkT[\frac{1}{1 - trC/b}C - I] \tag{13}$$

with τ_a the stress caused by amorphous phase, n the number of dumbbells, and k the Boltzmann constant.

The molecular chains in semi-crystalline phase can be treated as the rigid dumbbell model, i.e., two beads connected by a rigid rod. This rigid dumbbell cannot be stretched but can be oriented. Through the force analysis, the orientation equation of rigid dumbbell can be obtained. Substitution of orientation equation into continuity equation of configurational distribution function leads to the well-known Fokker–Planck equation [8,27]. Here, we also use macroscopic method to solve it. By moment operation of Fokker–Planck equation, evolution equation of orientation tensor is obtained [8]:

$$\overset{\nabla}{< RR >} = -\frac{1}{\lambda_{sc}(\alpha, T)}(< RR > -\frac{I}{3}) - \dot{\gamma} :< RRRR > \tag{14}$$

where $< RR >$ is the second-order orientation tensor, $\lambda_{sc}(\alpha, T)$ is the time constant of the rigid dumbbell, and $\dot{\gamma}$ is the shear rate tensor. Time constant of the rigid dumbbell $\lambda_{sc}(\alpha, T)$ is related with the relaxation time of fluid $\lambda_a(T)$ by the following empirical form [4,8]

$$\frac{\lambda_{sc}(\alpha, T)}{\lambda_a(T)} = \frac{(\alpha/A)^{\beta_1}}{(1 - \alpha/A)^{\beta}} \qquad \alpha < A \tag{15}$$

where A, β, β_1 are the empirical parameters. Note that the fourth-order orientation tensor appears in Equation (14). In order to find the solution of second-order orientation $< RR >$, the closure approximation is needed. Different closure approximations are reported, including Linear, Quadratic, Hybrid, Invariant Based Orthotropic Fitted closure (IBOF), Eigenvalue Based Orthotropic Fitted closure (EBOF), etc. [28]. Here, we adopt the Quadratic closure approximation which is given by

$$< RRRR >_{ijkl} = < RR >_{ij}< RR >_{kl} \tag{16}$$

Stress caused by semi-crystalline phase τ_{sc} is written as follows [8]

$$\tau_{sc} = \frac{\eta_{sc}(\alpha, T)}{\lambda_{sc}(\alpha, T)}(< RR > +\lambda_{sc}(\alpha, T)\dot{\gamma} :< RRRR >) \tag{17}$$

where $\eta_{sc}(\alpha, T)$ is the viscosity of semi-crystalline phase which has the following relation with the viscosity of amorphous phase $\eta_a(T)$ [4,8]

$$\frac{\eta_{sc}(\alpha, T)}{\eta_a(T)} = \frac{(\alpha/A)^{\beta_1}}{(1 - \alpha/A)^{\beta}} \qquad \alpha < A \qquad (18)$$

Hence, the total stress of the crystallizing system is

$$\tau = \tau_a + \tau_{sc} \qquad (19)$$

which contains the contribution of both amorphous phase and semi-crystalline phase. The first normal stress difference in Equation (10) which is considered as the driving force for nucleation of shish-kebabs, as calculated by Equation (19).

2.3. Numerical Method

Monte Carlo method and finite difference method are used to capture the evolution of crystal morphology and to compute the evolution equation of amorphous phase and semi-crystalline phase, respectively.

2.3.1. Monte Carlo Method

Monte Carlo method is introduced here to capture the nucleation-growth-impingement of spherulites and shish-kebabs. We consider the polymer in a small spatial region, $[0, 1]mm \times [0, 1]mm \times [0, 1]mm$. The investigation is carried out under a certain temperature, shear rate and shear time. The nucleation density N_s and growth rate of spherulites G_s are given in Equations (4) and (5), respectively. The nucleation density of shish-kebabs N_{s-k} is listed in Equation (10). Length growth rate $G_{s-k,l}$ and radius growth rate of shish-kebabs $G_{s-k,r}$ are presented in Equations (8) and (9), respectively.

Figure 1 shows the Monte Carlo method we used in the simulation. To better implement this method, we refer to our work [24] for more details. Here, we briefly present the important techniques and parameters. Firstly, spatial region is divided into a large array of equally sized cubic cells and in our simulation this number is set as 10^7. Secondly, different crystals are distinguished by different colors. Different colors are assigned to the different nuclei and the spatial cells covered by growth are assigned to the same color with the corresponding crystal. Thirdly, relative crystallinity α is transferred to the volume fraction of crystals, which is calculated by the cells that have been transformed to the crystals with the total spatial number.

The main advantages of Monte Carlo method are that it can avoid the use of crystallization kinetics model and it can also capture the detailed morphology evolution.

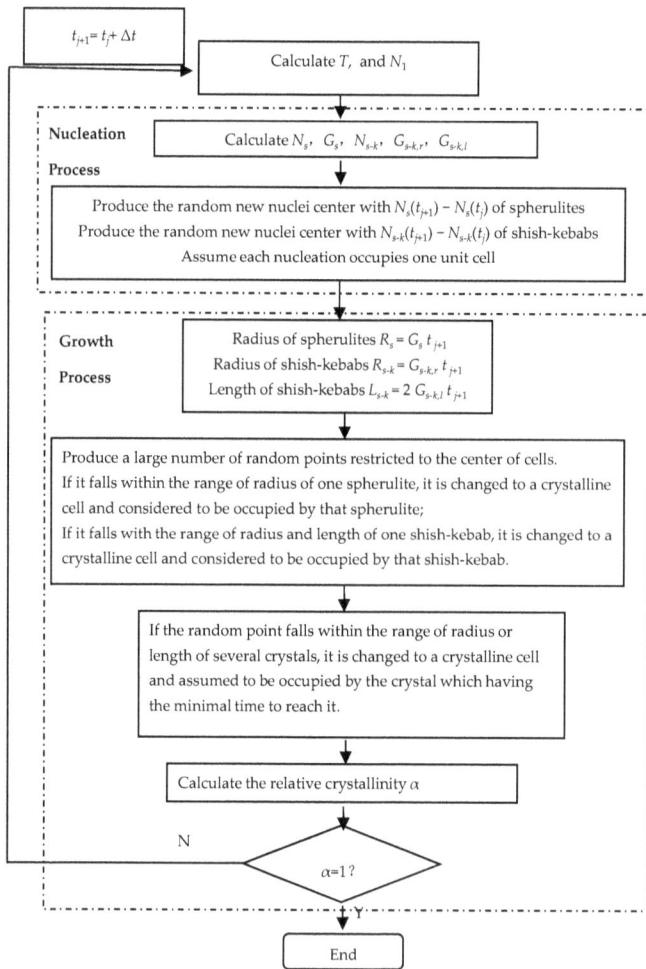

Figure 1. Flow chart for the Monte Carlo method in the simulation.

2.3.2. Finite Difference Method

Finite difference method is introduced to compute the equations in amorphous phase and semi-crystalline phase. Evolution of conformation tensor Equation (11) and orientation tensor Equation (14) are discretized by the first-order forward in time:

$$\frac{C^{n+1} - C^n}{\Delta t} = -\frac{1}{\lambda_a(T)}\left[\frac{1}{1 - trC^n/b}C^n - I\right] + (\nabla u)^T \cdot C^n + C^n \cdot (\nabla u) \tag{20}$$

$$\frac{<RR>^{n+1} - <RR>^n}{\Delta t} = -\frac{1}{\lambda_{sc}(\alpha,T)}\left(< RR >^n - \frac{I}{3}\right) - \dot{\gamma} :< RRRR > \\ +(\nabla u)^T \cdot < RR >^n + < RR >^n \cdot (\nabla u) \tag{21}$$

with the initial condition $C^0 = \frac{I}{3}$, $< RR >^0 = \frac{I}{3}$.

3. Results and Discussion

3.1. Parameters

The polymer we used here is the polyethylene. Material data and the parameters are listed in Table 1. Parameters for crystal morphology can be found in [7,10], and the parameters in amorphous phase and the semi-crystalline phase can be found in [8,29].

Table 1. Material data and input parameters.

Variables	Definition	Values	Variables	Definition	Values
$\tilde{a}$	Equation (4)	1.56×10^{-1}	$\lambda_{a,0}$	Equation (12)	4.00×10^{-2}s
$\tilde{b}$	Equation (4)	1.51×10^1	T_0	Equation (12)	476.15K
G_0	Equation (5)	2.83×10^2m/s	E_a/R_g	Equation (12)	5.602×10^3K
U^*/R_g	Equation (5)	755 K	b	Equations (11) and (13)	5
K_g	Equation (5)	5.5×10^5K^2	n	Equation (13)	$1.26 \times 10^{26}/m^3$
T_m^0	Equation (5)	483 K	k	Equation (13)	1.38×10^{-23}
T_g	Equation (5)	269 K	β	Equations (15) and (18)	9.2
$g_l/\dot{\gamma}_l^2$	Equation (8)	2.69×10^{-8}	β_1	Equations (15) and (18)	0.05
C	Equation (10)	10^6Pa$^{-1}\cdot$s$^{-1}\cdot$ m^{-1}	A	Equations (15) and (18)	0.44

3.2. Validity of the Simulation

To show the validity of our algorithm, results of relative crystallinity simulated by Monte Carlo method are compared with the data predicted by the Avrami model which are descripted in Figure 2. Here, we assume the nucleation of spherulites and shish-kebabs occur instantaneously with the density $N_s = 10^{12}/m^3$ and $N_{s-k} = 10^{12}/m^3$, respectively, spherulites growing with the rate $G_s = 10^{-6}$m/s and shish-kebabs growing with the length rate $G_{s-k,l} = 10^{-5}$m/s and radius rate $G_{s-k,r} = 10^{-6}$m/s, respectively. As can be seen in Figure 2, the simulation data show agreement with the Avrami model. Hence, the Monte Carlo method used is efficient and reliable.

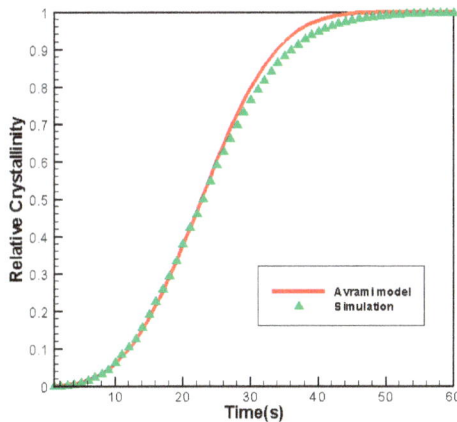

Figure 2. Comparison of simulation result with Avrami model.

We now show the reliability of our model. Simulations are carried out in 3D simple shear flow. Figure 3 displays the shear rate with the half crystallization time when the polymer suffers a constant shear time of 10 s. Results are compared with the experimental data obtained by Koscher and Fulchiron [7]. Our model predictions are in qualitative agreement with the experimental results. Therefore, the model we built is valid.

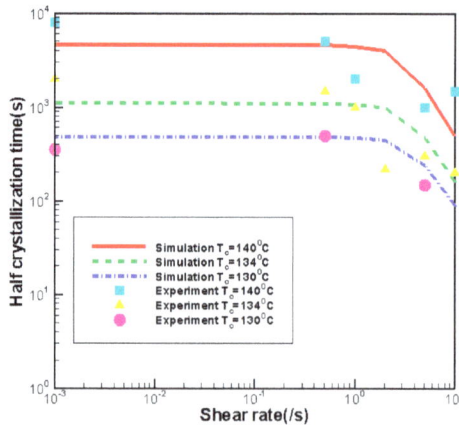

Figure 3. Comparison of simulation result with the experimental result [7].

3.3. Effects of Shear Rate

In this section, we show the effects of shear rate on the crystallization and rheology. Here, we set the shear time as $t_s = 10\,\text{s}$ and the temperature as $T_c = 140\,^\circ\text{C}$.

3.3.1. Effects of Shear Rate on Crystallization

Figure 4 gives the number of shish-kebabs with the shear rate $\dot{\gamma} = 0/\text{s}, 1/\text{s}, 2/\text{s}, 5/\text{s}, 10/\text{s}$. The case $\dot{\gamma} = 0/\text{s}$ represents the quiescent case. It is clear that the number of shish-kebabs increases rapidly with the increase of shear rate. After the cessation of shear, the number of shish-kebabs keeps constant.

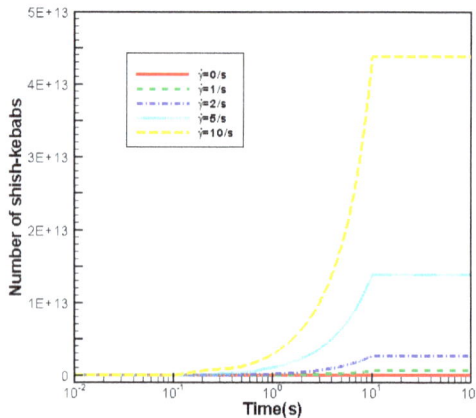

Figure 4. Number of shish-kebabs with different shear rates.

Figure 5 shows the relative crystallinity with the different shear rates. It is evident that crystallization rate is quicker in the case of considering the shearing effect. This accelerating effect is mainly contributed by the shear induced shish-kebabs. Due to the shear effect in the flow field, the nucleation and the length growth rate of shish-kebabs are provided. These promise the growth of shish-kebabs and contribute to the acceleration of crystallization process. As can be seen in Figure 5, increase of shear rate significantly increases the crystallization rate. This trend is consistent with the simulation results of Zheng et al. [8], Boutaous et al. [11] and Rong et al. [30].

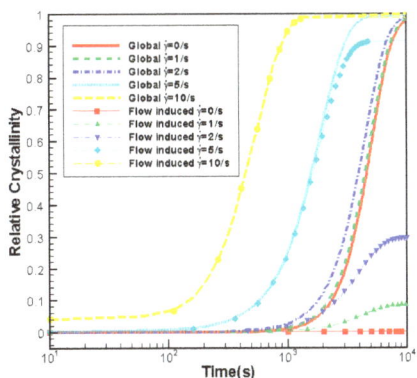

Figure 5. Relative crystallinity comparison with different shear rates.

Figure 6 plots the predicted crystal morphology with the shear rate $\dot{\gamma} = 0/\text{s}, 5/\text{s}, 10/\text{s}$ when $\alpha \approx 0.5$. In the quiescent case ($\dot{\gamma} = 0/\text{s}$), the crystal structure is spherulite, while in the shearing case ($\dot{\gamma} = 5/\text{s}, 10/\text{s}$), the crystal structures are spherulite and shish-kebab. As we can see, with the increase of shear rate, the global number of crystals increases obviously. In the $\dot{\gamma} = 5/\text{s}$ case, the shish-kebab structure is not notable. However, in the $\dot{\gamma} = 10/\text{s}$ case, shish-kebab structure is apparent. Thus, we can conclude that the increase of shear rate will lead to a higher anisotropy of the shish-kebab structure and also more impact of shish-kebab on the global crystal morphology.

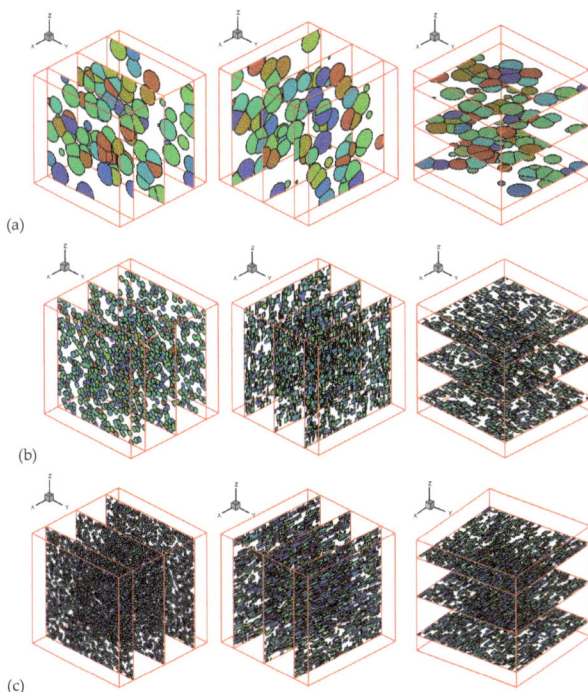

Figure 6. Morphology comparison with different shear rates ($\alpha \approx 0.5$). (**a**) $\dot{\gamma} = 0\,\text{s}^{-1}$; (**b**) $\dot{\gamma} = 5\,\text{s}^{-1}$; and (**c**) $\dot{\gamma} = 10\,\text{s}^{-1}$.

3.3.2. Effects of Shear Rate on Rheology

Figure 7 shows the evolution of viscosity in the system with different shear rates. It is obvious that the viscosity increases slowly with time before it reaches the critical value; however, when it reaches the critical value, the viscosity increases dramatically. This is caused by the crystallization. As is shown in Equation (18), the viscosity of semi-crystalline phase is calculated as $\eta_{sc}(\alpha, T) = (\alpha/A)^{\beta_1}\eta_a(T)/(1-\alpha/A)^{\beta}$; As $\alpha \to A$, $\eta_{sc} \to \infty$. Thus, the viscosity of system changes dramatically as $\alpha \to A$. Besides, the higher shear rate leads to an earlier sudden increase in viscosity. This is also in agreement with the work by Zheng et al. [8].

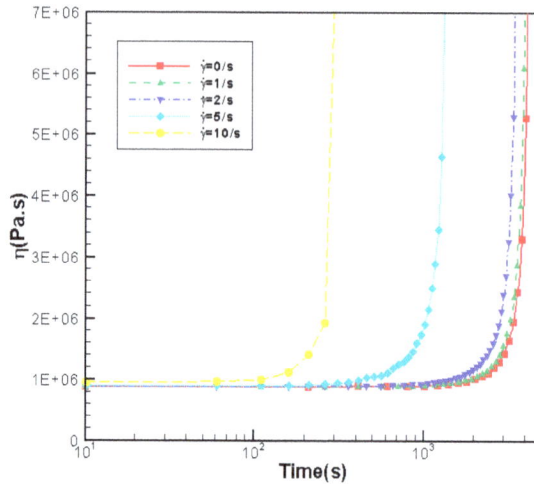

Figure 7. Evolution of viscosity with different shear rate.

3.4. Effects of Shear Time

In this section, we discuss the effects of shear time on the crystallization and rheology. The shear rate is set as $\dot{\gamma} = 10/s$ and the temperature is set as $T_c = 140\,°C$.

3.4.1. Effects of Shear Time on Crystallization

Relative crystallinity with shear time $t_s = 0\,s, 1\,s, 2\,s, 5\,s, 10\,s$ is shown in Figure 8. Crystallization rate in the shear flow increases more noticeably than in the quiescent condition (shear time $t_s = 0\,s$). Additionally, crystallization rate increases rapidly with the increase of shear time. This acceleration effect is also caused by the flow induced shish-kebabs. As can be seen in Figure 8, the contribution of relative crystallinity induced by flow increases as the shear time increases. Results here are also in consist with the work by Zheng et al. [8], Boutaous et al. [11] and Rong et al. [30].

Crystal morphology when $\alpha \approx 0.5$ with the shear time $t_s = 5\,s, 10\,s, 15\,s$ is plotted in Figure 9. As expected, shish-kebab structure is more apparent in the case with longer shear time. The morphology obtained here is similar to the experimental results by Koscher and Fulchiron [7].

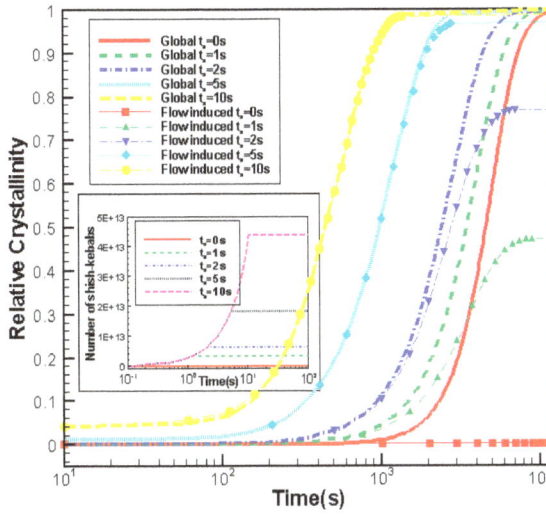

Figure 8. Relative crystallinity comparison with different shear time.

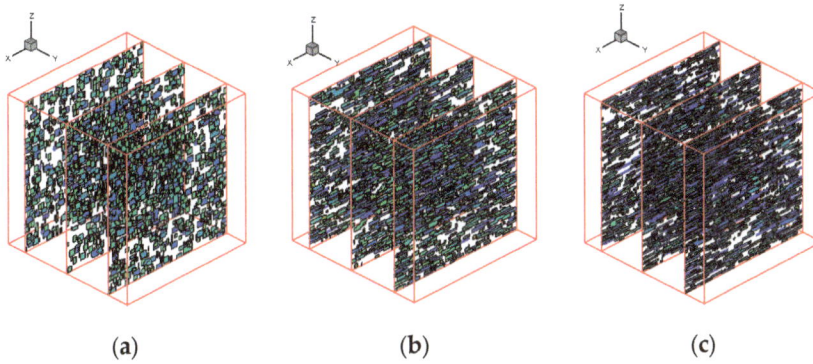

Figure 9. Morphology comparison with different shear time ($\alpha \approx 0.5$): (**a**) $t_s = 5$ s; (**b**) $t_s = 10$ s; and (**c**) $t_s = 15$ s.

3.4.2. Effects of Shear Time on Rheology

In Figure 10, the evolution of viscosity with different shear time is given. The viscosity of the system changes slowly at first, but becomes suddenly very large when time reaches a certain critical value. Besides, the longer the shear time, the earlier occurring of the sudden increase in viscosity. This is also caused by the crystallization process, which we explained in Section 3.3.2.

3.5. Effects of Shear Strain

In this section, we discuss the effects of shear strain. We set the temperature as $T_c = 140\,^\circ$C.

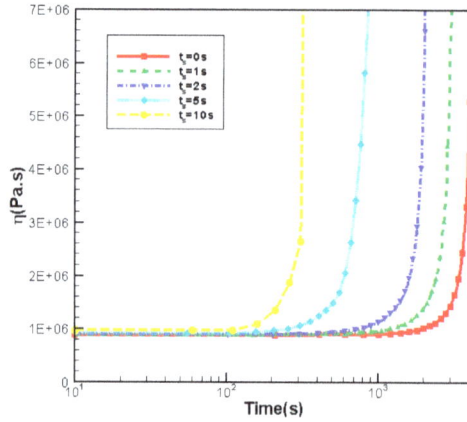

Figure 10. Evolution of viscosity with different shear time.

3.5.1. Effects of Shear Strain on Crystallization

Figure 11 plots the evolution of relative crystallinity with time at the total shear strain equal to 50 in three cases: shear rate $\dot{\gamma} = 2\,\mathrm{s}^{-1}$ and shear time $t_s = 25\,\mathrm{s}$, shear rate $\dot{\gamma} = 5\,\mathrm{s}^{-1}$ and shear time $t_s = 10\,s$, and shear rate $\dot{\gamma} = 10\,\mathrm{s}^{-1}$ and shear time $t_s = 5\,\mathrm{s}$. As shown in Figure 11, the case with higher shear rate and shorter shear time ($\dot{\gamma} = 10\ \mathrm{s}^{-1}$, $t_s = 5\,\mathrm{s}$) obtains the quickest crystallization rate. This is mainly due to the following two reasons: (1) as seen in Figure 11, the number of shish-kebabs in this case is the largest; and (2) the length of shish-kebabs L_{s-k} is the product of the length growth rate $G_{s-k,l}$ and the growing time $\tilde{t}$ (begin with the nucleation of the shish-kebab and end with the shear time), which can be approximated with $G_{s-k,l}t_s$. With the help of Equation (8), we know that L_{s-k} is a function of $\dot{\gamma}^2 t_s$; when we keep the shear stain $\gamma = \dot{\gamma}t_s$ as constant, the length of shish-kebabs L_{s-k} is larger in the case with higher shear rate $\dot{\gamma}$. Thus, in the higher shear rate and short shear time case ($\dot{\gamma} = 10\,\mathrm{s}^{-1}$, $t_s = 5\,\mathrm{s}$), the contribution of shish-kebabs is larger. This also agrees with the numerical work by Zheng et al. [8].

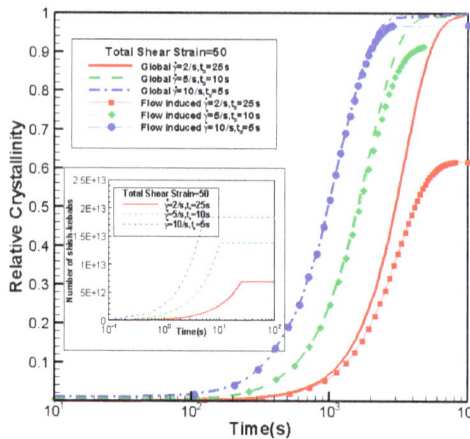

Figure 11. Relative crystallinity comparison with same shear strain but different shear rate and shear time.

3.5.2. Effects of Shear Strain on Rheology

Figure 12 shows the evolution of viscosity at three different conditions. As expected, the case with higher shear rate and short shear time ($\dot{\gamma} = 10\,\mathrm{s}^{-1}$, $t_s = 5\,\mathrm{s}$) has the earliest occurring sudden increase.

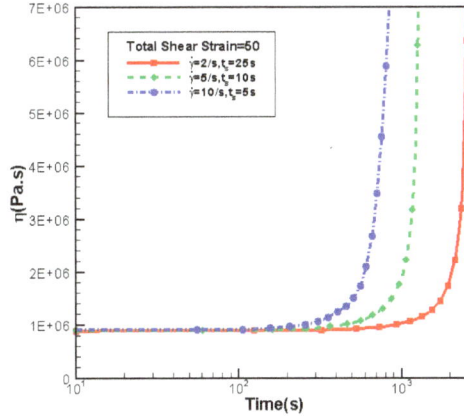

Figure 12. Evolution of viscosity with same shear strain but different shear rate and shear time.

4. Conclusions

A morphological Monte Carlo simulation is carried out to calculate the crystallization kinetics and capture the crystal morphology in 3D simple shear flow. Effects of shear rate, shear time and the shear strain on crystallization kinetics, crystal morphology and rheology of the system are discussed. The conclusions are drawn as follows.

(1) The evolution model and Monte Carlo method established are effective and reliable. With the evolution model and Monte Carlo algorithm, we obtain reliable crystallization kinetics and detailed crystal morphology.

(2) Effects of shear rate, shear time and shear strain on crystallization and rheology obtained here is in agreement with other numerical work and experimental results. We show the great influence of shear rate and shear time on the crystallization kinetics, crystal morphology and rheology of the system. In a higher shear rate or longer shear time case, the contribution of shish-kebabs to both crystallization kinetics and morphology becomes more significant and the sudden increase of viscosity occurs earlier. Under the same shear strain, the case with higher shear rate and shorter shear time can lead to a quicker crystallization rate and an earlier occurrence of sudden increase of viscosity.

Acknowledgments: The financial supports provided by the Natural Sciences Foundation of China (Nos. 11402078, 51375148, and U1304521) and the Scientific and Technological Research Project of Henan Province (Nos. 122102210198 and 14B110020) were fully acknowledged.

Conflicts of Interest: The authors declare no conflict of interest.

References

1. Eder, G.; Janeschitz-Kriegl, H. *Materials Science and Technology*; Wiley: New York, NY, USA, 1997.
2. Zheng, R.; Tanner, R.I.; Fan, X.J. *Injection Molding: Integration of Theory and Modeling Methods*; Springer: Berlin, Germany, 2011.
3. Pantanin, R.; Coccorullo, I.; Speranza, V.; Titomanlio, G. Modeling of morphology evolution in the injection molding process of thermoplastic polymers. *Prog. Polym. Sci.* **2005**, *30*, 1185–1222. [CrossRef]

4. Doufas, A.K.; Mchugh, A.J.; Miller, C. Simulation of melt spinning including flow-induced crystallization. Part 1. Model development and predictions. *J. Non-Newtonian Fluid Mech.* **2000**, *92*, 27–66. [CrossRef]

5. Tanner, R.I. A suspension model for low shear rate polymer solidification. *J. Non-Newtonian Fluid Mech.* **2002**, *102*, 397–408. [CrossRef]

6. Ziabicki, A.; Janecki, L.; Sorrentino, A. The role of flow-induced crystallization in melt spinning. *e-Ploymers* **2004**, *4*, 823–836.

7. Koscher, E.; Fulchiron, R. Influence of shear on polypropylene crystallization: Morphology development and kinetics. *Polymer* **2002**, *43*, 6931–6942. [CrossRef]

8. Zheng, R.; Kennedy, P.K. A model for post-flow induced crystallization: General equations and predictions. *J. Rheol.* **2004**, *48*, 823–842. [CrossRef]

9. Schneider, W.; Koppl, A.; Berger, J. Non-isothermal crystallization of polymers. *Int. Polym. Process* **1988**, *3*, 151–154.

10. Zuidema, H.; Peters, G.W.M.; Meijer, H.E.H. Development and validation of a recoverable strain-based model for flow induced crystallization of polymers. *Macromol. Theory Simul.* **2001**, *10*, 447–460. [CrossRef]

11. Boutaous, M.; Bourgin, P.; Zinet, M. Thermally and flow induced crystallization of polymers at low shear rate. *J. Non-Newtonian Fluid Mech.* **2010**, *165*, 227–237. [CrossRef]

12. Raabe, D. Mesoscale simulation of spherulite growth during polymer crystallization by use of a cellular automaton. *Acta Mater.* **2004**, *52*, 2653–2664. [CrossRef]

13. Raabe, D.; Godara, A. Mesoscale simulation of the kinetics and topology of spherulite growth during crystallization of isotactic polypropylene (iPP) by using a cellular automaton. *Model. Simul. Mater. Sci. Eng.* **2005**, *13*, 733–751. [CrossRef]

14. Raabe, D. Simulation of spherulite growth during polymer crystallization by use of a cellular automaton. *Mater. Sci. Forum* **2004**, *467*, 603–608. [CrossRef]

15. Lin, J.X.; Wang, C.Y.; Zheng, Y.Y. Prediction of isothermal crystallization parameters in monomer cast nylon 6. *Comput. Chem. Eng.* **2008**, *32*, 3023–3029. [CrossRef]

16. Spina, R.; Spekowius, M.; Hopmann, C. Multi-scale thermal simulation of polymer crystallization. *Int. J. Mater. Form* **2015**, *8*, 497–504. [CrossRef]

17. Spina, R.; Spekowius, M.; Hopmann, C. Multiphysics simulation of thermoplatic polymer crystallization. *Mater. Des.* **2016**, *95*, 455–469.

18. Liu, Z.J.; Ouyang, J.; Zhou, W.; Wang, X.D. Simulation of polymer crystallization under isothermal and temperature gradient conditions using praticle level set method. *Crystals* **2016**, *6*, 90. [CrossRef]

19. Liu, Z.J.; Ouyang, J.; Ruan, C.; Liu, Q.S. Numerical simulation of the polymer crystallization during cooling stage by using level set method. *Comput. Mater. Sci* **2015**, *97*, 245–253. [CrossRef]

20. Micheletti, A.; Burger, M. Stochastic and deterministic simulation of nonisothermal crystallization of polymers. *J. Math. Chem.* **2001**, *30*, 169–193. [CrossRef]

21. Ruan, C. Multiscale numerical study of 3D polymer crystallization during cooling stage. *Math. Probl. Eng.* **2012**, *2012*, 802420. [CrossRef]

22. Ruan, C.; Ouyang, J.; Liu, S. Computer modeling of isothermal crystallization in short fiber reinforced composites. *Comput. Chem. Eng.* **2011**, *35*, 2306–2317. [CrossRef]

23. Ketdee, S.; Anantawaraskul, S. Simulation of crystallization kinetics and morphological development during isothermal crystallization of polymers: Effect of number of nuclei and growth rate. *Chem. Eng. Commun.* **2008**, *195*, 1315–1327. [CrossRef]

24. Ruan, C.; Liu, C.; Zheng, G. Monte Carlo simulation for the morphology and kinetics of spherulites and shish-kebabs in isothermal polymer crystallization. *Math. Probl. Eng.* **2015**, *2015*, 50624. [CrossRef]

25. Godara, A.; Raabe, D.; Van Puyvelde, P.; Moldenaers, P. Influence of flow on the global crystallization kinetics of iso-tactic polypropylene. *Polym. Test.* **2006**, *25*, 460–469. [CrossRef]

26. Hoffman, J.D.; Miller, R.L. Kinetics of crystallization from the melt and chain folding in polyethylene fractions revisited: theory and experiment. *Polymer* **1997**, *38*, 3151–3212. [CrossRef]

27. Owens, R.G.; Phillips, T.N. *Computational Rheology*; Imperial College Press: London, UK, 2002.

28. Chung, D.H.; Kwon, T.H. Invariant-based optimal fitting closure approximation for the numerical prediction of flow-induced fiber orientation. *J. Rheol.* **2002**, *46*, 169–194. [CrossRef]

29. Tanner, R.I.; Qi, F. A comparison of some models for describing polymer crystallization at low deformation rates. *J. Non-Newtonian Fluid Mech.* **2005**, *127*, 131–141. [CrossRef]
30. Rong, Y.; He, H.P.; Cao, W.; Shen, C.Y.; Chen, J.B. Multi-scale molding and numerical simulation of the flow-induced crystallization. *Comput. Mater. Sci* **2013**, *67*, 35–39. [CrossRef]

© 2017 by the author. Licensee MDPI, Basel, Switzerland. This article is an open access article distributed under the terms and conditions of the Creative Commons Attribution (CC BY) license (http://creativecommons.org/licenses/by/4.0/).

![crystals logo] *crystals*

MDPI

Article

Atomistic Modelling of Si Nanoparticles Synthesis

Giovanni Barcaro [1], Susanna Monti [2], Luca Sementa [1] and Vincenzo Carravetta [1,*]

[1]   CNR-IPCF, National Research Council—Institute of Chemical and Physical Processes, via G. Moruzzi 1, I-56124 Pisa, Italy; giovanni.barcaro@pi.ipcf.cnr.it (G.B.); luca.sementa@pi.ipcf.cnr.it (L.S.)
[2]   CNR-ICCOM, National Research Council—Institute of Chemistry of Organometallic Compounds, via G. Moruzzi 1, I-56124 Pisa, Italy; sapeptides@gmail.com
*   Correspondence: carravetta@ipcf.cnr.it

Academic Editor: Hiroki Nada
Received: 12 January 2017; Accepted: 8 February 2017; Published: 13 February 2017

Abstract: Silicon remains the most important material for electronic technology. Presently, some efforts are focused on the use of Si nanoparticles—not only for saving material, but also for improving the efficiency of optical and electronic devices, for instance, in the case of solar cells coated with a film of Si nanoparticles. The synthesis by a bottom-up approach based on condensation from low temperature plasma is a promising technique for the massive production of such nanoparticles, but the knowledge of the basic processes occurring at the atomistic level is still very limited. In this perspective, numerical simulations can provide fundamental information of the nucleation and growth mechanisms ruling the bottom-up formation of Si nanoclusters. We propose to model the low temperature plasma by classical molecular dynamics by using the reactive force field (ReaxFF) proposed by van Duin, which can properly describe bond forming and breaking. In our approach, first-principles quantum calculations are used on a set of small Si clusters in order to collect all the necessary energetic and structural information to optimize the parameters of the reactive force-field for the present application. We describe in detail the procedure used for the determination of the force field and the following molecular dynamics simulations of model systems of Si gas at temperatures in the range 2000–3000 K. The results of the dynamics provide valuable information on nucleation rate, nanoparticle size distribution, and growth rate that are the basic quantities for developing a following mesoscale model.

Keywords: Si nanoparticle; plasma synthesis; theoretical model; reactive force field; molecular dynamics

1. Introduction

The synthesis of nanoparticles in the gas phase by a bottom-up process is emerging as a promising technology compared to traditional production techniques. This is because, even though the wet phase synthesis has a better control of the nanoparticle morphology, it requires solvents and cannot be easily used for industrial production. In contrast, the milling techniques are relatively simple and can be adapted to large scale productions, but do not provide a good control of size and morphology of the nanoparticles. Furthermore, some bottom-up processes do not use solvents and are therefore more eco-friendly. In the case of hard materials that are stable at high temperatures, the synthesis from gaseous phase in plasma reactors [1], flame reactors [2], and hot-wall reactors [3] is especially interesting. The operating conditions—in terms of density and high temperature (2000–3000 K) of the gas—in these types of reactors make the environment in which the synthesis occurs extremely complex, difficult to monitor and investigate with experimental techniques. This is due to the non-equilibrium thermodynamic processes, the kinetics that may develop in the range of ns or even of ps, and the chains of chemical reactions between the precursors and any contaminant.

For such complex systems, computational modelling might in principle provide valuable insights into the basic mechanisms that affect the synthesis process. These are, for example, nucleation, growth of the embryo-particles, and their surface chemical activity. Of course, numerical modelling has its objective difficulties too, due to the size of the models and the time scale of the simulations required to obtain results comparable to experiments. The basic approach in this case is to define—when possible—different modeling regions of spatial and temporal dimensions that are contiguous and partially superimposed. In the specific case of the plasma synthesis, the modelling range extends from the atomic size of the primary species up to the macroscopic continuous reactor and from the femtoseconds of molecular dynamics up to the milliseconds of diffusion and turbulence. Chemical engineers have developed and already employ numerical models on the macroscopic scale of the reactors, based on empirical parameters which, however, are not easily measurable (as in the case of the reactors here considered). This has led to the development of meso-scale modelling for a better understanding of the systems at a microscopic level. However, such models also depend on a set of basic parameters derived from thermodynamical data obtained on bulk material that can generally have properties very different from that of a nanoparticle. Atomistic modelling can provide insight about the basic processes that occur on the scale of nanometers and picoseconds, and then drive the subsequent models on larger scales to a better description of the system.

Atomistic modelling is the subject of the present report concerning the bottom-up synthesis of nanoparticles from Si low temperature plasma. The main goal is to estimate—by means of first-principle and molecular mechanics calculations with force fields (FF)–the morphology of embryo-particles (small size clusters) that form through homogeneous nucleation of the monoatomic gas and their growth during the first steps of the nucleation process. The numerical procedures described here can be easily extended to other hard materials. They are based on density functional theory (DFT), basin hopping [4] (DFT-BH) calculations, and molecular dynamics simulations based on reactive force fields. These force fields can describe bond breaking and formation during the evolution of the systems, which is fundamental in the cases examined in this study.

After a brief overview of the features of the reactive force field [5] and of the procedure to optimize its parameters for the case of Si nanoparticles, the results of the dynamics in terms of inception time of the nanoparticles as a function of their size and of the environmental conditions (temperature and density of the primary gas) will be presented. These results will be critically discussed and compared to the predictions provided by the Classical Nucleation Theory (CNT) [6], which, in different forms, is largely used to describe homogeneous nucleation.

2. Model and Computational Details

The atomistic model adopted to simulate the early stages of the gas phase condensation of Si nanoparticles consists of classical molecular dynamics simulations based on a reactive force field (ReaxFF) originally developed by van Duin [5], where the parameters were recently optimized [7] for Si-containing systems. ReaxFF is flexible, computationally efficient, and has been successfully applied to a large variety of material modelling problems [8–15]. For these reasons, it has been selected to model the early stages of the plasma-assisted growth of Si nanostructures.

For the simulation of this kind of process in Si-containing materials, two formally simpler potentials—namely, Tersoff and Stillinger–Weber potentials—have been frequently employed. Among the most recent applications there are the computational studies for the simulation of the condensation, growth, and crystallization of Si nanoclusters obtained by magnetron sputtering [16–18]. In this technique, the fragments of Si (mainly embryo clusters but also isolated atoms) produced by sputtering at very high temperatures (even 6000 K) pass through a condensation region occupied by an inert gas (Ar) at a much lower temperature. The sample is obviously very far from equilibrium, and has been simulated by molecular dynamics for an ensemble with constant number of atoms, volume and total energy (NVE) with a time gradient of the temperature that is taken as large as 30 K per ns. In our case we intend to refer to environmental conditions much closer to equilibrium and then better represented by simulations for an ensemble with constant number of atoms, volume and average temperature (NVT) with large fluctuations induced by the Berendsen's thermostat adopted. These conditions correspond locally to those that are present in the thermal plasma reactors of industrial type employed for large-scale production of nanoparticles.

ReaxFF is expressed by an extended set of analytical functions that reflect the complex picture of the atomic interactions and depend on a number of different parameters. These parameters have been derived, in the present case, from quantum chemistry calculations on model systems representative of the species present in the Si plasma. The complex analytic form of the force field is, anyway, much less computationally expensive than the Hamiltonian formulation of the quantum molecular dynamics (QMD) methods, and as a consequence, classical simulations are the only practical choice for describing large aggregates containing up to one million atoms. In comparison to most of the force fields available in the literature, which essentially provide a *physical picture* of the system, ReaxFF allows the simulation of bond breaking and formation. This aspect is very relevant in order to properly simulate not only interactions, but also reactions, and then dynamically characterize the various processes that determine and regulate the early formation of nanomaterial aggregates from the plasma. The potential energy in ReaxFF is calculated as the sum of a number of energy terms depending on the bond distance/bond order and bond order/bond energy relationships. The algorithm appropriately describes under- and over-coordination and dismisses the energy terms depending on bond distance, bond, and dihedral angles upon bond breaking. Non-bonded interactions (Coulomb and van der Waals forces) are calculated between atom pairs with a shielding factor [19] to dampen close-range interactions; atomic charges are calculated using the electronegativity equalization method (EEM) [20] and long-range Coulomb interactions are calculated by means of a taper function with an outer cut-off radius.

2.1. Optimization of the ReaxFF Parameters

Starting from a parametrization of the FF for Si based materials (VD) available in the literature [7] and designed to describe the bulk structure of SiC and SiO_2, we re-optimized the FF for the specific systems under study in order to consider surface effects of small Si clusters. These effects can be fundamental in the nucleation and growth processes, leading to nanoparticles with maximum diameter of about 1 nm. The FF parameters have been optimized in order to reproduce a training-set (*t*-set) of properties (i.e., energies and geometries) derived from quantum chemical (QC) calculations on selected model systems. Few experimental measurements of dissociation/ionization energies and mobilities of the ionized structures are available in the literature [21–24], whereas geometries and electronic structures of small silicon clusters have been studied extensively by QC approaches over the last few years [25–38].

The first-principle quantum method we have adopted is based on DFT [39], which is widely used for studying medium-large molecular systems. Calculations have been performed by the ESPRESSO package [40] using ultrasoft pseudo potentials [41] for an approximate description of the core electrons that do not directly participate in the formation of molecular bonds, whereas the electronic structure of the valence electrons (4 for Si) is described by the GGA-PBE [42] functional. Energy spin-restricted optimizations were performed in a cubic box with side between 10 an 20 Å (depending on the size of the cluster) and periodic boundary conditions in order to effectively project the wave function on a set of plane waves with cut-offs on the wave-function/electronic density of about 140/1100 eV, respectively. Other details of the DFT calculations are: gaussian smearing for the occupation of the Kohn–Sham single-particle states of 0.03 eV; Brillouin zone explored at the Gamma point only; single local minimization with a threshold on energy/forces of $10^{-3}/10^{-4}$ au.

A small Si cluster may adopt various conformations corresponding to slightly different energy values; at the plasma temperatures (>2000 K), a great number of such conformations is accessible. For the optimization of the force field, we selected the most populated minimum energy configurations of the clusters that have been identified by a specific algorithm based on geometry optimization and Basin-Hopping [4]. This is one of the several global optimization methods that are used to find the minimum of a multi-variable Potential Energy Surface (PES). As any other similar method, this approach does not guarantee a successful identification of the absolute energy minimum in a finite time, but provides an efficient algorithm for exploring the PES and accumulating a voluminous structural database. In our approach, for each cluster size, we performed a search for the minimum by running several BH calculations (from three to five), composed of 500 steps each. At each step, the cluster geometry is slightly modified randomly and a DFT geometry optimization is performed; the step is accepted or refused according to the Monte Carlo/Metropolis criterion [43] with a kT value in the range 0.5–1.0 eV [44]. The lowest-energy structures resulting from this search are summarized in Figure 1 for Si clusters composed of an even number of atoms with a size in the range 6 to 36. These structures were considered as model systems for computing the benchmark data forming the training set.

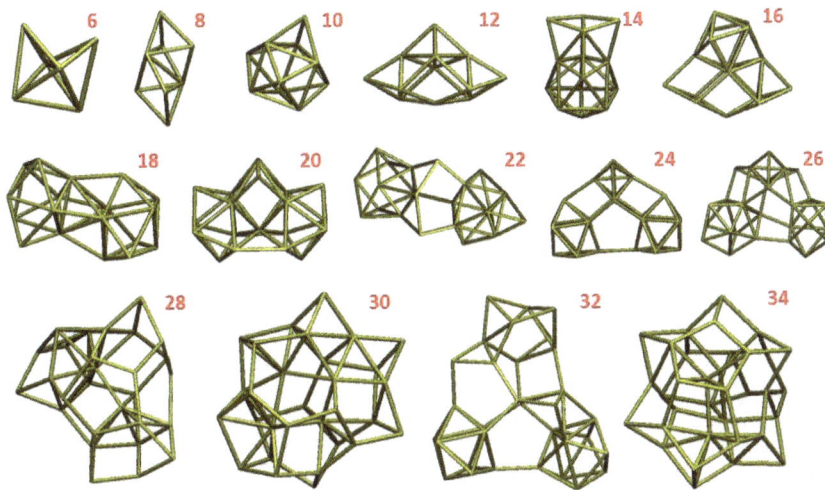

Figure 1. Lowest-energy structures of Si clusters with an even number of atoms and size in the range 6–34.

To refine the VD FF, we adopted the ReaxFF standalone code (version 2.0). The *t*-set adopted was given by geometries and energies of a number of model systems selected among the structures optimized by the DFT-BH searches, namely: Si_2, Si_3, Si_4, Si_5, Si_6, Si_8, Si_{10}, Si_{12} (two structures), see Figure 2, to which the basic unit of silicon diamond cubic crystal was added. Starting from the original parametrization (VD) [7], we optimized only the relevant parameters responsible for the flexibility of the Si clusters at high temperature against our *t*-set: (i) Si-Si bond (14 parameters); (ii) Si-Si-Si valence angle (5 parameters); (iii) Si-Si-Si-Si torsion angle (5 parameters). These latest parameters were not present in the original parametrization, but were added in this new force field to get a better agreement with the *t*-set data. Starting values of the parameters were taken from carbon in VD.

The ReaxFF code adopts a sequential optimization procedure of the parameters by the minimization of an error function (EF) which is calculated in the following way:

$$EF = \sum_i \frac{|f_{FF,i} - f_{QC,i}|}{w(i)} \tag{1}$$

where $f_{FF,i}$ and $f_{QC,i}$ are, respectively, the running FF and the benchmark QC values of the *i*-th entry of the *t*-set and $w(i)$ is its weight. The sum is extended to all the geometrical/energetic entries of the *t*-set. The ReaxFF optimization procedure works in the following way: at each step, a single parameter is considered and its starting value is varied by adding and subtracting a given quantity; for these three values of the considered parameter, the total EF is estimated and a parabolic extrapolation is achieved by finding (in correspondence of the minimum of the parabolic expression) the optimal value of the parameter within the chosen window; if this new predicted value corresponds to a lower value of the EF, the change is accepted and the algorithm moves to the successive parameter. As the parameters are strongly entangled, several sequential optimization cycles have to be repeated in order to achieve a real reduction of the EF against the *t*-set. It should be added that at each step the QC geometries optimized by the DFT-BH procedure are compared to those optimized by using the running FF; this is done in view of using the FF for an effective and realistic search of the different stable conformations of the clusters.

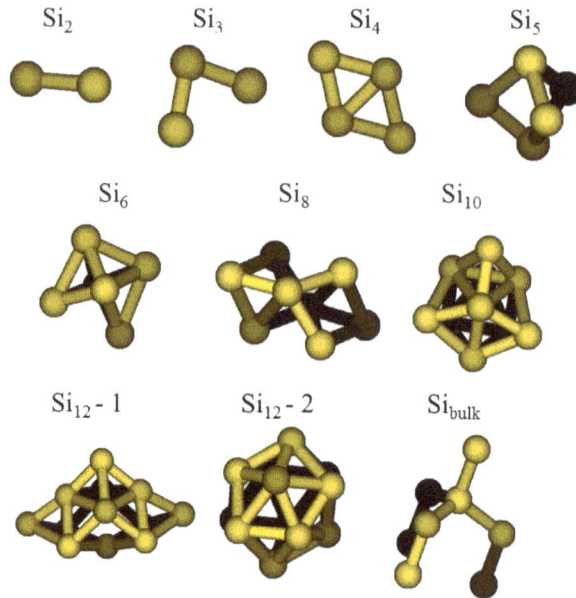

Figure 2. Si clusters included in the training set used in the FF optimization.

From the point of view of this performance it is convenient to analyze the predicted lowest-energy structures by using the two synthetic structural descriptors: R and θ. The first one refers to an average value of the interatomic distances:

$$r_i = \sum_j r_{i,j}/n_j \qquad\qquad R = \sum_j r_j/N \qquad\qquad (2)$$

where N is the number of atoms in the cluster, each $i-$atom has n_i first-neighbours, and $r_{i,j}$ is the distance between atoms i and j; the second one refers to an average value of the molecular angles:

$$\theta_i = \sum_j \theta_{i,j}/n_j \qquad\qquad \theta = \sum_j \theta_j/N \qquad\qquad (3)$$

2.2. Reactive Molecular Dynamics

Classical Molecular Dynamics simulations of the Si + Ar plasma have been carried out by using the LAMMPS code [45] with the presently optimized reactive force field (in the following named ND) in the NVT ensemble with the Berendsen's thermostat in order to have temperature fluctuations around a constant average temperature. In our approach, such simulations—performed at different compositions and temperatures—intend to model the environment present in thermal plasma reactors employed for large-scale production of Si nanoparticles. The results are employed to investigate nucleation processes and estimate specific parameters and mathematical expressions that can be used to feed mesoscale models for the simulation of much larger systems.

On the basis of the good performance of the ND FF on the exploration of the PES at 0 K, extensive MD simulations have been carried out by using this FF; nevertheless, a comparison with VD on a specific set of thermodynamic conditions will be discussed in the next section as well, in order to show how the two different parameterizations influence the nucleation and growth processes.

The computational details of the nucleation simulations are the following:

- *Simulation box:* an appropriate value of the length of the cubic simulation box has been chosen in order to reduce the effect of the periodic boundary conditions. The number of events that occur on the border becomes less and less relevant than the number of total events when the size of the box is increased. In the limit of a box with an infinite size, the biasing constraint of the translational periodicity disappears. According to a series of simulations in which the volume of a cubic unit cell has been progressively increased (from a minimum of 64 $\times 10^3$ nm^3 to a maximum of 13824 $\times$ 10^3 nm^3), the volume chosen was 4096 $\times 10^3$ nm^3.
- *Temperature:* each simulation has been carried out at fixed average temperature. The temperature range between 1800 and 2400 K has been explored by increasing the temperature by steps of 100 K in each simulation.
- *Atom density:* the number of atoms within the simulation box has been chosen on the basis of particular experimental conditions: total gas pressure within the plasma reactor around 1 atmosphere (standard ambient pressure). On the basis of this latter value, the density of the simulation gas (which is composed of Si and Ar atoms) is about 3.6×10^{24} atoms/m^3, which corresponds—in the chosen cubic box—to an ensemble of about 12,000 atoms.
- *Composition of the mixture:* the content of the (Si, Ar) mixture has been varied according to the following percentages: (i) 25% Si and 75% Ar; (ii) 50% Si and 50% Ar; and (iii) 75% Si and 25% Ar.
- *Time step:* The time step of the MD simulations has been set to 2 fs, having verified that a smaller time step of 0.5 fs reproduces the same results at both the qualitative and quantitative levels.

We underline that the adopted models refer to environmental conditions close to equilibrium corresponding, locally, to those that are present in the thermal plasma reactors of industrial type for large-scale production of nanoparticles.

3. Results and Discussion

From the point of view of assessing the performance of the reactive FF in exploring the PES of the small Si clusters, it is interesting to compare the structures predicted by the VD and the presently refined parametrization ND of the reactive FF against DFT results. In Figure 3, the lowest-energy structures predicted at sizes 12, 20, 30, and 36 are reported, together with a structural and energetic analysis of the conformations obtained. It should be noted that the lowest-energy structures shown in Figure 3 have been obtained by adopting the same BH search protocol described above, with energies and forces calculated at the FF level, but the energy of such structures has been then computed at the DFT level for a correct comparison with the DFT-BH results.

From the structural point of view, we can observe two main features that characterize the geometries of small Si clusters:

(i)   a compact arrangement of atoms with coordination numbers and coordination angles much different from bulk silicon, where the coordination number is strictly four and the angles are perfectly tetrahedral; in the DFT-BH optimized structures, instead, the number of nearest neighbours can be as large as 6 and even 7, and the bond angles are remarkably different from tetrahedral;

(ii) cluster surfaces are not smooth, but sharp with the presence of convex re-entrances, leaving the surface atoms with a very low coordination of one, or maximum two, neighbours.

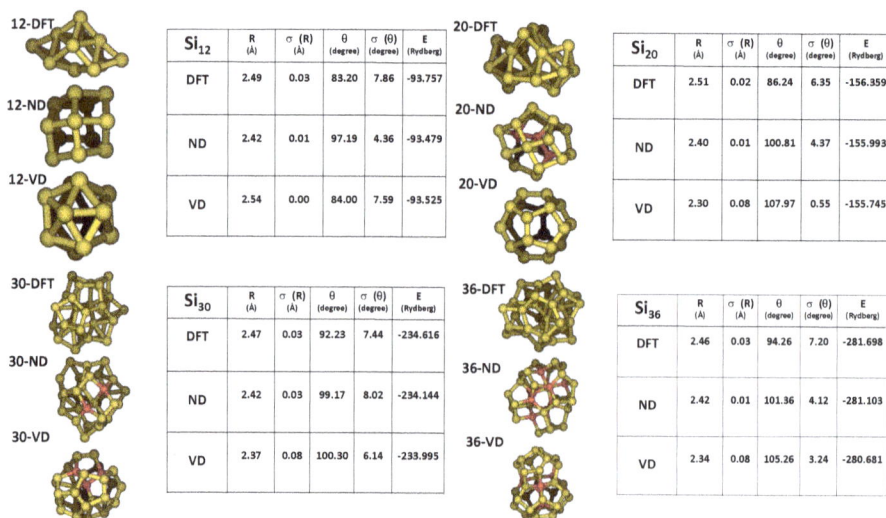

Si_{12}	R (Å)	σ (R) (Å)	θ (degree)	σ (θ) (degree)	E (Rydberg)
DFT	2.49	0.03	83.20	7.86	-93.757
ND	2.42	0.01	97.19	4.36	-93.479
VD	2.54	0.00	84.00	7.59	-93.525

Si_{20}	R (Å)	σ (R) (Å)	θ (degree)	σ (θ) (degree)	E (Rydberg)
DFT	2.51	0.02	86.24	6.35	-156.359
ND	2.40	0.01	100.81	4.37	-155.993
VD	2.30	0.08	107.97	0.55	-155.745

Si_{30}	R (Å)	σ (R) (Å)	θ (degree)	σ (θ) (degree)	E (Rydberg)
DFT	2.47	0.03	92.23	7.44	-234.616
ND	2.42	0.03	99.17	8.02	-234.144
VD	2.37	0.08	100.30	6.14	-233.995

Si_{36}	R (Å)	σ (R) (Å)	θ (degree)	σ (θ) (degree)	E (Rydberg)
DFT	2.46	0.03	94.26	7.20	-281.698
ND	2.42	0.01	101.36	4.12	-281.103
VD	2.34	0.08	105.26	3.24	-280.681

Figure 3. Structural and energetic results obtained by using different parametrizations of ReaxFF for Si clusters of size 12, 20, 30, and 36. The structural data R and θ are defined in Equations (2) and (3), respectively, while σ refers to their root-mean-square dispersion. DFT: density functional theory; ND: present optimization of ReaxFF; VD: parametrization for Si-based materials available in the literature [7].

By inspecting the results of the two FFs, we can see that the presently optimized parametrization ND using the *t*-set shown in Figure 1 better reproduces both the structural and energetic DFT results. Indeed, the two structural features discussed above can be found in the geometries predicted by the ND FF, and the relative FF energies are closer—compared to the VD values—to the benchmark DFT ones. The limitations exhibited in the present application by the VD FF could be due to its apparent tendency to favor Si atoms with "bulk-character". As a consequence, in the structures predicted for the small clusters, the average coordination number is about four, as evidenced by the high number of red atoms in Figure 3, which are tetrahedrally shaped, and by very smooth surfaces; moreover, the DFT energy of the VD FF optimized structures is higher than that predicted by the ND FF [44].

A direct comparison of our present MD results with those available in the literature [16–18]—using the Tersoff or Stillinger–Weber potentials in order to evaluate the performance of different force fields—would be biased by the large differences in the model systems adopted and in the characteristics of the MDs employed. An accurate comparison would require calculations in which the features of the simulations are more homogeneous, which is outside the scope of the present work.

Two typical results of the nucleation MD simulations at different temperatures are shown in Figure 4. On a short time scale, the dynamics is dominated by the formation of smaller aggregates (mainly Si dimers, which are quickly formed); the concentration of the smaller clusters reaches a plateau and then starts decreasing due to the formation of the larger aggregates. As can be noted in Figure 4a, a further peak is found at size 8, which corresponds to a remarkably stable structure of the Si system. As the simulation proceeds, larger aggregates start to form. At lower temperature (Figure 4a), several Si_{30} clusters can be found at the end of the run (about 300 ns), whereas at higher temperature (Figure 4b), the particles growth is slowed down due to the larger thermal agitation and only few aggregates of size around 30 can be obtained at the end of the simulation. This picture is different from that provided by the classical nucleation theory, in which, out of nothing, a nanoparticle of critical mass is produced without a previous history. In our opinion, this phenomenological model does not apply to the cases where—as for Si—the interaction between atomic components leads to the formation of covalent bonds with bond energy much higher than the available average thermal energy.

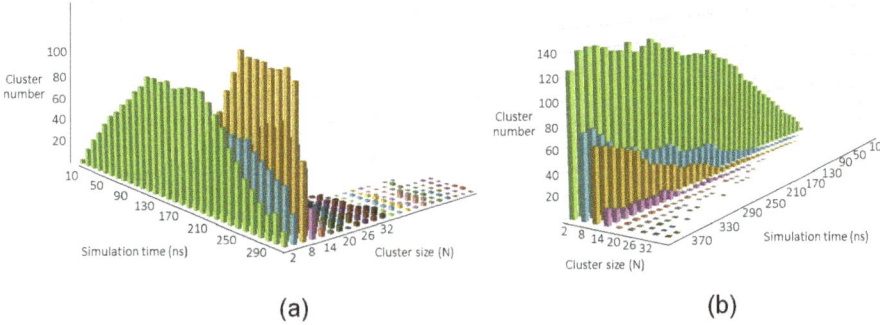

(a)   (b)

Figure 4. Nucleation and growth picture at (**a**) 1800 K and (**b**) 2400 K.

In Table 1, the characteristic inception times for particles of size 20, 30, and 40 are reported as a function of the temperature for the three mixture compositions specified above. The inception time of a particle composed of N atoms is estimated as the time when the first particle of that size appears in the simulation box. For a specific nuclearity, the inception time increases when increasing the temperature due to the higher thermal agitation that slows down the aggregation process. As could be expected, the inception time decreases at increasing Si content when comparing the same particle size and temperature. The Ar gas present in industrial thermal plasma reactors provides a more effective coupling with the electromagnetic field applied for the formation and maintenance of the plasma; the case of a pure Si gas is practically not realized. Our simulations (not included in this study) of such a system show that the processes of condensation and growth of Si nanoparticles remain substantially unchanged; the presence of Ar is not necessary for their implementation.

Table 1. Inception times (ns) for Si particles of 20, 30, and 40 atoms.

System	Si 25% N = 20	Si 25% N = 30	Si 25% N = 40	Si 50% N = 20	Si 50% N= 30	Si 50% N = 40	Si 75% N = 20	Si 75% N = 30	Si 75% N = 40
T = 1800 K	165	176	229	89	91	103	47	61	71
T = 1900 K	156	219	266	70	106	132	54	77	84
T = 2000 K	205	249	273	90	123	142	58	77	89
T = 2100 K	219	271	305	89	115	156	67	78	83
T = 2200 K	199	314	332	121	137	169	78	96	112
T = 2300 K	247	340	366	125	143	208	67	95	114
T = 2400 K	270	391		126	198		86	119	

When using the force field VD, the results of the nucleation and growth processes are completely different: at T = 2000 K with a plasma made of 50% Si and 50% Ar (within a time scale of around 160 ns), only a very low percentage of atoms form aggregates, as it can be evinced by inspecting Figure 5a; moreover, only dimers are formed (about 50 during the simulation, see Figure 5b) with the sporadic appearance of trimers, which soon dissociates again in dimers and single atoms (see Figure 5c). This behaviour could be due to an instability of the trimer that originates a bottleneck in the growth process [44], as this nuclearity is a mandatory step in the growth of larger aggregates, due to the fact that the formation of a tetramer due to the aggreagtion of two dimers is a very unlikely event in the present conditions.

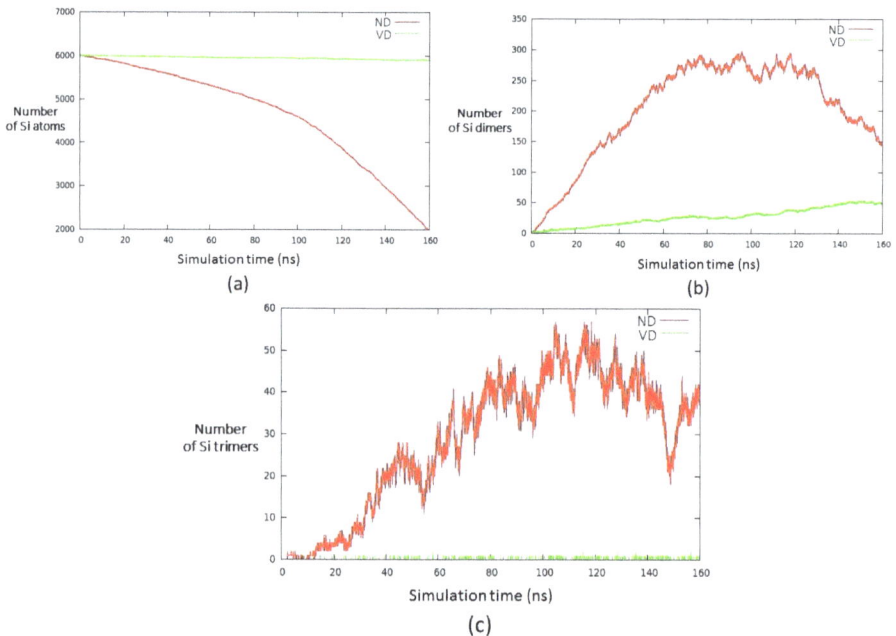

Figure 5. Number of (**a**) unreacted Si atoms; (**b**) dimers; and (**c**) trimers as a function of time at T = 2000 K. The results from the ND and VD FFs are shown in red and green, respectively.

To conclude, we compare the results of the present MD simulations with the ND FF and the only—to our knowledge—work in the literature reporting the values of partial pressures of small Si_N clusters (N = 1–6) in the vapour phase in equilibrium with the Si liquid phase calculated on the basis of classical thermodynamics [46]. In order to compare with the reported data, in Figure 6 we have plotted the ratio between the numbers of dimers and single atoms vs. the ratio between the number of trimers and dimers as a function of the simulation time. In this case, an NVT dynamics calculation has been performed at a total pressure of 1.22×10^{-2} MPa (1 atm = 0.1 MPa) and a temperature of T = 3000 K, matching one of the conditions considered in the experiment. The choice to operate at a higher temperature (with respect to the previous simulations) has been done to work with a higher value of Si pressure, and hence collect more statistical information about the investigated system. The simulation box contained 2860 atoms, and time-step has been set to 2.0 fs; the Berendsen's thermostat has been employed, in agreement with the methodological procedure applied for the estimation of the nucleation rates. According to the results reported in Figure 6, it can be seen that the dimers/atoms ratio is quite stable and reaches a plateau around the value 0.05; on the other side, the trimers/dimers ratio is less regular and is characterized by remarkable oscillations, mainly due to statistical reasons (lower absolute number of trimers in the simulation box during the dynamics); nevertheless, we can estimate that the latter ratio oscillates around the value 0.10 during the simulation. The obtained values of the ratios are in qualitative agreement with the data of Reference [46], which reports a dimers/atoms ratio of about 0.11 and a trimers/dimers ratio of about 0.25 at the considered values of temperature and pressure.

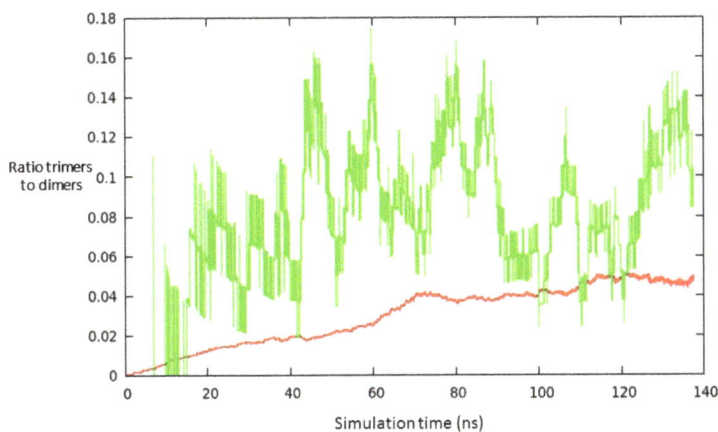

Figure 6. Ratio between the number of dimers and unreacted atoms (in red) vs. ratio between trimers and dimers (in green) for a set of thermodynamic variables specified in the main text.

4. Conclusions

We have proposed an theoretical approach to the difficult problem of modelling the nucleation and growth of embryo- particles in a low temperature plasma, with application to the Si gas. The approach is based on reactive molecular dynamics with a FF specifically optimized using DFT calculations as benchmark for a set of small Si clusters. The presently proposed (ND) FF—optimized starting from the one (VD) available in the literature—shows appreciable improvements in the description of Si nanoparticles. The results obtained by exploring the PES at 0 K evidenced the tendency of VD FF to over-estimate the bulk behavior in Si clusters. On the contrary, ND FF was capable of more confidently reproducing small clusters and favoring the formation of compact aggregates. The results of the molecular dynamics simulations performed with the ND FF are in reasonable agreement with the very few experimental data available for this kind of system, although the experimental environment corresponding to Si vapor in equilibrium with the liquid phase is only roughly represented by our simulation model. Unfortunately, an extension of the model including a representation of the liquid phase would lead to too-CPU-demanding calculations. Our results can be viewed as describing a state of quasi-equilibrium only in the sense that the obtained ratios are quite stable in a wide time window; consequently, the agreement of theory and experiment can be expected only at a qualitative level. The results obtained show that the proposed theoretical model can be satisfactorily applied to systems where the experimental studies are very difficult to perform and an accurate description of the basic processes at the atomistic level is presently missing.

Acknowledgments: This work has been developed as part of the European project Nanodome that has received funding form the European Union's Horizon 2020 Research and Innovation Programme, under Grant Agreement number 646121.

Author Contributions: Susanna Monti, Vincenzo Carravetta and Giovanni Barcaro conceived and designed the theoretical approach; Giovanni Barcaro and Luca Sementa performed the calculations; Giovanni Barcaro, Vincenzo Carravetta and Luca Sementa analysed the results; Vincenzo Carravetta, Susanna Monti, Giovanni Barcaro and Luca Sementa wrote the paper.

Conflicts of Interest: The authors declare no conflict of interest.

References

1. Shigeta, M.; Murphys, A.B. Thermal plasmas for nanofabrication. *J. Phys. D Appl. Phys.* **2011**, *44*, 174025–174041.

2. Koirala, R.; Pratsinisa, S.E.; Baiker, A. Synthesis of catalytic materials in flames: Opportunities and challenges. *Chem. Soc. Rev.* **2016**, *45*, 3053–3068.
3. Schwade, B.; Roth, P. Simulation of nano-particle formation in a wall-heated aerosol reactor including coalescence. *J. Aerosol Sci.* **2003**, *34*, 339–357.
4. Wales, D.; Doye, P.K. Global Optimization by Basin-Hopping and the Lowest Energy Structures of Lennard-Jones Clusters Containing up to 110 Atoms. *J. Phys. Chem. A* **1997**, *101*, 5111–5116.
5. Van Duin, A.C.T.; Dasgupta, S.; Lorant, F.; Goddard, W.A. ReaxFF: A Reactive Force Field for Hydrocarbons. *J. Phys. Chem. A* **2001**, *105*, 9396–9409.
6. Kashchiev, D. *Nucleation: Basic Theory with Applications*; Butterworth-Heinemann: Oxford, UK, 2000.
7. Newsome, D.A.; Sengupta, D.; Foroutan, H.; Russo, M.F.; van Duin, A.C. Oxidation of Silicon Carbide by O_2 and H_2O: A ReaxFF Reactive Molecular Dynamics Study, Part I. *J. Phys. Chem. C* **2012**, *116*, 16111–16121.
8. Han, Y.; Jiang, D.; Zhang, J.; Li, W.; Gan, Z.; Gu, J. Development, applications and challenges of ReaxFF reactive force field in molecular simulations. *Front. Chem. Sci. Eng.* **2016**, *10*, 16–38.
9. Li, C.; Monti, S.; Carravetta, V. Journey toward the Surface: How Glycine Adsorbs on Titania in Water Solution. *J. Phys. Chem. C* **2012**, *116*, 18318–18326.
10. Monti, S.; Li, C.; Carravetta, V. Reactive dynamics simulation of monolayer and multilayer adsorption of glycine on Cu(110). *J. Phys. Chem. C* **2013**, *117*, 5221–5228.
11. Carravetta, V.; Monti, S.; Li, C.; Ågren, H. Theoretical simulations of structure and X-ray photoelectron spectra of glycine and diglycine adsorbed on Cu(110). *Langmuir* **2013**, *29*, 10194–10204.
12. Li, C.; Monti, S.; Ågren, H.; Carravetta, V. Cysteine on TiO2(110): A theoretical study by reactive dynamics and photoemission spectra simulation. *Langmuir* **2014**, *30*, 8819–8828.
13. Monti, S.; Li, C.; Ågren, H.; Carravetta, V. Dropping a Droplet of Cysteine Molecules on a Rutile (110) Interface: Reactive versus Nonreactive Classical Molecular Dynamics Simulations. *J. Phys. Chem. C* **2015**, *119*, 6703–6712.
14. Monti, S.; Carravetta, V.; Ågren, H. Simulation of Gold Functionalization with Cysteine by Reactive Molecular Dynamics. *J. Phys. Chem. Lett.* **2016**, *7*, 272–276.
15. Da Silva, Á.M.; Mocellin, A.; Monti, S.; Li, C.; Marinho, R.R.T.; Medina, A.; Agren, H.; Carravetta, V.; de Brito, A.N. Surface-Altered Protonation Studied by Photoelectron Spectroscopy and Reactive Dynamics Simulations. *J. Phys. Chem. Lett.* **2015**, *6*, 807–811.
16. Zhao, J.; Singh, V.; Grammatikopoulos, P.; Cassidy, C.; Kengo, A.; Sowwan, M.; Nordlund, K.; Djurabekova, F. Crystallization of silicon nanoclusters with inert gas temperature control. *Phys. Rev. B* **2015**, *91*, 35419–35431.
17. Zhao, J.; Baibuz, E.; Vernieres, J.; Grammatikopoulos, P.; Jansson, V.; Nagel, M.; Steinhauer, S.; Sowwan, M.; Kuronen, A.; Nordlund, K.; et al. Formation Mechanism of Fe Nanocubes by Magnetron Sputtering Inert Gas Condensation. *ACS Nano* **2016**, *10*, 4684–4694.
18. Singh, V.; Cassidy, C.; Grammatikopoulos, P.; Djurabekova, F.; Nordlund, K.; Sowwan, M. Heterogeneous Gas-Phase Synthesis and Molecular Dynamics Modeling of Janus and Core-Satellite Si-Ag Nanoparticles. *J. Phys. Chem. C* **2014**, *118*, 13869–13875.
19. Janssens, G.O.A.; Baekelandt, B.G.; Toufar, H.; Mortier, W.J.; Schoonheydt, R.A. Comparison of Cluster and Infinite Crystal Calculations on Zeolites with the Electronegativity Equalization Method (EEM). *J. Phys. Chem.* **1995**, *99*, 3251–3258.
20. Mortier, W.J.; Ghosh, S.K.; Shankar, S. Electronegativity-equalization method for the calculation of atomic charges in molecules. *J. Am. Chem. Soc.* **1986**, *108*, 4315–4320.
21. Liu, B.; Lu, Z.Y.; Pan, B.; Wang, C.Z.; Ho, K.M.; Shvartsburg, A.A.; Jarrold, M.F. Ionization of medium-sized silicon clusters and the geometries of the cations. *J. Chem. Phys.* **1998**, *109*, 9401–9409.
22. Hudgins, R.R.; Imai, M.; Jarrold, M.F.; Dugourd, P. High-resolution ion mobility measurements for silicon cluster anions and cations. *J. Chem. Phys.* **1999**, *111*, 7865.
23. Bachels, T.; Schäfer, R. Binding energies of neutral silicon clusters. *Chem. Phys. Lett.* **2000**, *324*, 365–372.
24. Muller, J.; Liu, B.; Shvartsburg, A.A.; Ogut, S.; Chelikowsky, J.R.; Siu, K.W.M.; Ho, K.M.; Gantefor, G. Spectroscopic evidence for the tricapped trigonal prism structure of semiconductor clusters. *Phys. Rev. Lett.* **2000**, *85*, 1666–1669.
25. Kaxiras, E.; Jackson, K. Shape of Small Silicon Clusters. *Phys. Rev. Lett.* **1993**, *71*, 727–730.
26. Ho, K.M.; Shvartsburg, A.A.; Pan, B.; Lu, Z.Y.; Wang, C.Z.; Wacker, J.G.; Fye, J.L.; Jarrold, M.F. Structures of medium-sized silicon clusters. *Nature* **1998**, *392*, 582–585.

27. Mitas, L.; Grossman, J.; Stich, I.; Tobik, J. Silicon Clusters of Intermediate Size: Energetics, Dynamics, and Thermal Effects. *Phys. Rev. Lett.* **2000**, *84*, 1479–1482.
28. Rata, I.; Shvartsburg, A.A.; Horoi, M.; Frauenheim, T.; Siu, K.W.M.; Jackson, K.A. Single-parent evolution algorithm and the optimization of Si clusters. *Phys. Rev. Lett.* **2000**, *85*, 546–549.
29. Wang, J.; Wang, G.; Ding, F.; Lee, H.; Shen, W.; Zhao, J. Structural transition of Si clusters and their thermodynamics. *Chem. Phys. Lett.* **2001**, *341*, 529–534.
30. Sun, Q.; Wang, Q.; Jena, P.; Rao, B.K.; Kawazoe, Y. Stabilization of Si_{60} Cage Structure. *Phys. Rev. Lett.* **2003**, *90*, 135503.
31. Yoo, S.; Zeng, X.C. Global geometry optimization of silicon clusters described by three empirical potentials. *J. Chem. Phys.* **2003**, *119*, 1442–1450.
32. Jackson, K.A.; Horoi, M.; Chaudhuri, I.; Frauenheim, T.; Shvartsburg, A.A. Unraveling the shape transformation in silicon clusters. *Phys. Rev. Lett.* **2004**, *93*, 013401.
33. Li, B.X. Stability of medium-sized neutral and charged silicon clusters. *Phys. Rev. B Condens. Matter Mater. Phys.* **2005**, *71*, 235311.
34. Wang, J.; Zhou, X.; Wang, G.; Zhao, J. Optimally stuffed fullerene structures of silicon nanoclusters. *Phys. Rev. B Condens. Matter Mater. Phys.* **2005**, *71*, 113412.
35. Ona, O.; Bazterra, V.E.; Caputo, M.C.; Facelli, J.C.; Fuentealba, P.; Ferraro, M.B. Modified genetic algorithms to model cluster structures in medium-sized silicon clusters: Si18–Si60. *Phys. Rev. A* **2006**, *73*, 1–11.
36. Tereshchuk, P.L.; Khakimov, Z.M.; Umarova, F.T.; Swihart, M.T. Energetically competitive growth patterns of silicon clusters: Quasi-one-dimensional clusters versus diamond-like clusters. *Phys. Rev. B Condens. Matter Mater. Phys.* **2007**, *76*, 125418.
37. Zhou, R.L.; Pan, B.C. Low-lying isomers of Si_n^+ and Si_n^- (n = 31–50) clusters. *J. Chem. Phys.* **2008**, *128*, 234302.
38. Yoo, S.; Shao, N.; Zeng, X.C. Structures and relative stability of medium- and large-sized silicon clusters. VI. Fullerene cage motifs for low-lying clusters Si39, Si40, Si50, Si60, Si70, and Si80. *J. Chem. Phys.* **2008**, *128*, 104316.
39. Kohn, W.; Becke, A.D.; Parr, R.G. Density Functional Theory of Electronic Structure. *J. Phys. Chem.* **1996**, *100*, 12974–12980.
40. Giannozzi, P.; Baroni, S.; Bonini, N.; Calandra, M.; Car, R.; Cavazzoni, C.; Ceresoli, D.; Chiarotti, G.L.; Cococcioni, M.; Dabo, I.; et al. QUANTUM ESPRESSO: A modular and open-source software project for quantum simulations of materials. *J. Phys. Condens. Matter* **2009**, *21*, 395502.
41. Vanderbilt, D. Soft self-consistent pseudopotentials in a generalized eigenvalue formalism. *Phys. Rev. B* **1990**, *41*, 7892–7895.
42. Perdew, J.P.; Burke, K.; Ernzerhof, M. Generalized Gradient Approximation Made Simple. *Phys. Rev. Lett.* **1996**, *77*, 3865–3868.
43. Metropolis, N.; Ulam, S. The Monte Carlo method. *J. Am. Stat. Assoc.* **1949**, *44*, 335–341.
44. Barcaro, G.; Carravetta, V.; Monti, S.; Sementa, L. Force Field Parameters for Reactive Molecular Dynamics Simulation of Si Nanoparticles. *J. Chem. Theory Comput.* **2017**, in preparation.
45. LAMMPS Molecular Dynamics Simulator. Available online: http://lammps.sandia.gov (accessed on 12 February 2017).
46. Sevast'yanov, V.G.; Nosatenko, P.Y.; Gorskii, V.V.; Ezhov, Y.S.; Sevast'yanov, D.V.; Simonenko, E.P.; Kuznetsov, N.T. Experimental and theoretical determination of the saturation vapor pressure of silicon in a wide range of temperatures. *Russ. J. Inorg. Chem.* **2010**, *55*, 2073–2088.

© 2017 by the authors. Licensee MDPI, Basel, Switzerland. This article is an open access article distributed under the terms and conditions of the Creative Commons Attribution (CC BY) license (http://creativecommons.org/licenses/by/4.0/).

crystals

MDPI

Article

A Scheme for the Growth of Graphene Sheets Embedded with Nanocones

Yu-Peng Liu [1,2], Jing-Tian Li [1,2], Quan Song [1,2], Jun Zhuang [3] and Xi-Jing Ning [1,2,*]

[1] Institute of Modern Physics, Fudan University, Shanghai 200433, China; ypliu16@fudan.edu.cn (Y.-P.L.);
    12110200005@fudan.edu.cn (J.-T.L.); 13110200005@fudan.edu.cn (Q.S.)
[2] Applied Ion Beam Physics Laboratory, Fudan University, Shanghai 200433, China
[3] Department of Optical Science and Engineering, Fudan University, Shanghai 200433, China;
    junzhuang@fudan.edu.cn
*   Correspondence: xjning@fudan.edu.cn; Tel.: +86-21-65643119; Fax: +86-21-65642787

Academic Editor: Hiroki Nada
Received: 16 November 2016; Accepted: 18 January 2017; Published: 15 February 2017

Abstract: Based on the monolayer growth mode of graphene sheets (2D crystal) by chemical vapor deposition (CVD) on a Cu surface, it should be possible to grow the 2D crystal embedded with single wall carbon nanocones (SWCNC) if nano-conical pits are pre-fabricated on the surface. However, a previous experiment showed that the growing graphene sheet can cross grain boundaries without bending, which seems to invalidate this route for growing SWCNCs. The criterion of Gibbs free energy was applied in the present work to address this issue, showing that the sheet can grow into the valley of a boundary if the boundary has a slope instead of a quarter-turn shape, and SWCNCs can be obtained by this route as long as the lower diameter of the pre-fabricated pit is larger than 1.6 nm and the deposition temperature is higher than 750 K.

Keywords: graphene; carbon nanocones; molecular dynamics simulation

---

1. Introduction

As a 2D crystal, graphene has vast applications. It would be highly interesting if single wall carbon nanocones (SWCNCs) can be growth in graphene sheets since SWCNCs also have many extraordinary properties. Theoretical studies show that SWCNCs have good mechanical properties [1], a strong ability for electronic emission [2,3], heat flux rectification [4], and an electrical rectification effect [5]. Molecular dynamics (MD) simulations indicated that SWCNCs are stable even at 1500 K [6], and our previous work employing a statistical model showed that the cones can survive for 10^7 years at room temperature [7], which makes them potential candidates for future nanoscale devices. However, it remains a challenge to prepare SWCNCs without elaborately tailoring graphene sheets [7]. In the present work, we explored approaches to grow graphene sheets with desired structures, such as SWCNCs, carbon nanotubes, or cubic boxes, embedded in the sheets.

Recently, the chemical vapor deposition (CVD) method has been proven to be the most powerful approach to grow large-scale high-quality monolayer graphene [8], and the sheets are usually grown on a Cu substrate. Because of the low catalytic reactivity for dehydrogenation on top of graphene and the negligible carbon solubility in Cu, a graphene sheet with monolayer thickness grows larger and larger on the surface of a Cu substrate. Thus, if we pre-fabricated some nano-scale pits with conical or cylindrical shapes on the surface of a Cu substrate by physical or chemical methods (such as highly focused electron beam bombardment), then the growing sheet might occupy the pits and finally form some 3D structures embedded in the sheet. However, this seems impossible because of an experimental observation that the graphene sheet can grow continuously over grain boundaries without bending [9].

In this work, we first performed MD simulations of the growth of a monolayer graphene sheet on the grain boundaries of a Cu substrate, showing that on the quarter-turn boundary of a Cu crystal grain, the graphene sheet always maintain a flat structure, which is why the sheet can grow continuously across the grain boundaries [9]. However, if the quarter-turn boundary is replaced with a slope, the simulations show that the structure is a bent sheet which is stuck to the surface. Because MD simulation can only cover a very short time period (<1 μs), the criterion of Gibbs free energy (GFE) was applied to address this issue, showing that the bent graphene sheets are of lower GFE on a slope than on the flat surfaces, indicating that the bent graphene sheet can be obtained on the sloped boundary. On this basis, the GFE criterion was applied to explore the conditions for growing graphene sheets embedded with nano-conical structures by pre-fabricating conical pits on the surface of Cu substrates, showing that SWCNCs with cone angles larger than 90° can be achieved as long as the lower diameter of the pit is larger than 1.6 nm and the growth temperature is higher than 750 K.

2. Growth on the Cu Grain Boundaries

Regarding atomistic sizes, most crystal grain boundaries on metal surfaces should be of a quarter-turn shape, as shown in Figure 1. Therefore, we performed MD simulations on the relaxation of a piece of graphene sheet initially stuck to a quarter-turn boundary. Specifically, a Cu substrate of nine atomic layers with (111) surfaces was constructed according to the lattice constant 3.615 Å, and the motion of a graphene sheet containing 290 carbon atoms placed on the quarter-turn boundary was simulated by MD. In the simulation, the carbon atoms of the left-most two rows as well as all the Cu atoms were fixed, and the periodic conditions were applied in the Y-direction. Throughout this work, the standard Verlet algorithm was used to integrate the Newton equation with a time step of 0.2 fs, and the interactions between C-C and Cu-Cu were described by the Brenner function [10] and the tight-binding potential [11], respectively, while the interaction between C and Cu atoms was described by the Lennard-Jones (L-J) potential [12].

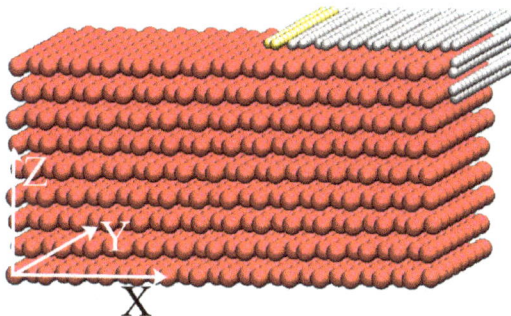

Figure 1. Schematic of a grain boundary with a quarter-turn shape covered by a piece of graphene sheet containing 290 C atoms.

Considering that the temperature for graphene growth by CVD is about 1200 K, the C atoms, except for the ones of the left-most two rows (the yellow balls in Figure 1), were initially assigned velocities of Maxwell distribution at 1200 K, and were then allowed to relax freely. After about 50 ps, the bent graphene sheet became flat with a little vibration. The initial temperature has little effect on the process. For example, when the mobile C atoms were initially assigned velocities of Maxwell distribution at 300 K, the bent sheet also turned into a flat structure in about 50 ps. However, things changed when the quarter-turn boundary was replaced with a slope of bent angles of 15°, 30°, and 45°, as shown in Figure 2. After the mobile C atoms were assigned the velocities of Maxwell distribution at 1200 K, the graphene sheet kept the curved shape until the end of the MD simulation, at 200 ps.

Nevertheless, we do not know what will happen if the MD simulation can cover a period of more than 1 s, so we have to apply the criterion of GFE to approach this problem.

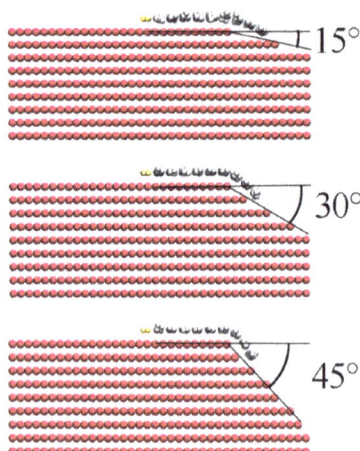

Figure 2. The structure of a piece of graphene sheet consisting of 290 C atoms on different slopes of the grain boundary after 200 ps relaxation at 1200 K.

For zero pressure, the GFE is

$$G = U - TS, \tag{1}$$

where U is the internal energy, and the entropy S is calculated by

$$S = S_0 + \int_{T_0}^{T} \frac{C_p}{T} dT, \tag{2}$$

where S_0 is the entropy for $T_0 = 0.01$ K, and was neglected in this work. The heat capacity C_p was calculated by $C_p = dE/dT$, where the internal energy E as the function of temperature T was obtained by MD simulations. Specifically, the mobile C atoms of the graphene sheet of 290 atoms were first cooled below 0.01 K, and were then assigned velocities of the Maxwell velocity distribution at the temperature T every 0.1 ps until the system reached the prescribed temperature T. Then the system was allowed to relax freely for about 2 ps, during which the internal energy and temperature was recorded every 20 fs to produce the average value. Setting the temperature T from 10 K–40 K with an interval of 10 K, we obtained the relationship between the internal energy and temperature. The obtained GFE of the graphene sheets stuck on the slopes of different angles are shown in Figure 3a, where the GFE of a flat sheet on a quarter-turn boundary is also presented, indicating that the GFE of the curved graphene on the slopes is significantly smaller than the flat graphene on the quarter-turn boundary, and the smaller the slope angle, the smaller the GFE. For a given slope, the GFE of the flat sheet (G_f) is significantly larger than for the bent sheet (G_b) if the slope angle is smaller than 60° (Figure 3b), indicating that in the process of graphene growth on the Cu surface by CVD, the sheet would stick to the surface of the slopes to form a bent structure as long as the angle is smaller than 60°, and an angle of 30° should be favorable for growth of the bent sheet.

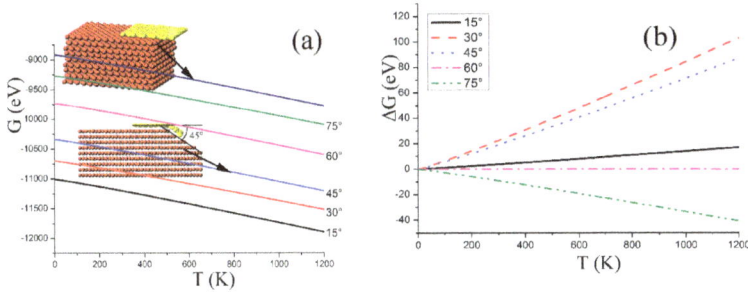

Figure 3. (a) The GFE of a piece of graphene sheet on boundaries of different slopes as a function of the temperature, and (b) the difference defined as G_f-G_b for a given slope.

3. SWCNC Growth by CVD Methods

Based on the discussion in the previous section, it is not impossible to grow SWCNCs embedded in graphene sheets by pre-fabricating conical pits on the surface of Cu substrates, and it is necessary to explore the specific conditions for growth. MD simulation seems to be an ideal method to display the growth process and therefore can determine under what conditions SWCNCs may grow on Cu substrates. However, the realistic growth rate of graphene by CVD is too slow for MD to simulate, and we have to resort to the GFE criterion that states for an isobaric process, such as CVD growth, the system tends to a structure of lower GFE. Therefore, we calculated the GFE of a piece of graphene sheet crossing over a pit on the surface (Figure 4a), or bending into the pit (Figure 4b), to see which geometry is of smaller GFE.

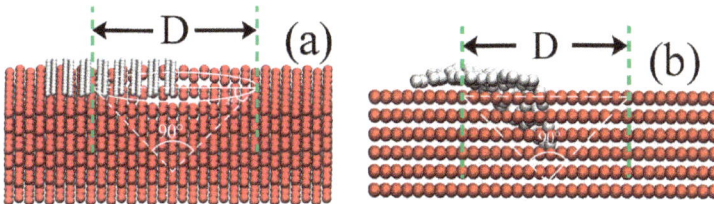

Figure 4. (a) Schematic of a piece of flat graphene sheet covering half of a conical pit with an angle of 90°; or (b) the sheet bending into the pit.

A technique of time-going-backwards [13,14] in the MD simulation was employed to obtain the probable structure of a piece of graphene sheet bent into the pit (Figure 4b). Specifically, a graphene sheet containing 176 C atoms was initially placed over a conical pit with a 45° angle slope (Figure 4a), corresponding to a conical angle of 90°, and then the C atoms of the leftmost two rows and all the Cu atoms were kept fixed while the Newton's equations for the other mobile C atoms were integrated by a negative step of −0.2 fs, during which every mobile C atom was assigned a velocity every 0.16 ps by

$$v_i = \left[v_i^0 - \theta^{1/2} v^T(\xi) \right] / (1-\theta)^{1/2}, \tag{3}$$

where v_i^0 and $v^T(\xi)$ are the present velocity of atom i and the velocity randomly chosen from the Maxwell distribution at 300 K, respectively, while θ is a random number ($1 < \theta < 1$). In this process, the temperature increased step by step, and when the temperature increased to 4000 K, the Newton's equations were integrated by a positive step of 0.2 fs and the mobile C atoms were assigned a velocity every 0.16 ps by

$$v_i = (1-\theta)^{1/2} v_i^0 + \theta^{1/2} v^T(\xi) \tag{4}$$

When the system arrived at about 300 K, we cooled the system down to 0.01 K by a damping method [13], and recorded the geometrical structure as well as the potential energy.

The above heating and cooling cycle ran 100 times for each of the conical pits with lower diameters of 1.2, 1.6 or 2.0 nm, producing 100 geometrical structures for each pit. As listed in Table 1, among the 100 configurations there are 41, 25 and 5 structures extending into the pits with diameters of 2.0, 1.6, and 1.2 nm, respectively, indicating that the larger pits are more favorable to the growth of the bent structures. The potential energy of the resultant structures changes in the range of 30–40 eV, and some of the bent structures have lower potential but others do not, implying that the potential energy is insufficient to judge the dynamic probability for growth of the bent structure.

Table 1. Number of bent structures appearing in 100 rounds of heating and cooling cycles.

Substrate	Diameter of Pit Bottom (nm)	Number of Bent Structures
2	2.0	41
4	1.6	25
26	1.2	5

We calculated the GFE of the above bent structures (G_b) and of the flat structure (G_f) with the same number of C atoms. The calculations showed, generally, that the lower the potential of the bent structure, the lower the GFE. So only the GFE of the bent structures with lower potential energies need to be inspected. For the pit with a lower diameter of 2.0 nm, three bent structures of lower potential energy denoted by S1, S2 and S3 are shown in the inset in Figure 5a, where the temperature dependence of the GFE difference, $\Delta G = G_f - G_b$, is presented for each of the structures. Below room temperatures, the GFE of the flat sheet is significantly smaller than the bent sheets, while as the temperature increases above 470 K, the GFE of the bent structure S1 gets smaller than the flat sheets. Among the three structures, the GFE of S1 is significantly lower than the others in the entire temperature range (0–1200 K), so at about 1200 K, the CVD growth of graphene on Cu surface should probably be formed. For the pit with a lower diameter of 1.6 nm, two bent structures of lower potential energy, denoted by C1 and C2 in Figure 5b, have GFEs smaller than the flat structures when the temperature is above 950 K, and the GFE of C1 is significantly smaller in the entire temperature range. Therefore, the bent structure C1 would be more probable in the CVD growth. However, for the pit with a lower diameter of 1.2 nm, the GFEs of all the bent structures are always larger than the flat structures in the temperature range from 0–1200 K, implying that the bent structures can not be formed in this pit.

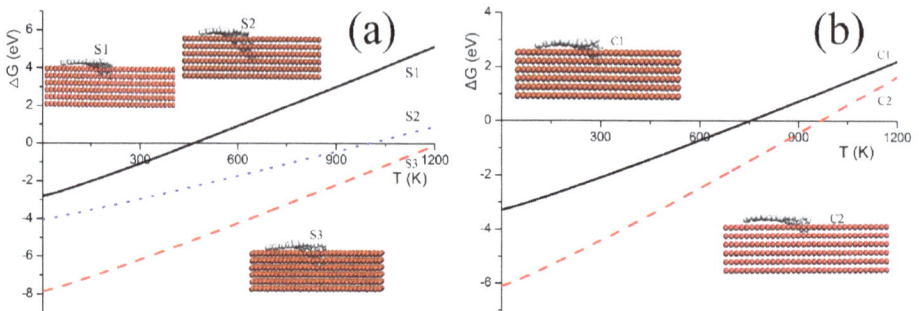

Figure 5. (a) Temperature dependence of the GFE difference $\Delta G = G_f - G_b$ for structures S1, S2 and S3 bent into the pit with a lower diameter of 2.0 nm, and (b) for structure C1 and C2 bent into the pit with a lower diameter of 1.6 nm.

4. Conclusions

Based on the GFE criterion, it is shown that single wall carbon nanocones embedded in single layer graphene sheets can be obtained by CVD on a Cu surface with pre-existing conical pits. For a 90° conical angle, the growth requires that the lower diameter of the pit must be larger than 1.6 nm and the temperature of the substrate should be higher than 750 K, which is below the temperature required for the CVD growth of graphene. Certainly, it is necessary to understand the dynamics of how the carbon atoms move into the pit to form the nanocone, but the process is too slow to be simulated by MD, so the specific pathway or mechanism for the nanocone growth and the exact structures of the bent sheet still escape prediction.

Acknowledgments: This work was supported by Specialized Research Fund for the Doctoral Program of Higher Education under Grant No. 20130071110018 and the National Natural Science Foundation of China under Grant No. 11274073.

Author Contributions: Yu-Peng Liu performed the simulation and prepared the manuscript. Jing-Tian Li and Quan Song edited the manuscript. All authors contributed to the discussions. Xi-Jing Ning designed and supervised this project.

Conflicts of Interest: The authors declare no conflict of interest.

References

1. Jordan, S.P.; Crespi, V.H. Theory of carbon nanocones: Mechanical chiral inversion of a micron-scale three-dimensional object. *Phys. Rev. Lett.* **2004**, *93*, 255504. [CrossRef] [PubMed]
2. Charlier, J.C.; Rignanese, G.M. Electronic structure of carbon nanocones. *Phys. Rev. Lett.* **2001**, *86*, 5970–5973. [CrossRef] [PubMed]
3. Compernolle, S.; Kiran, B.; Chibotaru, L.F.; Nguyen, M.T.; Ceulemans, A. Ab initio study of small graphitic cones with triangle, square, and pentagon apex. *J. Chem. Phys.* **2004**, *121*, 2326–2336. [CrossRef] [PubMed]
4. Yang, N.; Zhang, G.; Li, B.W. Carbon nanocone: A promising thermal rectifier. *Appl. Phys. Lett.* **2008**, *93*, 3. [CrossRef]
5. Ming, C.; Lin, Z.Z.; Zhuang, J.; Ning, X.J. Electronic rectification devices from carbon nanocones. *Appl. Phys. Lett.* **2012**, *100*, 4. [CrossRef]
6. Tsai, P.C.; Fang, T.H. A molecular dynamics study of the nucleation, thermal stability and nanomechanics of carbon nanocones. *Nanotechnology* **2007**, *18*. [CrossRef]
7. Ming, C.; Lin, Z.Z.; Cao, R.G.; Yu, W.F.; Ning, X.J. A scheme for fabricating single wall carbon nanocones standing on metal surfaces and an evaluation of their stability. *Carbon* **2012**, *50*, 2651–2656. [CrossRef]
8. Munoz, R.; Gomez-Aleixandre, C. Review of cvd synthesis of graphene. *Chem. Vap. Depos.* **2013**, *19*, 297–322. [CrossRef]
9. Yu, Q.; Jauregui, L.A.; Wu, W.; Colby, R.; Tian, J.; Su, Z.; Cao, H.; Liu, Z.; Pandey, D.; Wei, D.; et al. Control and characterization of individual grains and grain boundaries in graphene grown by chemical vapour deposition. *Nat. Mater.* **2011**, *10*, 443–449. [CrossRef] [PubMed]
10. Brenner, D.W. Empirical potential for hydrocarbons for use in simulating the chemical vapor-deposition of diamond films. *Phys. Rev. B* **1990**, *42*, 9458–9471. [CrossRef]
11. Cleri, F.; Rosato, V. Tight-binding potentials for transition-metals and alloys. *Phys. Rev. B* **1993**, *48*, 22–33. [CrossRef]
12. Shi, X.H.; Yin, Q.F.; Wei, Y.J. A theoretical analysis of the surface dependent binding, peeling and folding of graphene on single crystal copper. *Carbon* **2012**, *50*, 3055–3063. [CrossRef]
13. Ning, X.J.; Qin, Q.Z. A new molecular dynamics method for simulating trapping site structures in cryogenic matrices. *J. Chem. Phys.* **1999**, *110*, 4920–4928. [CrossRef]
14. Ning, X.-J.; Qin, Q.-Z. Trapping site structures of O3 isolated in argon matrices. *J. Chem. Phys.* **1999**, *111*, 7047–7052. [CrossRef]

© 2017 by the authors. Licensee MDPI, Basel, Switzerland. This article is an open access article distributed under the terms and conditions of the Creative Commons Attribution (CC BY) license (http://creativecommons.org/licenses/by/4.0/).

crystals

MDPI

Article

Strength of Alkane–Fluid Attraction Determines the Interfacial Orientation of Liquid Alkanes and Their Crystallization through Heterogeneous or Homogeneous Mechanisms

Yuqing Qiu and Valeria Molinero *

Department of Chemistry, The University of Utah, 315 South 1400 East, Salt Lake City, UT 84112-0850, USA
* Correspondence: Valeria.Molinero@utah.edu; Tel.: +1-801-585-9618

Academic Editor: Hiroki Nada
Received: 2 February 2017; Accepted: 9 March 2017; Published: 15 March 2017

Abstract: Alkanes are important building blocks of organics, polymers and biomolecules. The conditions that lead to ordering of alkanes at interfaces, and whether interfacial ordering of the molecules leads to heterogeneous crystal nucleation of alkanes or surface freezing, have not yet been elucidated. Here we use molecular simulations with the united-atom OPLS and PYS alkane models and the mW water model to determine what properties of the surface control the interfacial orientation of alkane molecules, and under which conditions interfacial ordering results in homogeneous or heterogeneous nucleation of alkane crystals, or surface freezing above the melting point. We find that liquid alkanes present a preference towards being perpendicular to the alkane–vapor interface and more parallel to the alkane–water interface. The orientational order in the liquid is short-ranged, decaying over ~1 nm of the surface, and can be reversed by tuning the strength of the attractions between alkane and the molecules in the other fluid. We show that the strength of the alkane–fluid interaction also controls the mechanism of crystallization and the face of the alkane crystal exposed to the fluid: fluids that interact weakly with alkanes promote heterogeneous crystallization and result in crystals in which the alkane molecules orient perpendicular to the interface, while crystallization of alkanes in the presence of fluids, such as water, that interact more strongly with alkanes is homogeneous and results in crystals with the molecules oriented parallel to the interface. We conclude that the orientation of the alkanes at the crystal interfaces mirrors that in the liquid, albeit more pronounced and long-ranged. We show that the sign of the binding free energy of the alkane crystal to the surface, ΔG_{bind}, determines whether the crystal nucleation is homogeneous ($\Delta G_{\text{bind}} \geq 0$) or heterogeneous ($\Delta G_{\text{bind}} < 0$). Our analysis indicates that water does not promote heterogeneous crystallization of the alkanes because water stabilizes more the liquid than the crystal phase of the alkane, resulting in $\Delta G_{\text{bind}} > 0$. While $\Delta G_{\text{bind}} < 0$ suffices to produce heterogeneous nucleation, the condition for surface freezing is more stringent, $\Delta G_{\text{bind}} < -2 \, \gamma_{\text{xl}}$, where γ_{xl} is the surface tension of the liquid–crystal interface of alkanes. Surface freezing of alkanes is favored by their small value of γ_{xl}. Our findings are of relevance to understanding surface freezing in alkanes and to develop strategies for controlling the assembly of chain-like molecules at fluid interfaces.

Keywords: surface ordering; chain molecules; homogeneous nucleation; heterogeneous nucleation; surface freezing; complete wetting; assembly; crystallization

---

1. Introduction

Alkanes are simple organic molecules and the main building block of complex organic compounds, including surfactants, polymers and lipids. One of the unique properties of alkanes is the surface freezing effect [1–3]. Linear alkanes with 16 to 50 carbons form a crystalline monolayer with the molecules perpendicular to the alkane–vapor interface at temperatures up to three Kelvin above their bulk equilibrium melting temperatures [2]. A recent experimental and simulation study of supercooled alkane droplets demonstrates that alkanes crystallize heterogeneously at the alkane–vapor interface [4]. A monolayer of alkanes perpendicular to the interface is formed before the interior of the droplet crystallizes [4], even for alkanes that are too short to present surface freezing.

There is no surface freezing of alkanes at the alkane–water interface [5], although alkanes interact more strongly with water than with alkane vapor. The order of alkanes at the alkane–water interface has been studied with total internal reflection second-harmonic generation spectroscopy [6–8]. The calculated effective second-order susceptibilities from simulations of configurations having the alkane chains parallel to the alkane–water interface result in better agreement with experimental measurements than the ones calculated from configurations with the alkane chains perpendicular to the interface [9]. That result is consistent with previous molecular simulations of alkane–water slabs that found the alkane to be more parallel at the interface than in the bulk of the liquid [10,11]. The first goal of this study is to identify which property controls the interfacial ordering of alkanes.

A recent simulation study of the heterogeneous crystallization of alkanes by silicon-like templating model surfaces indicates that the rate of crystallization increases as the alkane–surface interaction potential becomes more attractive [12]. As vacuum, that has no attraction to alkanes, promotes the heterogeneous crystallization of alkanes, this poses the question of whether the same rule applies to crystallization of alkanes by fluid interfaces, which cannot act as a template to order the alkane crystal. To the best of our knowledge, it has not been demonstrated whether alkanes crystallize at the water interface, with which they experience dispersion interactions. Our second goal is to determine the crystallization mechanism—homogeneous or heterogeneous—of alkanes in the presence of water and other fluids.

A commonality of surface freezing and heterogeneous crystallization of alkanes is that a layer of crystalline alkane perpendicular to the interface precedes the bulk crystallization [2,4]. The third goal of this study is to interpret the conditions that lead to homogeneous or heterogeneous nucleation of supercooled alkanes, and the existence of surface freezing above their melting point. We have recently demonstrated that the condition for heterogeneous nucleation is that the binding free energy of the crystal to the nucleating surface is negative, $\Delta G_{\text{bind}} < 0$, and that complete wetting of the surface by the crystal is attained when the binding free energy is not only negative but less than twice the liquid–crystal surface tension, $\Delta G_{\text{bind}} < -2\,\gamma_{\text{xl}}$ [13]. Here we use the framework of ref [13] to interpret the mechanism of crystallization of alkanes at fluid interfaces, and to explain why—for a particular combination of surface and alkane—heterogeneous nucleation can occur without surface freezing but surface freezing cannot occur without heterogeneous nucleation.

2. Methods

Force fields. Alkanes are modeled with two different united-atom (UA) force fields: PYS [14–16] and OPLS [17,18]. These force fields have been widely used to investigate the structure, interfacial properties and phase transitions of alkanes [4,12,19–37]. Water is modeled with the monatomic water model, mW [38], which has been extensively used to study the structure, thermodynamics, interfacial properties, and phase transitions of water [38–83]. All the force fields in this study are united-atom force fields and all interactions are short-ranged. We model the interaction of methyl and methylene groups with water through Lennard-Jones potentials, and assume that both alkane moieties, which we here call C, interact identically with water. The C–C and C–mW interactions are cut off at 1.2 nm with a long-range van der Waals tail correction to the energy and pressure [84] (implemented in LAMMPS [85] through the pair_modify tail command) as recommended for PYS alkanes in [22]. The strength and size of the mW water and OPLS nonane interaction were parameterized in ref [37]. The size of water–carbon interaction was taken as the average of SPC/E water–water distance [86] and OPLS methylene–methylene distance [18], $\sigma_{\text{wc}} = 0.35$ nm, and the strength of the water–carbon interaction,

$\varepsilon_{wc} = 0.17$ kcal·mol^{-1}, was parameterized to reproduce the experimental liquid water–liquid nonane surface tension γ_{lw} [37]. Here we follow the same procedure to parameterize ε_{wc} for mW water and PYS alkanes by matching the liquid–liquid surface tension γ_{lw} of nonane–water, hexadecane–water and eicosane–water to their experimental counterparts [87,88], assuming that $\sigma_{wc} = 0.35$ nm. The optimized strength of interactions between mW water and PYS alkanes ε_{wc} are 0.22, 0.20, 0.19 kcal·mol^{-1} for nonane (C9), hexadecane (C16) and eicosane (C20), respectively.

Simulations settings. We perform molecular dynamic simulations using LAMMPS [85]. The equations of motions are integrated with the velocity Verlet algorithm with timestep of 5 fs. The simulation cell is periodic in the three directions. Except when otherwise is indicated, we control the temperature and pressure using the Nose-Hoover [89,90] thermostat and barostat with time constant of 0.5 ps and 1.25 ps, respectively.

Properties. The melting temperatures of alkanes, T_m, are determined through phase coexistence in the *NpT* ensemble, following the procedure of ref [22,23]. The simulation cells contain 960 nonane molecules (C9), or 1024 hexadecane molecules (C16), or 1024 eicosane molecules (C20). We start with crystalline alkane structures from the Cambridge Structural Database [91], and we equilibrate the alkane crystals at 10 K below their corresponding T_m in the models for over 2 ns in the *NpT* ensemble. We then melt half of the simulation cell, exposing to the liquid the (100) or (001) faces of the crystals. The error bar of T_m is determined as the gap between the highest temperature at which the simulation cell crystallizes and the lowest temperature at which it melts.

The enthalpy of fusion of nonane modeled with OPLS and PYS at their corresponding melting temperature T_m, is calculated as the enthalpy difference between liquid and crystalline alkanes, $\Delta H_m = H_{liquid} - H_{crystal}$, computed from one-phase simulations of cells with 960 nonane molecules (C9), or 1024 hexadecane molecules (C16), or 1024 eicosane molecules (C20), at their corresponding T_m. The entropy of fusion ΔS_m at T_m is calculated from the enthalpy of fusion and the melting temperature, $\Delta S_m = \Delta H_m / T_m$. The enthalpies of vaporization, ΔH_{vap}, at 298 K, are calculated as the enthalpy difference between gas and liquid alkanes, $\Delta H_{vap} = H_{gas} - H_{liquid}$, where the enthalpy of the gas was computed from simulation of one gas molecule in a 7 nm $\times$ 7 nm $\times$ 14 nm simulation cell in the *NVT* ensemble with the a Langevin thermostat [92], with a damping constant of 1 ps.

The surface tension of the liquid alkane–vacuum interface, γ_{lv}, and liquid alkane–liquid water, γ_{lw}, are calculated through the mechanical route [93], as $\gamma = (L_z/2)(<p_n> - <p_t>)$, where $<p_n>$ and $<p_t>$ are the pressure tensors normal and tangential to the interface, averaged over 50 ns of *NVT* simulations, and L_z is the length of the simulation cell in the direction perpendicular to the interfaces. The simulation cells used for calculating γ_{lv} contain slabs of 114 nonane molecules, or 64 hexadecane molecules, or 120 eicosane molecules. The simulation cells used for calculating γ_{lw} contain 114 nonanes + 1024 waters, or 64 hexadecanes + 1024 waters, or 120 eicosanes + 2048 waters. The *xyz* dimensions of the simulation cells for the alkane/vacuum and alkane/water two-phase systems are identical: 3.5 nm $\times$ 3.5 nm $\times$ 7 nm for nonane, 6 nm $\times$ 6 nm $\times$ 6 nm for hexadecane and 4 nm $\times$ 4 nm $\times$ 8 nm for eicosane. The alkane molecules occupy about half of the cell, with the alkane–vacuum interface parallel to the *xy* plane. To verify that the simulation cells are sufficiently large, we also compute the surface tension of the nonane–vapor and hexadecane–vapor interfaces using 8 times larger simulation cells (with twice thicker slabs of alkane), finding identical results.

To measure the orientation of liquid alkanes at their liquid–vapor interface and liquid alkane–water interface, we run simulations in the *NVT* ensemble of OPLS nonane/mW water or PYS nonane/mW water with various ε_{wc} above the melting temperature of their corresponding alkane model. The simulation cells contain 912 nonanes (all in the liquid phase) or 912 nonanes plus 4096 waters (each in its respective liquid phase). Each periodic cell has two equivalent alkane/vapor or alkane/water interfaces. We define an end-to-end (methyl-to-methyl) vector for each nonane molecule and measure the angle θ it forms with respect to the normal to the interface. The use of the end-to-end vector to characterize the orientation of nonane assumes that the chains are extended and unbent, which is the case for nonane, although it may not be the case for long alkanes. We locate the vector at the center of the chain. We divide the box along the surface normal into 0.1 nm wide bins, and average the θ in each bin over 20 ns simulations. For OPLS nonane we compute the orientation of the liquid at 310 K and for PYS nonane at 240 K, which correspond to 3 and 20 K above the corresponding melting points when exposing the (100) face of the crystal to the liquid (see Results section). We cannot run the simulation with PYS nonane closer to its melting point because the dynamics of the alkane become too slow to allow ergodic sampling of the molecular orientations in tens of nanoseconds. To assess the effect of water–alkane interaction on the interfacial ordering of alkanes, we perform two-phase simulations of water–alkane in which we tune the strength of ε_{wc} while keeping the original water–water and alkane–alkane interactions unchanged.

Crystallization of alkanes. We run simulations of the liquid alkane–vacuum and liquid alkane–water two-phase systems under supercooled conditions in the *NpT* ensemble to crystallize the alkanes. Crystallization of PYS hexadecane is simulated in a slab of 1024 hexadecanes in contact with vacuum or two-phase 1024 hexadecanes/8192 waters at 240 K. Crystallization of OPLS nonane is simulated in a slab of 114 nonanes in contact with vacuum or two-phase 114 nonanes/1024 waters at 270 K. The maximum waiting time for crystallization in each simulation is 100 ns. To assess the effect of water–alkane attraction on the mechanism of crystallization of alkanes, we perform simulations of crystallization, same as described above for water–alkane but in which we tune the strength of ε_{wc} while keeping the original water–water and alkane–alkane interactions. We apply the local bond order parameter q_6 [94] to distinguish crystalline from liquid hexadecane, following ref [21]. We select the largest crystalline cluster by applying the criteria that a molecule belongs to the crystal if $q_6 > 0.3$ for any six consecutive atoms in a single alkane molecule.

3. Results and Discussion

3.1. Thermodynamic Properties of the Models

3.1.1. Thermodynamics of Pure Alkanes

We first characterize the thermodynamic properties of alkanes modeled with OPLS and PYS force fields. We find that PYS reproduces the experimental melting temperature of nonane (C9), hexadecane (C16), and eicosene (C20) exposing the (100) face to the liquid (Table 1), in agreement with refs [22,23]. The (001) face of alkanes, which exposes the methyl ends of the molecules to the liquid, has lower surface energy than the (001) face, which exposes the side of the molecule to the liquid phase [22,23]. In principle, the melting temperature is a bulk property independent of the crystal face exposed, however in finite size simulations T_m can depend on the face exposed to the liquid [95]. PYS alkanes exposing the (001) face grow and dissolve at a slower rate than when exposing the (100) face, making it challenging to determine the melting points of PYS alkanes in cells exposing the (001) interface. We compute the melting temperatures of OPLS alkanes, which have not been previously reported, and find them to be much higher than their experimental counterparts. Moreover, the OPLS model fails to predict the rapid increase in melting temperature with chain length observed in experiments. T_m of OPLS nonane is almost 90 K above the experimental value (Table 2). T_m of OPLS hexadecane is 325 K, which is over 30 K higher than their experimental T_m. Because of the high overestimation of the melting temperatures of nonane by OPLS, we did not use this force field to compute melting temperatures for longer alkanes. We find that T_m computed for OPLS alkanes exposing the (001) face is at least 30 K higher than T_m determined with the (100) face exposed, due to finite size effects in the simulations. Larger simulation cells would be required for an accurate determination of the melting temperatures of alkanes models.

Table 1. Comparison of thermodynamic properties of PYS alkanes and their interfaces with mW water with the experimental ('exp.') counterparts.

Alkane	T_m (K) [a]	Exp. T_m (K)	γ_{lv} (mJ·m^{-2})	Exp. γ_{lv} (mJ·m^{-2})	γ_{lw} (mJ·m^{-2})	Exp. γ_{lw} (mJ·m^{-2})
Nonane	219 ± 2	219.5 ± 0.5 [b]	14 ± 1 [c]	22.70 [e]	54 ± 1 [c]	52.4 [e]
Hexadecane	289 ± 2	291 ± 1 [b]	18 ± 1 [c]	26.26 [f]	55 ± 1 [c]	55.2 [e]
Eicosane	309 ± 2	310 ± 1 [b]	19 ± 1 [d]	27.62 [f]	58 ± 1 [c]	56.7 [g]

[a] computed for cells exposing the (100) plane; [b] ref [96]; [c] at T = 295 K; [d] T = 313.15 K; [e] at T = 295 K, from ref [87]; [f] T = 313.15 K, from ref [97]; [g] at T = 295 K, from ref [88].

Table 2. Comparison of thermodynamic properties of pure PYS and OPLS nonane.

Nonane Model	T_m (K)	γ_{lv} (mJ·m^{-2})	ΔH_{vap} (kcal·mol^{-1})	ΔH_m (kcal·mol^{-1})	ΔS_m (cal K^{-1}·mol^{-1})
OPLS	307 ± 2 [a]	23 ± 1 [c]	12.70 [e]	4.25 [g]	13.9
PYS	219 ± 2 [a]	14 ± 1 [c]	11.19 [e]	3.50 [h]	15.9
Experiment	219.5 ± 0.5 [b]	22.7 [d]	11.16 [f]	3.59 [i]	16.4

[a] T_m determined with (100) interface; [b] ref [96]; [c] at T = 295 K; [d] at T = 295 K from ref [87]; [e] T = 298 K; [f] at T = 299 K from ref [98]; [g] T = 307 K; [h] T = 219 K; [i] at T = 219.5 K from ref [99].

Both the OPLS and PYS models underestimate the entropy of melting of the alkanes. The ability of PYS to reproduce the experimental ΔH_m and ΔS_m may be dependent on whether the alkane has an odd or even number of carbons, because although PYS underestimates ΔH_m and ΔS_m of octane by 40% [22], we find it reproduces quite well these properties for nonane (Table 2). OPLS overestimates ΔH_m of nonane by 18%. However, due to the high melting temperature of this model it underestimates ΔS_m by at least 15% compared to the experiment (Table 2). Table 2 also shows that PYS reproduces well the vaporization enthalpy ΔH_{vap} of nonane, while OPLS overestimates it.

PYS consistently underestimates the liquid–vacuum surface tension γ_{lv} of nonane, hexadecane, and eicosane by more than 30% (Table 2). OPLS overestimates so much the melting temperatures, that it is not possible to measure the liquid–vacuum surface tensions of hexadecane and eicosane at the same temperatures as in the experiments without spontaneous crystallization of the alkanes. For OPLS nonane, γ_{lv} can be measured and reproduces well the experimental value (Table 2). However, we note that OPLS overestimates the liquid–vacuum surface tension of ethane by 20% [100], indicating that the agreement is not transferable along chain lengths. We conclude that the OPLS and PYS united-atom force fields either reproduce the liquid–crystal phase equilibrium or the liquid–vacuum surface properties of alkanes, but not both.

3.1.2. Thermodynamics of Alkane–Water Systems

The interaction between the united atom methylene and methyl groups of OPLS nonane with mW water was parameterized in [37] to reproduce the experimental surface tension of the alkane/water interface, resulting in ε_{wc} = 0.17 kcal·mol^{-1} and σ_{wc} = 0.35 nm. Here we keep σ_{wc} and follow the same strategy to parameterize ε_{wc} of PYS alkanes with mW water. We find that the strength of water–methylene (or water–methyl) interactions ε_{wc} needed to reproduce the experimental alkane–water surface tension decreases slightly (within 0.03 kcal·mol^{-1}) with the length of the PYS alkane chains (Table 1). To assess the sensitivity of the liquid-alkane-water surface tension γ_{lw} to ε_{wc}, we apply the ε_{wc} of nonane–water to the other alkanes, and find that γ_{lw} deviates from the experimental values by less than 3 mJ·m^{-2}. We do not assess here whether ε_{wc} is transferable over the chain length for OPLS alkanes, as we only model nonane with that force field. Use of different strength of water–methyl and water–methylene interactions may allow the use of a single set of parameters for all the PYS alkane–water interactions.

Although OPLS and PYS alkanes have different force field parameters and thermodynamic properties, in what follows we show that they display the same trends in the orientational order of alkanes and in the mechanism of crystallization in the presence of water or vacuum, indicating that the results of this study are robust and independent on the details of the force fields.

3.2. Interfacial Orientation of Liquid Alkanes is Controlled by the Strength of Attraction to the Other Phase

Before investigating the crystallization mechanisms of alkanes, we characterize the orientational ordering of liquid alkanes in contact with vacuum and water. We address the effect of alkane–water attraction ε_{wc} on the interfacial orientation and surface tension of the latter interface. To identify the position of the interfaces, we compute the density profile of the center of mass of the alkanes for slabs of nonane in contact with vacuum or water (Figure 1a,b) and find the Gibbs dividing surface, defined as the plane where the density reaches half the bulk value. Figure 1 shows that the density of liquid alkane peaks at about 0.5 nm from that interface. The existence of interfacial density peaks has been previously reported for other alkanes [101,102]. The density peaks are sharper at the nonane–water interface than at the nonane–vacuum interface.

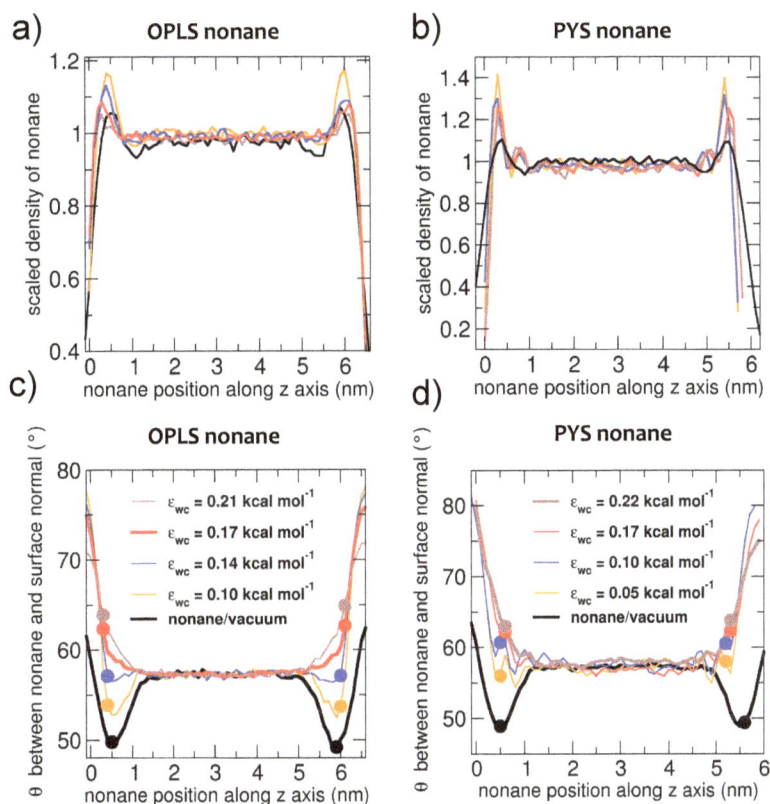

Figure 1. Profiles of the density (**a,b** panels) and average orientation (**c,d** panels) of liquid nonane in contact with vacuum and water, for which the strength ε_{wc} of the water–carbon interaction is tuned. The ordering of the liquid by the interface is short-ranged, about the length of a nonane molecule, and turns from leaning parallel to the surface to leaning more perpendicularly to the surface on decreasing the strength of the coupling between alkane and the other fluid phase (water or vacuum). The parameters that reproduce the experimental water–nonane surface tension are shown with thick lines: brown for PYS (ε_{wc} = 0.22 kcal·mol^{-1}) and red for OPLS (ε_{wc} = 0.17 kcal·mol^{-1}). The densities are presented scaled with respect to the bulk values; the orientations are computed with respect to the surface normal, as shown in Figure 2. Densities and orientations are computed at 310 K for OPLS nonane and at 240 K for PYS nonane. Solid circles signal the average orientation θ measured 0.5 nm from the Gibbs dividing surface.

To characterize the orientation of the molecules, we measure the angle θ between the methyl-to-methyl vector that each molecule forms with the normal to the interface (Figure 2): $\theta = 90°$ means that the molecule is parallel to the interface and $\theta = 0°$ that it is perpendicular to the interface. The orientational order of nonane at the interface is short ranged (Figure 1c,d), decaying to the bulk value at distances beyond 1.0 nm (the length of a nonane molecule) from the Gibbs dividing surface. Liquid nonane has a preference towards being perpendicular to the alkane–vacuum interface, irrespective of the force field and in agreement with what was previously reported for the orientation of a slab of OPLS decane in contact with vacuum [103,104]. At the water interface, the alkanes are more parallel compared to their average orientation in the bulk. We find that the extent of orientational order is slightly more pronounced at the vacuum than the water interface. Our results are consistent with previous simulations of GROMOS decane in contact with SPC or SPC/E water [10] and a recent interpretation of sum frequency scattering experiments of the decane–water interface [9]. We conclude that, irrespective of the force fields used for the calculations, liquid alkanes have opposite orientational preferences at the alkane–water and the alkane–vacuum interface.

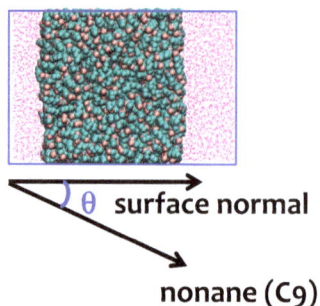

Figure 2. The orientation of the alkane molecules is characterized by the angle θ between the alkane molecules and surface normal. The upper panel shows a typical snapshot of the water–nonane simulation cell. Methyl groups are shown with pink beads and methylene groups with cyan beads. Water is represented by magenta points. Blue squares denote the periodic boundaries of the simulation box. The lower panel illustrates the relation between the angle θ and the orientation of nonane.

To understand how the strength of attraction between alkane and the other phase (which we below call solvent) impacts the interfacial orientation of liquid alkanes, we perform simulations of alkane in contact with water varying the strength of the water–carbon attraction, ε_{wc}, to span the range of the interactions from water to vacuum (Figure 1c,d). We find that the orientational preference of the alkanes at the interface evolves from parallel to perpendicular with decreasing strength of water–carbon coupling ε_{wc}. Decreasing ε_{wc} also results in a linear increase in the liquid alkane–water surface tension γ_{lw} (Figure 3). We note that in the limit of null interaction between water and the alkane, the interfacial tension of the water–alkane interface should be the sum of the water–vacuum and alkane–vacuum interfaces. In next section, we investigate whether the tuning of the interfacial ordering of the alkanes results in a change in mechanism from heterogeneous to homogeneous nucleation and distinct preference of interfacial orientation of alkanes in the crystal phase.

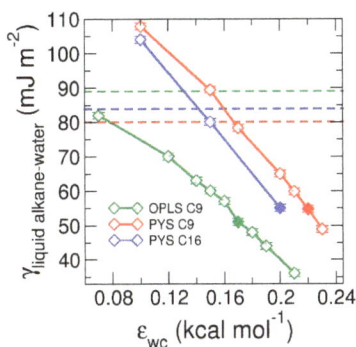

Figure 3. Surface tension of liquid alkane–water interface, γ_{lw}, as a function of the coupling between water and alkanes, ε_{wc}. Green, red, blue curves correspond to surface tensions of OPLS nonane (C9), PYS nonane (C9) and PYS hexadecane (C16), respectively. The surface tensions are computed at the same temperatures as in the experimental references: 295 K for nonane [87] and 313.15 K for hexadecane [97]. Solid squares represent the ε_{wc} at which experimental surface tension are reproduced. Note that in the limit where the water–alkane interactions become purely repulsive, the surface tension of the water–alkane interface can be larger than the sum of the surface tensions of non–interacting water–vacuum and alkane–vacuum interfaces. The latter is given by the sum of the surface tension of the water–vacuum interface (66 mJ·m^{-2} at 295 K [105]) and the liquid alkane–vacuum interfaces (Tables 1 and 2), $\gamma_{wv} + \gamma_{lv}$, and are shown with dashed lines.

3.3. Strength of the Alkane–Solvent Attraction Determines Whether Alkanes Crystallize through Heterogeneous or Homogeneous Nucleation

We study the crystallization mechanisms of nonane and hexadecane at the alkane–water and alkane–vacuum interfaces under highly supercooled conditions, at least 45 K below the corresponding T_m. Alkanes crystallize heterogeneously at the alkane–vacuum interface (Figure 4a), irrespective of the force field and in agreement with previous studies using the PYS model [4]. In the presence of water, the alkanes crystallize homogeneously, forming the critical crystal nucleus in the interior of the liquid phase (Figure 4b).

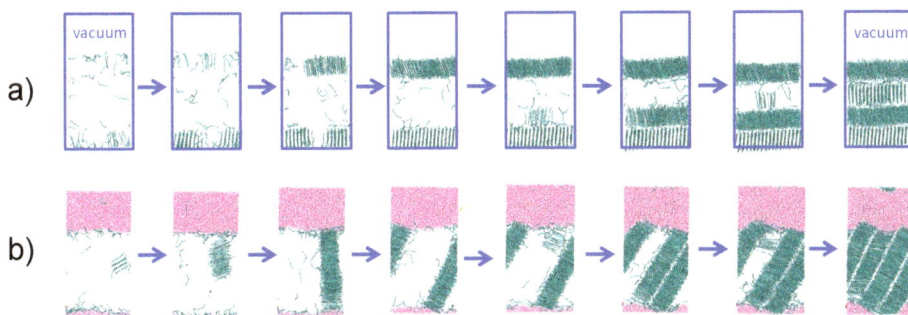

Figure 4. Snapshots of PYS hexadecane crystallizing (**a**) heterogeneously at the hexadecane–vacuum interface at 240 K and (**b**) homogeneously in the presence of water (ε_{wc} = 0.20 kcal·mol^{-1}) at 245 K. The induction time that precedes crystallization is not shown. Cyan lines indicate crystalline C16; the liquid alkane phase is hidden. Pink points represent water. Blue rectangles in (**a**) denote the boundaries of the periodic alkane/vacuum simulation box.

The nucleus of the alkane crystal has the same shape for heterogeneous and homogeneous nucleation: a cylindrical one-molecule thick bundle of partially aligned alkane chains [21,23] (Figure 4). In the presence of a vacuum interface, the crystal nucleus forms at the surface, with the molecules aligned perpendicular to the vacuum interface (Figure 4a). The crystal grows first in the direction parallel to the interface, before it nucleates and grows subsequent layers. The same mechanism has been reported for the crystallization of large PYS nonane droplets [4]. The simulations suggest that the alkane–vacuum interface stabilizes the chain-end rather than the side of the alkane molecules. The perpendicular orientation of the alkanes at the crystal–vacuum interface is consistent with the orientation in the liquid and in the one-layer-thick surface freezing in medium length alkanes.

The one-molecule thick bundle-like nucleus for the homogeneous crystallization in the presence of water grows by first adding alkanes on the side of the nucleus—which results in a one-molecule thick crystal layer—and only then growing a second crystal layer (Figure 4b). Same as in the growth in the presence of vapor, there is a separation of time scales for growth of the one-molecule thick crystal layer and the secondary nucleation and growth of subsequent layers. Irrespective of the initial orientation of the crystal nucleus, when the alkane crystal reaches the water interface, it reorients to expose the long side of the alkane molecules to water. We equilibrate this crystalline structure at the melting temperature, and find that alkanes align to maximize the hexadecane–water interface, producing a 'seesaw' shaped interface that exposes the (100) surface of the crystal to liquid water. We conclude that, same as for the vacuum interface, the orientational order of the molecules at the crystal–water interface mirrors the one in the liquid, albeit very pronounced and long ranged.

The orientations of crystalline alkanes at the alkane–water interface can be rationalized as follows. The strength of the interaction between water and CH_3 or CH_2 are the same, but the atom density on the side of the crystallized alkanes (i.e., the (100) surface) is higher than that on the end of the crystallized molecules (i.e., the (001) surface), which makes water stabilize the (denser) side face of crystalline alkanes more than their (less dense) CH_3 end. The 'seesaw' shape of the (100) interface occurs because the hexadecane molecules in the crystal are not parallel to the interface between alkane layers, which we confirm in the simulations of growth of PYS hexadecane. We note that PYS overestimates the tilt of octane in the crystal with respect to the experimental value, $120.0°$ in the model vs. $105.8°$ in the experimental crystal structure [22], and it may overestimate the tilt also for hexadecane.

An earlier simulation study of ultrathin films with less than three layers of alkanes on solid substrates showed that very attractive surfaces orient the first layer of alkanes parallel to the surface while weakly attractive surfaces orient them perpendicular [106]. We find the same trend for the surface orientation of bulk alkanes at fluid interfaces as we tune the strength of the solvent–alkane attraction ε_{wc} (Figure 3) between PYS hexadecane or OPLS nonane and water. Figure 5 shows representative snapshots of the simulation trajectories displaying crystalline alkanes for each ε_{wc} in the alkane/water systems. At low ε_{wc} the alkane molecules orient mostly perpendicular to the surface (Figure 1) and the crystallization is heterogeneous at the interface. We define as the 'neutral' ε_{wc} the one for which θ measured half a molecule length (0.5 nm) from the Gibbs dividing surface (solid circles in Figure 1c,d) is the same as in bulk. The neutral ε_{wc} depends on the force field and the length of the chain: 0.14 kcal·mol^{-1} for OPLS nonane, 0.05 kcal·mol^{-1} for PYS nonane and 0.15 kcal·mol^{-1} for PYS hexadecane. We find a transition from heterogeneous to homogeneous nucleation as ε_{wc} increases above the neutral orientation value. At the neutral ε_{wc} alkanes can either nucleate homogeneously or heterogeneously, suggesting that the crystalline nuclei have similar stabilities in the bulk and at the interface. We conclude that the attraction between alkanes and the fluid phase reverses the orientation of the crystalline alkanes with respect to the surface and controls the mechanism of crystallization.

A recent simulation study of crystallization of pentacontane (C50) finds that it crystallizes heterogeneously at solid silicon-like templating surfaces, resulting in crystals with the chains oriented parallel to the interface, and that the nucleation is faster for more strongly interacting surfaces [12,19,20]. Based on these results, it may be expected that increasing ε_{wc} between alkane and the water fluid may increase the ordering of the molecules parallel to the interface, and could result in a new region of heterogeneous nucleation at high ε_{wc}. However, we find that increasing ε_{wc} between PYS hexadecane and water to 0.22 kcal·mol^{-1} (10% over the value for mW water–PYS C16) does not result in heterogeneous nucleation but in mixing of the two components. This suggests that different from solid crystal-templating surfaces, liquids that interact with alkanes more strongly than water would rather dissolve the alkanes than result in heterogeneous crystallization.

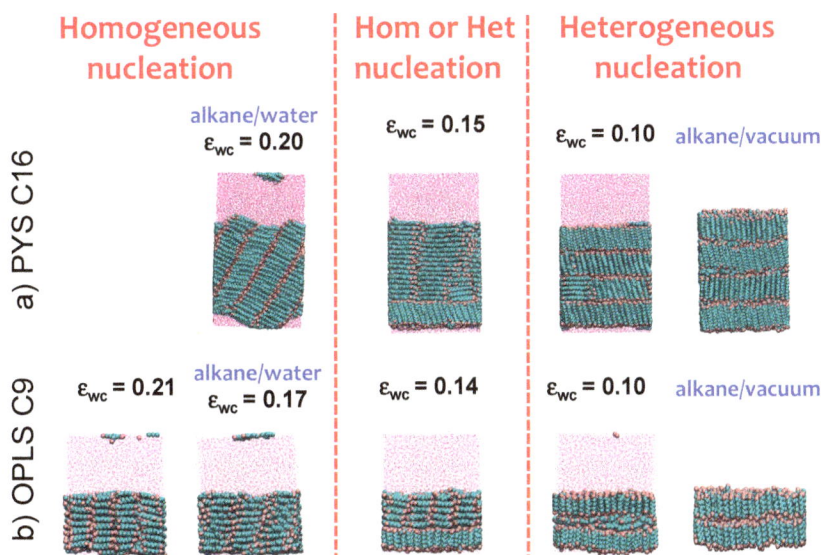

Figure 5. Interfacial orientation of (**a**) PYS hexadecane or (**b**) OPLS nonane crystallized in the presence of solvent (water or vacuum) is reversed from parallel to perpendicular by decreasing the strength of water–carbon interaction ε_{wc}. The trend does not depend on the alkane length or force field. Snapshots shown correspond to the end of the alkane crystallization in the presence of solvent with coupling ε_{wc} indicated. We highlight with blue labels the cells corresponding to the alkane/water parameters that reproduce the experimental surface tension and the alkane/vacuum systems. The change in interfacial orientation from parallel to perpendicular in the crystal coincides with the change from homogeneous to heterogeneous nucleation, and occurs for the neutral ε_{wc}, for which the orientation of the alkanes at the surface of the liquid is the same as in the bulk.

3.4. The Sign and Magnitude of the Binding Free Energy of the Alkane Crystal to the Surface Determine Whether the Crystal Nucleation is Homogeneous, Heterogeneous, or There Is Surface Freezing

In what follows we use classical nucleation theory [107] (CNT) to identify the conditions that lead to homogeneous crystal nucleation, heterogeneous crystal nucleation, and surface freezing above the equilibrium melting point. We then use that framework to rationalize why alkane crystals nucleate heterogeneously in contact with vacuum or weakly interacting fluids and homogeneously with more strongly interacting liquids, such as water.

The rate of crystal nucleation in CNT is $J = A \exp(-\Delta G^*/k_B T)$, in which the prefactor A depends mostly on the diffusion coefficient of the molecules in the liquid, ΔG^* is the free energy barrier of nucleation, k_B is the Boltzmann constant and T the temperature. The heterogeneous nucleation barrier in CNT is given by $\Delta G^*_{het} = N^* \Delta \mu + A_{xl}^* \gamma_{xl} + A_{xs}^* \Delta \gamma$, where N^* is the number of molecules in the critical nucleus, $\Delta \mu$ is the difference in chemical potential between liquid and crystal, and A_{xl}^* and A_{xs}^* are the areas of the liquid–crystal and surface–crystal interfaces of the critical nucleus, γ_{xl} is the surface tension of the liquid–crystal interface, and $\Delta \gamma = \gamma_{xs} - \gamma_{ls}$ is the difference between the surface tension of the crystal–surface and liquid–surface interfaces. The geometry of the nucleus is determined by

Young's equation, $\gamma_{xl} \cos\alpha + \Delta\gamma = 0$, where α is the contact angle of crystal nucleus on the surface [108]. Heterogeneous nucleation can only be induced when $\cos\alpha > -1$, which implies that the binding free energy of the crystal nucleus to the surface is negative, $\Delta G_{bind} = \gamma_{xs} - (\gamma_{ls} + \gamma_{xl}) < 0$ (Figure 6) [13]. Surface freezing can be considered a case of complete wetting of the surface by the crystal, which requires that $\Delta G_{bind} < -2\,\gamma_{xl}$ (Figure 6) if the line tension of the crystal–liquid–surface interface is neglected [13,109]. Heterogeneous nucleation at a surface can occur without surface freezing when $-2\,\gamma_{xl} < \Delta G_{bind} < 0$. If surface freezing occurs, then the bulk crystal will nucleate heterogeneously from the frozen interface at T_m.

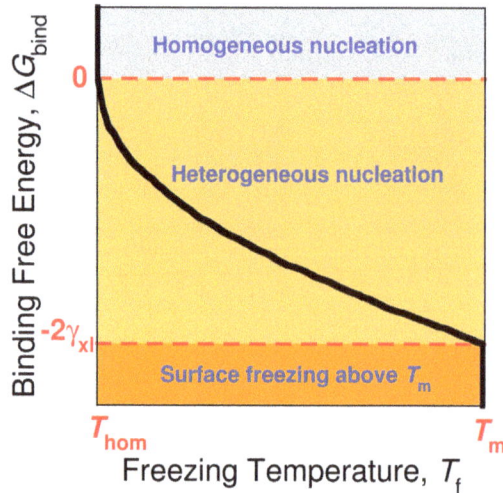

Figure 6. Sketch of the relation between the bulk freezing temperature T_f of the alkane crystal (black line) and the binding free energy ΔG_{bind} of the crystalline alkane to the nucleating surface. The curve corresponds to a constant nucleation rate, which determines the homogeneous nucleation temperature T_{hom} [13]. The sign and magnitude of ΔG_{bind} determines whether the surface can induce heterogeneous nucleation or promote surface freezing. If $\Delta G_{bind} > 0$, the crystal nucleates homogeneously at T_{hom}. For surfaces that produce $0 > \Delta G_{bind} > -2\,\gamma_{xl}$, the nucleation of the alkane crystal is heterogeneous but there is no surface freezing above the equilibrium melting temperature T_m. If $\Delta G_{bind} < -2\,\gamma_{xl}$, the surface induces surface freezing of the alkane above T_m, and the crystal nucleates heterogeneously from the frozen surface just below T_m.

Heterogeneous crystallization of the alkanes at the vapor interface implies that the crystallite is more stable at the vapor interface than in the bulk of the liquid. This happens when $\Delta G_{bind} = \gamma_{xv} - (\gamma_{lv} + \gamma_{xl}) < 0$, where γ_{lv}, γ_{xv}, and γ_{xl} are the are the surface tension of liquid–vapor, crystal–vapor and liquid–crystal interfaces for the alkanes. The liquid–crystal surface tension γ_{xl} of alkanes is in the order of a few mJ·m^{-2} in experiments and simulations [22,23]. The γ_{lv} of alkanes are 20 to 30 mJ·m^{-2}, as reported in Tables 1 and 2. The surface energy between crystal alkanes and vapor γ_{xv} has not been, to our knowledge, experimentally determined. γ_{xv} has been recently computed for PYS octane (C8) and Trappe nonadecane (C19) at their corresponding melting temperatures [30], and found to be ~40% lower for the longer alkane, 35 and 24 mJ·m^{-2} [30]. We note that for heterogeneous crystallization to occur at the vapor interface, the condition $\Delta G_{bind} < 0$ has to be satisfied at the non-equilibrium crystallization temperature, although it may not be satisfied at the melting point.

Surface freezing of alkanes at the vapor interface requires $\Delta G_{bind} = \gamma_{xv} - (\gamma_{lv} + \gamma_{xl}) < -2$ γ_{xl} [13,110,111], a requirement stronger than for hetergeneous nucleation. Since the surface tension of liquid–crystalline alkane γ_{xl} is small, a few mJ·m^{-2} [22,23], there is a narrow range of ΔG_{bind} for which heterogeneous crystallization of alkanes at the vapor interface occurs without surface freezing. C8 satisfy this condition, and does not present surface freezing although it heterogeneously crystallizes at the vacuum interface [30]. C16 to C50 display surface freezing [1–3]. Our analysis indicates that the small free energy cost of the liquid–crystal interface γ_{xl} in alkanes is key for the realization of surface freezing at surfaces as different as vacuum [1–3] and SiO_2 [112,113].

For the crystallization of the alkanes at the water interface, the binding free energy is $\Delta G_{bind} = \gamma_{xw}$ $- (\gamma_{lw} + \gamma_{xl})$, where γ_{xw}, and γ_{lw} are the surface tension of water in contact with crystalline and liquid alkane, respectively. While γ_{lw} is readily available from experiments or simulations (see Section 3.1), we are unable to find experimental data for the surface tension of the water–crystalline alkane interface, γ_{xw}. To interpret why the alkane–water interface cannot induce heterogeneous crystallization of alkanes, we draw a schematic diagram of the evolution of γ_{lw} and γ_{xw} with increasing ε_{wc} (Figure 7) based on the crystallization mechanism vs. ε_{wc} reported in Figure 5. We interpret that increasing ε_{wc} stabilizes the water–liquid alkane interface more than the water–crystal alkane interface, although both interfaces would be stabilized (i.e., γ_{lw} and γ_{xw} decrease) on increasing alkane–solvent attraction. This differential stabilization of γ_{lw} vs. γ_{xw} with ε_{wc} should result in a crossover between γ_{lw} and $\gamma_{xw} - \gamma_{lx}$ that, as we discussed above, signals the transition from heterogeneous to homogeneous nucleation when the interfacial liquid alkane has the same orientation as in the bulk. Our analysis indicates that liquid water does not induce heterogeneous crystal nucleation because it preferentially stabilizes the liquid phase of the alkane.

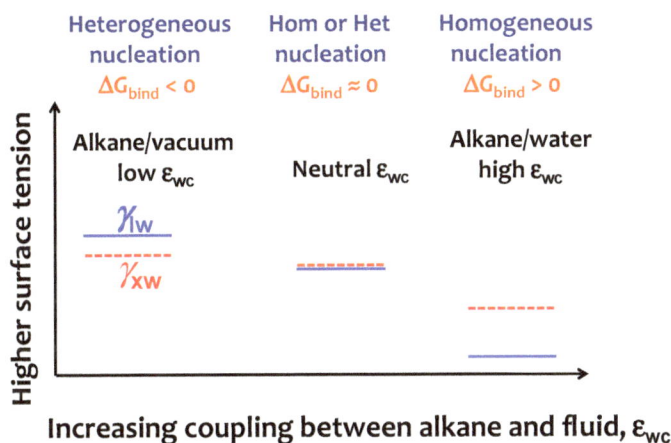

Figure 7. Scheme to interpret the change from heterogeneous to homogeneous nucleation mechanism on increasing the strength of alkane–solvent attraction ε_{wc}. The simulation trends of Figure 5 indicate that on increasing ε_{wc} the liquid alkane–water interfacial tension γ_{lw} (blue line) decreases more than the crystalline alkane–water surface tension γ_{xw} (red line), resulting in an increase of the binding free energy ΔG_{bind}. The neutral ε_{wc} signals the transition from heterogeneous to homogeneous at $\Delta G_{bind} = 0$, which coincides with the lack of preference for parallel or perpendicular orientation for the interfacial liquid alkane.

4. Conclusions

We use molecular dynamic simulations to investigate the relation between the interfacial orientation of alkanes in the liquid phase and their mechanism of crystallization. In agreement with previous simulation results [10,103,104] and the interpretation of experiments [9], we find that alkane molecules in the liquid orient in opposite directions at the vacuum and water interfaces: they preferentially align perpendicular to the vacuum interface and parallel to the water interface. However, we note that the orientation is not very pronounced: eight degrees below the bulk average for vacuum and six degrees above the bulk average for water for PYS alkanes, and even less for the OPLS model. We demonstrate that the interfacial orientation of the alkanes in the liquid can be tuned through the strength of the fluid solvent (water) and alkane attraction. Increasing the solvent–alkane attraction results in an orientation of the alkanes more parallel to the interface.

Although both water and vacuum orient liquid alkanes at the interface, only the vacuum interface promotes heterogeneous nucleation of alkane crystals. Heterogeneous crystallization of alkanes at the vacuum interface results in crystals with molecules oriented perpendicular to the surface, same as in the critical crystal nucleus from which they grow. Crystallization of alkanes in the presence of water occurs through homogeneous nucleation in the bulk of the alkane liquid phase (Figure 4a). Although the homogeneous critical nucleus forms with an arbitrary orientation in the bulk liquid with respect to the alkane–solvent interface, as the crystal grows and reaches the water interface, the nucleus rotates and the alkane molecules in the crystal end up aligned parallel to the surface (Figure 4b), maximizing the attractive interactions between alkane and water. We conclude that the preferential orientation of alkane molecules in the crystal with respect to the fluid surface mirrors the one in liquid alkanes, although the order in the crystal is long-ranged, in the liquid it involves only the first monolayer at the interface.

Tuning the interactions of the liquid alkane–solvent interface from non-interacting vacuum-like to water values results in a change from heterogeneous to homogeneous nucleation as the orientation of the molecules in the interfacial liquid turn from leaning perpendicular to leaning parallel to the surface. Our analysis indicates that the change in mechanism results from a preferential stabilization of the liquid alkane–solvent interface compared to the crystalline alkane–solvent interface on increasing the alkane–solvent attraction.

We use Classical Nucleation Theory to explain the transition from heterogeneous to homogeneous nucleation on increasing the strength of the alkane–solvent interactions and the distinct conditions that lead to heterogeneous nucleation from the supercooled liquid and surface freezing above the equilibrium melting temperature. Heterogeneous nucleation occurs when the binding free energy of the crystal (immersed in its melt) to the nucleating surface is negative ($\Delta G_{bind} < 0$), while surface freezing requires a more stringent condition: that the binding free energy is less than minus twice the liquid–crystal surface tension of the alkane, ($\Delta G_{bind} < -2\,\gamma_{xl}$) [13,109]. This implies that there is a range of binding free energies for which heterogeneous nucleation can occur without surface freezing. This must be the case for alkanes with fewer than 16 carbons at the vacuum interface. Interestingly, silica-coated Si (100) surfaces produce surface freezing in which the alkane molecules are oriented parallel to the surface [113], opposite to the order they present at the vacuum interface [1–3]. These results stress that surface freezing can be attained by solid surfaces that interact very weakly and strongly with the alkanes. Likewise, experiments and simulations indicate that crystalline surfaces that interact strongly with alkanes can induce heterogeneous nucleation with the chains ordered parallel to the surface [12,112,113]. This region of $\Delta G_{bind} < 0$ for highly interacting surfaces cannot be accessed with a fluid interface, as we observe that increase in interaction strength results in mixing instead of heterogeneous nucleation. This distinction should be important in developing strategies to control the assembly of alkane-containing organic and biological molecules at fluid and solid interfaces.

Acknowledgments: This work was supported by the National Science Foundation through award CHE-1305427 'Center for Aerosol Impacts on Climate and the Environment'. We thank the Center for High Performance Computing at The University of Utah for technical support and a grant of computer time.

Author Contributions: Yuqing Qiu and Valeria Molinero conceived the work and wrote the paper. Yuqing Qiu parameterized the models, performed the simulations and analyzed the data.

Conflicts of Interest: The authors declare no conflict of interest.

References

1. Wu, X.; Sirota, E.; Sinha, S.; Ocko, B.; Deutsch, M. Surface crystallization of liquid normal-alkanes. *Phys. Rev. Lett.* **1993**, *70*, 958–961. [CrossRef] [PubMed]

2. Ocko, B.M.; Wu, X.Z.; Sirota, E.B.; Sinha, S.K.; Gang, O.; Deutsch, M. Surface freezing in chain molecules: Normal alkanes. *Phys. Rev. E* **1997**, *55*, 3164–3182. [CrossRef]

3. Tkachenko, A.V.; Rabin, Y. Fluctuation-Stabilized Surface Freezing of Chain Molecules. *Phys. Rev. Lett.* **1996**, *76*, 2527–2530. [CrossRef] [PubMed]

4. Modak, V.P.; Pathak, H.; Thayer, M.; Singer, S.J.; Wyslouzil, B.E. Experimental evidence for surface freezing in supercooled n-alkane nanodroplets. *Phys. Chem. Chem. Phys.* **2013**, *15*, 6783–6795. [CrossRef] [PubMed]

5. Tikhonov, A.M.; Mitrinovic, D.M.; Li, M.; Huang, Z.; Schlossman, M.L. An X-ray reflectivity study of the water-docosane interface. *J. Phys. Chem. B* **2000**, *104*, 6336–6339. [CrossRef]

6. Conboy, J.; Daschbach, J.; Richmond, G. Studies of Alkane/Water Interfaces by Total Internal Reflection Second Harmonic Generation. *J. Phys. Chem.* **1994**, *98*, 9688–9692. [CrossRef]

7. Conboy, J.C.; Daschbach, J.L.; Richmond, G.L. Total internal reflection second-harmonic generation: Probing the alkane water interface. *Appl. Phys. A* **1994**, *59*, 623–629. [CrossRef]

8. Mitrinović, D.M.; Tikhonov, A.M.; Li, M.; Huang, Z.; Schlossman, M.L. Noncapillary-Wave Structure at the Water-Alkane Interface. *Phys. Rev. Lett.* **2000**, *85*, 582–585. [CrossRef] [PubMed]

9. Vácha, R.; Roke, S. Sodium Dodecyl Sulfate at Water–Hydrophobic Interfaces: A Simulation Study. *J. Phys. Chem. B* **2012**, *116*, 11936–11942. [CrossRef] [PubMed]

10. Van Buuren, A.R.; Marrink, S.J.; Berendsen, H.J.C. A molecular dynamics study of the decane/water interface. *J. Phys. Chem.* **1993**, *97*, 9206–9212. [CrossRef]

11. Xiao, H.; Zhen, Z.; Sun, H.; Cao, X.; Li, Z.; Song, X.; Cui, X.; Liu, X. Molecular dynamics study of the water/n-alkane interface. *Sci. China Chem.* **2010**, *53*, 945–949. [CrossRef]

12. Bourque, A.J.; Locker, C.R.; Rutledge, G.C. Heterogeneous Nucleation of an n-Alkane on Tetrahedrally Coordinated Crystals. *J. Phys. Chem. B* **2017**, *121*, 904–911. [CrossRef] [PubMed]

13. Qiu, Y.; Odendahl, N.; Hudait, A.; Mason, R.H.; Bertram, A.K.; Paesani, F.; de Mott, P.J.; Molinero, V. Ice nucleation efficiency of hydroxylated organic surfaces is controlled by their structural fluctuations and mismatch to ice. *J. Am. Chem. Soc.* **2017**, *139*, 3052–3064. [CrossRef] [PubMed]

14. Waheed, N.; Ko, M.J.; Rutledge, G.C. Molecular simulation of crystal growth in long alkanes. *Polymer* **2005**, *46*, 8689–8702. [CrossRef]

15. Waheed, N.; Lavine, M.S.; Rutledge, G.C. Molecular simulation of crystal growth in n-eicosane. *J. Chem. Phys.* **2002**, *116*, 2301–2309. [CrossRef]

16. Paul, W.; Yoon, D.Y.; Smith, G.D. An optimized united atom model for simulations of polymethylene melts. *J. Chem. Phys* **1995**, *103*, 1702–1709. [CrossRef]

17. Jorgensen, W.L.; Tirado-Rives, J. The OPLS [optimized potentials for liquid simulations] potential functions for proteins, energy minimizations for crystals of cyclic peptides and crambin. *J. Am. Chem. Soc.* **1988**, *110*, 1657–1666. [CrossRef] [PubMed]

18. Jorgensen, W.L.; Madura, J.D.; Swenson, C.J. Optimized intermolecular potential functions for liquid hydrocarbons. *J. Am. Chem. Soc.* **1984**, *106*, 6638–6646. [CrossRef]

19. Bourque, A.J.; Rutledge, G.C. Kinetic Model for Layer-by-Layer Crystal Growth in Chain Molecules. *Macromolecules* **2016**, *49*, 3956–3964. [CrossRef]

20. Bourque, A.; Locker, C.R.; Rutledge, G.C. Molecular Dynamics Simulation of Surface Nucleation during Growth of an Alkane Crystal. *Macromolecules* **2016**, *49*, 3619–3629. [CrossRef]

21. Anwar, M.; Turci, F.; Schilling, T. Crystallization mechanism in melts of short n-alkane chains. *J. Chem. Phys.* **2013**, *139*, 214904. [CrossRef] [PubMed]

22. Yi, P.; Rutledge, G.C. Molecular simulation of crystal nucleation in n-octane melts. *J. Chem. Phys.* **2009**, *131*, 134902. [PubMed]

23. Yi, P.; Rutledge, G.C. Molecular simulation of bundle-like crystal nucleation from n-eicosane melts. *J. Chem. Phys* **2011**, *135*, 024903. [PubMed]

24. Hrahsheh, F.; Wilemski, G. Fluctuating structure of aqueous organic nanodroplets. *AIP Conf. Proc.* **2013**, *1527*, 63–66.

25. Yang, H.; Zhao, X.J.; Sun, M. Induced crystallization of single-chain polyethylene on a graphite surface: Molecular dynamics simulation. *Phys. Rev. B* **2011**, *84*, 011803. [CrossRef] [PubMed]

26. Kaner, P.; Ruiz-Orta, C.; Boz, E.; Wagener, K.B.; Tasaki, M.; Tashiro, K.; Alamo, R.G. Kinetic control of chlorine packing in crystals of a precisely substituted polyethylene. Toward advanced polyolefin materials. *Macromolecules* **2013**, *47*, 236–245. [CrossRef]

27. Wang, J.; Zhu, X.; Lu, X.; Zhou, Z.; Wang, G. On structures and properties of polyethylene during heating and cooling processes based on molecular dynamics simulations. *Comp. Theor. Chem.* **2015**, *1052*, 26–34. [CrossRef]

28. Yamamoto, T. Molecular dynamics of crystallization in n-alkane mixtures; texture, compatibility, and diffusion in crystals. *Polymer* **2016**, *99*, 721–733. [CrossRef]

29. Yi, P.; Locker, C.R.; Rutledge, G.C. Molecular dynamics simulation of homogeneous crystal nucleation in polyethylene. *Macromolecules* **2013**, *46*, 4723–4733. [CrossRef]

30. Modak, V.P.; Wyslouzil, B.E.; Singer, S.J. On the determination of the crystal-vapor surface free energy, and why a Gaussian expression can be accurate for a system far from Gaussian. *J. Chem. Phys.* **2016**, *145*, 054710. [CrossRef] [PubMed]

31. Obeidat, A.; Hrahsheh, F.; Wilemski, G. Scattering Form Factors for Russian Doll Aerosol Droplet Models. *J. Phys. Chem. B* **2014**, *119*, 9304–9311. [CrossRef] [PubMed]

32. Anwar, M.; Schilling, T. Crystallization of polyethylene: A molecular dynamics simulation study of the nucleation and growth mechanisms. *Polymer* **2015**, *76*, 307–312. [CrossRef]

33. Wang, E.; Escobedo, F.A. Mechanical Properties of Tetrapolyethylene and Tetrapoly (Ethylene Oxide) Diamond Networks Via Molecular Dynamics Simulations. *Macromolecules* **2016**, *49*, 2375–2386. [CrossRef]

34. Romanos, N.A.; Theodorou, D.N. Crystallization and melting simulations of oligomeric α1 isotactic polypropylene. *Macromolecules* **2010**, *43*, 5455–5469. [CrossRef]

35. Lee, S.; Rutledge, G.C. Plastic deformation of semicrystalline polyethylene by molecular simulation. *Macromolecules* **2011**, *44*, 3096–3108. [CrossRef]

36. Yi, P.; Rutledge, G.C. Molecular origins of homogeneous crystal nucleation. *Annu. Rev. Chem. Biomol. Eng.* **2012**, *3*, 157–182. [CrossRef] [PubMed]

37. Qiu, Y.; Molinero, V. Morphology of Liquid–Liquid Phase Separated Aerosols. *J. Am. Chem. Soc.* **2015**, *137*, 10642–10651. [CrossRef] [PubMed]

38. Molinero, V.; Moore, E.B. Water modeled as an intermediate element between carbon and silicon. *J. Phys. Chem. B* **2009**, *113*, 4008–4016. [CrossRef] [PubMed]

39. Malkin, T.L.; Murray, B.J.; Salzmann, C.G.; Molinero, V.; Pickering, S.J.; Whale, T.F. Stacking disorder in ice I. *Phys. Chem. Chem. Phys.* **2015**, *17*, 60–76. [CrossRef] [PubMed]

40. Moore, E.B.; Molinero, V. Is it cubic? Ice crystallization from deeply supercooled water. *Phys. Chem. Chem. Phys.* **2011**, *13*, 20008–20016. [CrossRef] [PubMed]

41. Moore, E.B.; de la Llave, E.; Welke, K.; Scherlis, D.A.; Molinero, V. Freezing, melting and structure of ice in a hydrophilic nanopore. *Phys. Chem. Chem. Phys.* **2010**, *12*, 4124–4134. [CrossRef] [PubMed]

42. Solveyra, E.G.; Llave, E.D.L.; Molinero, V.; Soler-Illia, G.J. A.A.; Scherlis, D.A. Structure, Dynamics, and Phase Behavior of Water in TiO$_2$ Nanopores. *J. Phys. Chem. C* **2013**, *117*, 3330–3342. [CrossRef]

43. Hudait, A.; Molinero, V. Ice crystallization in ultrafine water-salt aerosols: Nucleation, ice-solution equilibrium, and internal structure. *J. Am. Chem. Soc.* **2014**, *136*, 8081–8093. [CrossRef] [PubMed]

44. Johnston, J.C.; Molinero, V. Crystallization, melting, and structure of water nanoparticles at atmospherically relevant temperatures. *J. Am. Chem. Soc.* **2012**, *134*, 6650–6659. [CrossRef] [PubMed]

45. Lu, J.; Qiu, Y.; Baron, R.; Molinero, V. Coarse-Graining of TIP4P/2005, TIP4P-Ew, SPC/E, and TIP3P to Monatomic Anisotropic Water Models Using Relative Entropy Minimization. *J. Chem. Theory Comput.* **2014**, *10*, 4104–4120. [CrossRef] [PubMed]

46. Lupi, L.; Hudait, A.; Molinero, V. Heterogeneous nucleation of ice on carbon surfaces. *J. Am. Chem. Soc.* **2014**, *136*, 3156–3164. [CrossRef] [PubMed]

47. Moore, E.B.; Molinero, V. Structural transformation in supercooled water controls the crystallization rate of ice. *Nature* **2011**, *479*, 506–508. [CrossRef] [PubMed]

48. Moore, E.B.; Allen, J.T.; Molinero, V. Liquid-ice coexistence below the melting temperature for water confined in hydrophilic and hydrophobic nanopores. *J. Phys. Chem. C* **2012**, *116*, 7507–7514. [CrossRef]

49. Moore, E.B.; Molinero, V. Ice crystallization in water's "no-man's land". *J. Chem. Phys.* **2010**, *132*, 244504. [CrossRef] [PubMed]

50. Moore, E.B.; Molinero, V. Growing correlation length in supercooled water. *J. Chem. Phys.* **2009**, *130*, 244505. [CrossRef] [PubMed]

51. Hujo, W.; Shadrack Jabes, B.; Rana, V.K.; Chakravarty, C.; Molinero, V. The Rise and Fall of Anomalies in Tetrahedral Liquids. *J. Stat. Phys.* **2011**, *145*, 293–312. [CrossRef]

52. Kastelowitz, N.; Johnston, J.C.; Molinero, V. The anomalously high melting temperature of bilayer ice. *J. Chem. Phys.* **2010**, *132*, 124511. [CrossRef] [PubMed]

53. Xu, L.; Molinero, V. Is There a Liquid–Liquid Transition in Confined Water? *J. Phys. Chem. B* **2011**, *115*, 14210–14216. [CrossRef] [PubMed]

54. Nguyen, A.H.; Molinero, V. Identification of Clathrate Hydrates, Hexagonal Ice, Cubic Ice, and Liquid Water in Simulations: The CHILL+ Algorithm. *J. Phys. Chem. B* **2015**, *119*, 9369–9376. [CrossRef] [PubMed]

55. Nguyen, A.H.; Koc, M.A.; Shepherd, T.D.; Molinero, V. Structure of the Ice–Clathrate Interface. *J. Phys. Chem. C* **2015**, *119*, 4104–4117. [CrossRef]

56. Holten, V.; Limmer, D.T.; Molinero, V.; Anisimov, M.A. Nature of the anomalies in the supercooled liquid state of the mW model of water. *J. Chem. Phys.* **2013**, *138*, 174501. [CrossRef] [PubMed]

57. Bullock, G.; Molinero, V. Low-density liquid water is the mother of ice: On the relation between mesostructure, thermodynamics and ice crystallization in solutions. *Faraday Discuss.* **2013**, *167*, 371–388. [CrossRef] [PubMed]

58. González Solveyra, E.A.; de la Llave, E.; Scherlis, D.A.; Molinero, V. Melting and crystallization of ice in partially filled nanopores. *J. Phys. Chem. B* **2011**, *115*, 14196–14204. [CrossRef] [PubMed]

59. Reinhardt, A.; Doye, J.P.K. Effects of surface interactions on heterogeneous ice nucleation for a monatomic water model. *J. Chem. Phys.* **2014**, *141*, 084501. [CrossRef] [PubMed]

60. Factorovich, M.H.; Molinero, V.; Scherlis, D.A. A simple grand canonical approach to compute the vapor pressure of bulk and finite size systems. *J. Chem. Phys.* **2014**, *140*, 064111. [CrossRef] [PubMed]

61. Factorovich, M.H.; Molinero, V.; Scherlis, D.A. Vapor pressure of water nanodroplets. *J. Am. Chem. Soc.* **2014**, *136*, 4508–4514. [CrossRef] [PubMed]

62. Limmer, D.T.; Chandler, D. The putative liquid-liquid transition is a liquid-solid transition in atomistic models of water. *J. Chem. Phys.* **2011**, *135*, 134503. [CrossRef] [PubMed]

63. Limmer, D.T.; Chandler, D. The putative liquid-liquid transition is a liquid-solid transition in atomistic models of water. II. *J. Chem. Phys.* **2013**, *138*, 214504. [CrossRef] [PubMed]

64. Espinosa, J.R.; Vega, C.; Valeriani, C.; Sanz, E. Seeding approach to crystal nucleation. *J. Chem. Phys.* **2016**, *144*, 034501. [CrossRef] [PubMed]

65. Zaragoza, A.; Conde, M.M.; Espinosa, J.R.; Valeriani, C.; Vega, C.; Sanz, E. Competition between ices Ih and Ic in homogeneous water freezing. *J. Chem. Phys.* **2015**, *143*, 134504. [CrossRef] [PubMed]

66. Espinosa, J.R.; Sanz, E.; Valeriani, C.; Vega, C. Homogeneous ice nucleation evaluated for several water models. *J. Chem. Phys.* **2014**, *141*, 18C529. [CrossRef] [PubMed]

67. Bi, Y.; Cabriolu, R.; Li, T. Heterogeneous Ice Nucleation Controlled by the Coupling of Surface Crystallinity and Surface Hydrophilicity. *J. Phys. Chem. B* **2016**, *120*, 1507–1514. [CrossRef]

68. Cabriolu, R.; Li, T. Ice nucleation on carbon surface supports the classical theory for heterogeneous nucleation. *Phys. Rev. E* **2015**, *91*, 052402. [CrossRef] [PubMed]

69. Li, T.; Donadio, D.; Russo, G.; Galli, G. Homogeneous ice nucleation from supercooled water. *Phys. Chem. Chem. Phys.* **2011**, *13*, 19807–19813. [CrossRef] [PubMed]

70. Jacobson, L.C.; Molinero, V. A Methane–Water Model for Coarse-Grained Simulations of Solutions and Clathrate Hydrates. *J. Phys. Chem. B* **2010**, *114*, 7302–7311. [CrossRef] [PubMed]

71. Dhabal, D.; Nguyen, A.H.; Singh, M.; Khatua, P.; Molinero, V.; Bandyopadhyay, S.; Chakravarty, C. Excess entropy and crystallization in Stillinger-Weber and Lennard-Jones fluids. *J. Chem. Phys.* **2015**, *143*, 164512. [CrossRef] [PubMed]

72. Singh, M.; Dhabal, D.; Nguyen, A.H.; Molinero, V.; Chakravarty, C. Triplet Correlations Dominate the Transition from Simple to Tetrahedral Liquids. *Phys. Rev. Lett.* **2014**, *112*, 147801. [CrossRef] [PubMed]

73. Hudait, A.; Qiu, S.; Lupi, L.; Molinero, V. Free energy contributions and structural characterization of stacking disordered ices. *Phys. Chem. Chem. Phys.* **2016**, *18*, 9544–9553. [CrossRef] [PubMed]

74. Cox, S.J.; Kathmann, S.M.; Slater, B.; Michaelides, A. Molecular simulations of heterogeneous ice nucleation. I. Controlling ice nucleation through surface hydrophilicity. *J. Chem. Phys.* **2015**, *142*, 184704. [CrossRef] [PubMed]

75. Cox, S.J.; Kathmann, S.M.; Slater, B.; Michaelides, A. Molecular simulations of heterogeneous ice nucleation. II. Peeling back the layers. *J. Chem. Phys.* **2015**, *142*, 184705. [CrossRef] [PubMed]

76. Lupi, L.; Kastelowitz, N.; Molinero, V. Vapor deposition of water on graphitic surfaces: Formation of amorphous ice, bilayer ice, ice I, and liquid water. *J. Chem. Phys.* **2014**, *141*, 18C508. [CrossRef] [PubMed]

77. Fitzner, M.; Sosso, G.C.; Cox, S.J.; Michaelides, A. The Many Faces of Heterogeneous Ice Nucleation: Interplay Between Surface Morphology and Hydrophobicity. *J. Am. Chem. Soc.* **2015**, *137*, 13658–13669. [CrossRef] [PubMed]

78. Li, T.; Donadio, D.; Galli, G. Ice nucleation at the nanoscale probes no man's land of water. *Nat. Commun.* **2013**, *4*, 1887. [CrossRef] [PubMed]

79. Lu, J.; Chakravarty, C.; Molinero, V. Relationship between the line of density anomaly and the lines of melting, crystallization, cavitation, and liquid spinodal in coarse-grained water models. *J. Chem. Phys.* **2016**, *144*, 234507. [CrossRef] [PubMed]

80. Jabes, B.S.; Nayar, D.; Dhabal, D.; Molinero, V.; Chakravarty, C. Water and other tetrahedral liquids: Order, anomalies and solvation. *J. Phys. Condens. Matter* **2012**, *24*, 284116. [CrossRef] [PubMed]

81. Factorovich, M.H.; Gonzalez Solveyra, E.; Molinero, V.; Scherlis, D.A. Sorption Isotherms of Water in Nanopores: Relationship Between Hydropohobicity, Adsorption Pressure, and Hysteresis. *J. Phys. Chem. C* **2014**, *118*, 16290–16300. [CrossRef]

82. Limmer, D.T.; Chandler, D. Phase diagram of supercooled water confined to hydrophilic nanopores. *J. Chem. Phys.* **2012**, *137*, 044509. [CrossRef] [PubMed]

83. Haji-Akbari, A.; de Fever, R.S.; Sarupria, S.; Debenedetti, P.G. Suppression of sub-surface freezing in free-standing thin films of a coarse-grained model of water. *Phys. Chem. Chem. Phys.* **2014**, *16*, 25916–25927. [CrossRef] [PubMed]

84. Sun, H. COMPASS: An ab initio force-field optimized for condensed-phase applications overview with details on alkane and benzene compounds. *J. Phys. Chem. B* **1998**, *102*, 7338–7364. [CrossRef]

85. Plimpton, S. Fast Parallel Algorithms for Short-Range Molecular Dynamics. *J. Comput. Phys.* **1995**, *117*, 1–19. [CrossRef]

86. Berendsen, H.J.C.; Grigera, J.R.; Straatsma, T.P. The missing term in effective pair potentials. *J. Phys. Chem.* **1987**, *91*, 6269–6271. [CrossRef]

87. Goebel, A.; Lunkenheimer, K. Interfacial tension of the water/n-alkane interface. *Langmuir* **1997**, *13*, 369–372. [CrossRef]

88. Zeppieri, S.; Rodríguez, J.; López de Ramos, A.L. Interfacial Tension of Alkane + Water Systems. *J. Chem. Eng. Data* **2001**, *46*, 1086–1088. [CrossRef]

89. Nosé, S. A unified formulation of the constant temperature molecular dynamics methods. *J. Chem. Phys.* **1984**, *81*, 511–519. [CrossRef]

90. Hoover, W.G. Canonical dynamics: Equilibrium phase-space distributions. *Phys. Rev. A* **1985**, *31*, 1695–1697. [CrossRef]

91. Groom, C.R.; Bruno, I.J.; Lightfoot, M.P.; Ward, S.C. The Cambridge Structural Database. *Acta Crystallogr. Sect. B* **2016**, *72*, 171–179. [CrossRef] [PubMed]

92. Schneider, T.; Stoll, E. Molecular-dynamics study of a three-dimensional one-component model for distortive phase transitions. *Phys. Rev. B* **1978**, *17*, 1302–1322. [CrossRef]

93. Gloor, G.J.; Jackson, G.; Blas, F.J.; de Miguel, E. Test-area simulation method for the direct determination of the interfacial tension of systems with continuous or discontinuous potentials. *J. Chem. Phys.* **2005**, *123*, 134703. [CrossRef] [PubMed]

94. Steinhardt, P.; Nelson, D.; Ronchetti, M. Bond-orientational order in liquids and glasses. *Phys. Rev. B* **1983**, *28*, 784–805. [CrossRef]

95. Rozmanov, D.; Kusalik, P.G. Anisotropy in the crystal growth of hexagonal ice, I(h). *J. Chem. Phys* **2012**, *137*, 094702. [CrossRef] [PubMed]

96. Acree, W.E.; Chickos, J.S. Phase Transition Enthalpy Measurements of Organic and Organometallic Compounds. In *NIST Chemistry WebBook, NIST Standard Reference Database Number 69*; Linstrom, P.J., Mallard, W.G., Eds.; National Institute of Standards and Technology: Gaithersburg MD, USA, 2016.

97. Rolo, L.I.; Caço, A.I.; Queimada, A.J.; Marrucho, I.M.; Coutinho, J.A.P. Surface tension of heptane, decane, hexadecane, eicosane, and some of their binary mixtures. *J. Chem. Eng. Data* **2002**, *47*, 1442–1445. [CrossRef]

98. Viton, C.; Chavret, M.; Jose, J. Enthalpies of Vaporization of Normal Alkanes from Nonane to Pentadecane at Temperatures from 298 to 359 K. *ELDATA Int. Electron. J. Phys. Chem. Data* **1996**, *2*, 103.

99. Mondieig, D.; Rajabalee, F.; Metivaud, V.; Oonk, H.; Cuevas-Diarte, M. n-Alkane binary molecular alloys. *Chem. Mater.* **2004**, *16*, 786–798. [CrossRef]

100. Benet, J.; MacDowell, L.G.; Menduiña, C. Liquid-vapor phase equilibria and surface tension of ethane as predicted by the TraPPE and OPLS models. *J. Chem. Eng. Data* **2010**, *55*, 5465–5470. [CrossRef]
101. Wick, C.D.; Siepmann, J.I.; Schure, M.R. Molecular Simulation of Concurrent Gas−Liquid Interfacial Adsorption and Partitioning in Gas−Liquid Chromatography. *Anal. Chem.* **2002**, *74*, 3518–3524. [CrossRef] [PubMed]
102. Haji-Akbari, A.; Debenedetti, P.G. Thermodynamic and kinetic anisotropies in octane thin films. *J. Chem. Phys.* **2015**, *143*, 214501. [CrossRef] [PubMed]
103. Harris, J.G. Liquid-vapor interfaces of alkane oligomers: Structure and thermodynamics from molecular dynamics simulations of chemically realistic models. *J. Phys. Chem.* **1992**, *96*, 5077–5086. [CrossRef]
104. Kawamata, M.; Yamamoto, T. Molecular dynamics simulation of surface ordering in liquid n-alkanes. *J. Phys. Soc. Jpn.* **1997**, *66*, 2350–2354. [CrossRef]
105. Baron, R.; Molinero, V. Water-driven cavity-ligand binding: Comparison of thermodynamic signatures from coarse-grained and atomic-level simulations. *J. Chem. Theory Comput.* **2012**, *8*, 3696–3704. [CrossRef] [PubMed]
106. Yamamoto, T.; Nozaki, K.; Yamaguchi, A.; Urakami, N. Molecular simulation of crystallization in n-alkane ultrathin films: Effects of film thickness and substrate attraction. *J. Chem. Phys.* **2007**, *127*, 154704. [CrossRef] [PubMed]
107. Turnbull, D.; Fisher, J.C. Rate of nucleation in condensed systems. *J. Chem. Phys.* **1949**, *17*, 71–73. [CrossRef]
108. Young, T. An essay on the cohesion of fluids. *Philos. Trans. R. Soc. Lond.* **1805**, *95*, 65–87. [CrossRef]
109. Iwamatsu, M. Heterogeneous Nucleation on a Completely Wettable Substrate. In *Advances in Contact Angle, Wettability and Adhesion*; John Wiley & Sons, Inc.: Salem, MA, USA, 2013; pp. 49–72.
110. Auer, S.; Frenkel, D. Line tension controls wall-induced crystal nucleation in hard-sphere colloids. *Phys. Rev. Lett.* **2003**, *91*, 015703. [CrossRef] [PubMed]
111. Sirota, E.; Wu, X.; Ocko, B.; Deutsch, M. What Drives the Surface Freezing in Alkanes? *Phys. Rev. Lett.* **1997**, *79*, 531. [CrossRef]
112. Merkl, C.; Pfohl, T.; Riegler, H. Influence of the molecular ordering on the wetting of SiO_2/air interfaces by alkanes. *Phys. Rev. Lett.* **1997**, *79*, 4625. [CrossRef]
113. Volkmann, U.G.; Pino, M.; Altamirano, L.; Taub, H.; Hansen, F.Y. High-resolution ellipsometric study of an n-alkane film, dotriacontane, adsorbed on a SiO_2 surface. *J. Chem. Phys.* **2002**, *116*, 2107–2115. [CrossRef]

© 2017 by the authors. Licensee MDPI, Basel, Switzerland. This article is an open access article distributed under the terms and conditions of the Creative Commons Attribution (CC BY) license (http://creativecommons.org/licenses/by/4.0/).

crystals

MDPI

Review

Computer Simulations of Crystal Growth Using a Hard-Sphere Model

Atsushi Mori

Graduate School of Science and Technology, Tokushima University, Tokushima 770-8506, Japan; atsushimori@tokushima-u.ac.jp

Academic Editors: Hiroki Nada and Helmut Cölfen
Received: 19 December 2016 ; Accepted: 29 March 2017 ; Published: 4 April 2017

Abstract: A review of computer simulation studies on crystal growth in hard-sphere systems is presented. A historical view on the crystallization of hard spheres, including colloidal crystallization, is given in the first section. Crystal phase transition in a system comprising particles without bonding is difficult to understand. In the early days, therefore, many researchers did not accept such crystalline structures as crystals that should be studied in the field of crystal growth. In the last few decades, however, colloidal crystallization has drawn attention because in situ observations of crystallization process has become possible. Next, simulation studies of the crystal/fluid interface of hard spheres are also reviewed. Although colloidal crystallization has now been recognized in the crystal growth field, the stability of the crystal–fluid coexistence state has still not been satisfactorily understood based on a bond-breaking picture, because of an infinite diffuseness of the interfaces in non-bonding systems derived from this picture. Studies of sedimentary colloidal crystallization and colloidal epitaxy using the hard-sphere model are lastly reviewed. An advantage of the colloidal epitaxy is also presented; it is shown that a template not only fixes the crystal growth direction, but also improves the colloidal crystallization. A new technique for reducing defects in colloidal crystals through the gravity effect is also proposed.

Keywords: hard spheres; crystal/fluid interface; colloidal crystals; sedimentation; colloidal epitaxy

---

1. Introduction

Bonds are commonly formed between various entities in solid materials. In contrast, a class of matter called "soft matter" does not have any bonding. A colloidal system is a typical example of soft matter. The presence of large entities is one of the characteristics of soft matter. Therefore, because of their large size, such entities have slow motion; unlike atomic systems, in situ observation of the crystal growth process at the particle level is possible in colloidal systems.

To get insight into the particle-level mechanism in atomic systems, researchers have to rely on computer simulations. One cannot only follow the rapid motion of atoms in reality and simultaneously look at the atoms in crystal growth processes. These difficulties are not faced with computer simulations such as molecular dynamics (MD) and Monte Carlo (MC) simulations. However, these simulation methods have a limitation in total computation time. For example, a simulation for atomic systems can be conducted for several microseconds at most. In contrast, because colloidal particles have slower motion than atoms, the simulation of colloidal particles can be performed for several days. In other words, simulations corresponding to real situations are possible for soft matter.

This paper is a review on computer simulations using a hard-sphere (HS) model for crystal growth. Information about the structure of the crystal/melt interface at the particle level is necessary to understand phenomena and develop techniques. For example, the relationship between the interface structure and the growth mode, and the colloidal crystallization defect behavior that depends on the interface structure are related to the quality of colloidal crystals. Computer simulations of the HS crystal/fluid interfaces were developed in the 1990s. To some crystal growth researchers, construction of a stable HS crystal/fluid interface in simulations is strikingly similar to the discovery of the crystalline phase in the HS system itself. If the difficulty of arranging HS particles at high number densities could be circumvented, an HS crystal–fluid coexistence state can be simulated starting from a dense HS configuration as an initial state. HS simulations have been further developed in the last few decades. The direction of studies on HS simulations coincides with that of colloidal crystals; a large number of studies have been conducted on controlling defects in colloidal crystals for functionalization.

After briefly introducing various aspects of the HS model, a description of the simulation methods is provided. Simulation of the HS crystal/fluid interface has been provided in Section 3. Simulations of HS systems in a gravitational field have also been discussed in Section 4. Crystal–fluid coexistence state under an external field is also a subject of this review. In the latter part of the review, the effect of different simulation techniques on colloidal crystallization has been extensively investigated.

1.1. Crystalline Phase in a Hard-Sphere System

In 1957, the crystalline phase of the HS system was reported by means of computer simulations [1,2]. Formation of face-centered cubic (fcc) structures by HSs has also been reported [3]. In the early days of the investigation of HS systems, researchers were surprised to see the occurrence of a crystalline phase in a system without any attractive interaction between the particles [4]. That is, it was natural for the researchers of the time to consider bond order as the reason for the occurrence of ordered phases. In general, competition between the configurational entropy and the potential energy of attractive interactions causes phase transition. The free energy of a disordered phase becomes small because of the entropic effect at high temperatures. In contrast, a crystal phase with bonding between atoms becomes energetically stable at low temperatures. This picture leads to the understanding that if there are no interparticle attractions, phase transitions into an ordered phase do not occur. Hence, bond formation can be considered to result from attractive interactions. Crystalline phase transition in the HS system is sometimes termed the Kirkwood–Alder–Wainwright transition (or simply as the Alder transition). Kirkwood was the first to predict this transition [5]. At present, we have an intuitive picture of the Alder transition based on the competition between the two entropic effects (i.e., configurational and vibrational entropies). At low particle number density, the effect of configurational entropy dominates that of vibrational entropy. In contrast, at high density, the vibrational entropy becomes predominant in the ordered phase of the system—even if the configurational entropy is lost, resulting in the total entropy increase. This picture is valid for systems comprising repulsive interactions only; i.e., the Alder transition in a wide sense can be understood in this manner.

Among the researchers who did not believe in the concept of the Alder transition, Onsager was an exception. He had already predicted the existence of a nematic phase in a hard-rod system [6]. He observed that ordering would more easily take place in systems with broken isotropy than in those with spherical particles. Nowadays, computer simulations of hard-rod systems for liquid crystals have been developed as an important branch of the soft matter field. In particular, different conclusions have been reported on the stability of columnar phase several times during the 1980s–1990s, which is both exploratory and interesting (e.g., [7]). The present author believes that the discussion on this development will help in understanding the role of the packing of particles in inducing crystalline order. However, this subject is not a topic of the present review.

Competition between attractive forces and thermal motion is responsible for the gas–liquid phase transition. Therefore, gas and liquid phases are indistinguishable in systems comprising repulsive interactions only, such as supercritical fluids. In terms of the crystal growth, the phase diagram of the HS system drawn on the density axis can be divided into three regions: the fluid region below ϕ_f, crystal region above ϕ_s, and a two-phase coexistence region between them; ϕ_f and ϕ_s are, respectively, the volume fractions of the coexisting fluid and crystal. Here, the density of the HS system is expressed in terms of its volume fraction, $\phi \equiv (\pi\sigma^3/6)(N/V)$, where σ is the HS diameter, N is the number of particles, and V is the total volume. The phase diagram in the P–T plane can be divided into two parts by a straight line: the region above $P/T = (P/T)_c$ is the crystal region, and the region below it is the fluid region. In 1968, Hoover and Ree applied the single-occupancy cell method and obtained the following: $\phi_f = 0.494$, $\phi_s = 0.545$, and $(P\sigma^3/k_BT)_c = 11.75$ (where k_B is Boltzmann's constant) [8]. These values were then revised by Davidchack and Laird in 1998 [9]. A direct two-phase coexistence simulation was performed, and the phase transition condition was corrected downward through minimization of stress in the crystal: $\phi_f = 0.491$, $\phi_s = 0.542$, and $(P\sigma^3/k_BT)_c = 11.55$. Recently, the following values have been obtained by constant-pressure MC simulations: $\phi_f = 0.492$, $\phi_s = 0.545$, and $(P\sigma^3/k_BT)_c = 11.576$ [10], while those obtained by MC simulations with a devised umbrella sampling technique are as follows: $\phi_f = 0.49188$, $\phi_s = 0.54312$, and$(P\sigma^3/k_BT)_c = 11.5727$ [11]. It should be noted that the phase transition densities are far lower than the close-pack density ($\phi_{cp} = \pi\sqrt{2}/6 \cong 0.74$).

Note also that Jackson's theory of the crystal/melt interface was presented in 1958 [12]. Although his model—the so-called two-level model of the crystal/melt interface—was a mean-field model based on different types of bonds (i.e. solid–solid, melt–melt, and solid–melt bonds), it was successfully applied to various types of materials. This indicates the versatility of the mean-field model. However, many researchers working on crystal growth believe that a bonding picture is essential [13].

1.2. Colloidal Crystals and Effective Hard-Sphere Model

Similar to HS systems, colloidal particles do not form any bonding between the particles. Therefore, crystal growth researchers face a conceptual difficulty when working on colloidal crystals. A colloidal crystal was first reported in 1954 [14]. However, this was not related with the HS phase transition because the colloidal crystals investigated were dried (i.e., they were essentially close-packed crystals). Lux et al. conducted a Bragg diffraction study in the visible light region for colloidal dispersions, the density of which corresponds to that obtained by the Alder transition [15]. A standard theory that is used to describe interparticle interactions between colloidal particles is the DLVO theory [16,17]. In the initial days of research on colloidal crystals, attempts were made to understand colloidal crystallization in the framework of the DLVO theory as a phenomenon in which the particles were trapped in the first minimum of interparticle potential. However, the distance to the first minimum was too small to explain the lattice spacing of colloidal crystals. In 1972, Wadach and Toda successfully explained colloidal crystallization by means of Alder transition [18]. They treated a colloidal sphere of diameter σ as an effective HS of diameter $\sigma_{eff} = \sigma + \alpha\kappa^{-1}$, with α being a factor of order unity, where κ was the Debye parameter. Parameter κ^{-1} is sometimes also referred to as the Debye screening length, which can be regarded as the thickness of the electric double layer around a colloidal particle. In this respect, the HS model was an idealized model for charged colloids. Experimental efforts have been made to realize HS colloids [19–21]. Nowadays, colloidal systems exhibiting the HS crystallization behavior are extensively studied using sterically stabilized colloids such as those reported in References [22–26].

A large number of studies on colloidal crystals have been conducted because colloidal crystals—in principle—possess properties similar to those of a photonic crystal. Structures with a specially ordered variation of dielectric constant—where the periodicity of the special order is on the same order as that of the wavelength of light—exhibit a photonic band [27–29]. Because the periodicity of colloidal crystals is on the same order as the wavelength of light, a photonic band structure arises which is similar to the band structure of electrons in semiconductors. In contrast with photonic crystals fabricated by nano-manufacturing, colloidal photonic crystals are more cost-effective, because the equipment for colloidal crystallization is cheaper than that for nano-manufacturing. In addition, since colloidal crystallization is a self-assembling phenomenon, it is less time-consuming to form a three-dimensional (3D) structure. However, a shortcoming of colloidal crystallization is the controlling of defects. Colloidal epitaxy is one of techniques that is used to reduce defects [22]. A key idea of the colloidal epitaxy is the uniqueness of stacking sequence in fcc (001) stacking. Growth direction is limited in fcc [001] through the use of a patterned substrate (i.e., a template). A recent development is the use of a template with a defect structure in order to introduce the desired structure in a colloidal crystal [26].

Some studies have also examined the effect of gravity on the defects for defect control. Zhu et al. performed a space shuttle experiment and found that a sediment of HS colloids in microgravity forms a random hexagonal close pack (rhcp) structure [23]. Traversing along the fcc ⟨111⟩ stacking sequence occurs in a three-fold manner in the regular fcc structure (where the fcc {111} planes can be classified into three with respect to the particle positions parallel to the plane—say, A, B, and C), while it is two-fold in the regular hexagonal close-packed (hcp) structure. That is, ABCABC··· stacking sequence corresponds to fcc, and ABAB··· corresponds to hcp. In turn, random sequences of A, B, and C correspond to rhcp. Zhu et al. also reported that the effect of gravity reduces the stacking disorder. Small free-energy differences between fcc and hcp crystals account for the difficulty in excluding the stacking disorder. Entropy difference between fcc and hcp crystals has been computed for HS systems, and it was found to be—at most—on the order of $10^{-3}k_B$ per particle [30–35]. In addition, fcc–hcp interfacial free energy was calculated to be on the order of $10^{-5}k_BT$ [33].

2. Simulation Methods

Unlike systems in which particles interact via interparticle potential of a continuous function, some difficulties are faced in hard-particle systems, because the potential function possesses some singular points. In this case, in an MD simulation, the equations of motion cannot be replaced with difference equations when performing the simulation. Differential of potential diverges at the contact of two hard particles. Section 2.1 describes the simulation algorithm for the HS systems. The MC simulation algorithm does not suffer too much from the singularity of potential. In Section 2.2, after a general explanation, the method used for MC simulation of HSs under gravity—on which the present author also worked—is described. Although Brownian dynamics simulation is more suitable for the simulation of suspended particles, this review does not explain it in detail because the present author has not worked on this method. Colloidal particles are subjected to random forces from the particles of dispersion media, in addition to the interparticle force between colloidal particles themselves. Therefore, the Langevin equation governs the time evolution of this system. The random forces should be treated together with the difficulty of potential singularity in the Brownian dynamics simulation.

2.1. MD Method for Hard Spheres in General

Alder and Wainwright presented the algorithm through which they tracked particle coordinates during time evolution by solving collision dynamics between HSs [36]. The review describes the collision dynamics of a system of non-identical HSs; the HS diameter and mass of a particle i are σ_i and m_i, respectively. Let $r_i(t)$ and $v_i(t)$ be the position and velocity of particle i at time t, respectively. For all pairs i and j, Equation (1)

$$|r_{ij} + v_{ij}t_{ij}| = \sigma_{ij}, \tag{1}$$

is solved by squaring both its sides to obtain a list $\{t_{ij}\}$ of times after which a collision between particles will occur. Here, $\sigma_{ij} \equiv (\sigma_i + \sigma_j)/2$ is the distance between particles i and j at collision. The following equation is obtained

$$t_{ij} = \frac{-b_{ij} - \sqrt{b_{ij}^2 - v_{ij}^2(r_{ij}^2 - \sigma_{ij}^2)}}{v_{ij}^2}, \tag{2}$$

where the relative position vector and relative velocity vector are expressed as $r_{ij}(t) \equiv r_i(t) - r_j(t)$ and $v_{ij}(t) \equiv v_i(t) - v_j(t)$, respectively, and $b_{ij} \equiv r_{ij} \cdot v_{ij}$. An inappropriate solution to Equation (1), $t_{ij} = [-b_{ij} + \sqrt{b_{ij}^2 - v_{ij}^2(r_{ij}^2 - \sigma_{ij}^2)}]/v_{ij}^2$, is excluded. In addition, positive b_{ij} corresponds to a collision that occurred in the past. After making the list $\{t_{ij}\}$ and sorting it, the time is evolved to $t' = t + t_{kl}$, where t_{kl} is the minimum value in the members of $\{t_{ij}\}$. All positions are updated as $r_i(t') = r_i(t) + v_i(t)t_{kl}$.

Velocities remain unchanged, except for particles k and l. From linear and angular momentum conservation at the collision, v_k and v_l are updated as

$$v_k(t') = v_k(t) - \frac{m_l}{m_k + m_l} \frac{b_{kl}}{\sigma_{kl}^2} r_{kl}(t'),$$

$$v_l(t') = v_l(t) + \frac{m_k}{m_k + m_l} \frac{b_{kl}}{\sigma_{kl}^2} r_{kl}(t'). \tag{3}$$

The list $\{t_{ij}\}$ is updated by subtracting t_{kl}, except for the members involved in the collided particles. The times to collision with particles k and/or i are recalculated using Equation (1). One can solve the time evolution by repeating this procedure.

However, this algorithm cannot be parallelized. Let us make a simple consideration. Consider a case where $\{t_{ij}^{(1)}\}$ is the list of times to collision for system 1 and $\{t_{ij}^{(2)}\}$ is the list for system 2; to obtain a list for a system comprising systems 1 and 2, the two lists should be combined. This list can be taken as the list for the composed system if the collision at the interface between the systems is neglected. The average number of collisions doubles until a specific event for the combined system occurs. In other words, the mean time between successive collisions would decrease by half. It can be said that for a large system, a collision would occur at almost any instant. Such a difficulty does not arise for continuous potential, for which the equation of motion is replaced with a difference equation with a non-zero constant time step.

The collision-by-collision method has been developed as an event-driven method (e.g., [37]). At a glance, computational cost for a system with pair-wise interaction is proportional to N^2, where N is the number of particles (or system size). However, when the interaction is of short range (e.g., r_{int}), the computational cost can be greatly reduced. One may determine interactions within r_{int}. A method for this purpose is the linked-cell method [38]. In a usual linked-cell method, the system is divided into different numbers of cells whose size is larger than r_{int}. For a particle, the interactions that need to be calculated are limited to within the cell this particle belongs to and its neighboring cells. For the HS system, the cell size is set small so as to include one particle at the most, and in addition to collisions between particles, events through which a particle crosses the cell boundaries are also investigated [39]. This method has been further improved by Isobe [40]. In contrast to Alder and Wainwright's algorithm, this scheme enables parallelization. An event-driven simulation package has been recently presented in [41].

2.2. MC Method Employed for Hard Spheres under Gravity

MC simulation for the HS system is very simple. If all interparticle distances are larger than or equal to the HS diameter ($r_{ij} \geqslant \sigma_{ij}$ for non-identical HSs) after an MC move, this MC attempt is accepted. If an overlap between HSs occurs, this attempt is rejected. The present author and coworkers employed the MC method for the simulation of HSs in gravity in order to investigate the sedimentation of HSs on a bottom wall. In this case, in addition to the interparticle interaction, interaction between particles and the wall should be considered. In an MD simulation, particle–wall collision event was also investigated. This is not a very difficult task. However, the MC method was employed because an algorithm to reject a configuration that a particle locates outside the wall was overwhelmingly simple compared to that for investigating collision events between a particle and the wall in the presence of gravity. In the presence of a gravitational field, the effect of gravitational energy should be incorporated. The Metropolis method was employed to investigate the change in the gravitational energy associated with the MC move.

3. Hard-Sphere Crystal/Fluid Interface

It is difficult to construct a crystal/fluid interface in HS systems because of the hard-core singularity. The particle density is relatively high in the simulations of a crystal/fluid interface or an interface between two condensed phases. In the HS system, the volume fraction of HSs in the coexistence region is about 0.5. When crystalline and fluid phases are prepared separately with appropriate coexistence densities and then joined together to create an interface, an overlap would occur between HSs because of the high density. To avoid such an overlap, crystal and fluid blocks should be kept slightly apart from each other or the overlapped particles should be removed. With both methods, the density of the entire system would be less than that of the coexistence one. The method to circumvent this difficulty is described in Section 3.1. Some results of interface simulations are shown in Section 3.2

3.1. Crystal–Fluid Coexistence in Equilibrium

In 1995, the present author and coworkers for the first time successfully performed an MD simulation of crystal–fluid coexistence states in an HS system [42]. It was also in 1995 that Kyrlidis and Brown performed a similar study by MC simulation [43]. To prepare a configuration at a high density, the simulation should be usually started with an fcc configuration, even if one aims at simulating a dense liquid. Overlap between HSs (or contact at a steep repulsive core, in general) occurs for a dense random distribution of particle centers. A crystalline configuration at a density corresponding to a dense liquid is well equilibrated to obtain a random configuration. Equilibration is also key in crystal–fluid coexistence simulations.

A direct method to avoid an overlap between HSs at the interface does not involve separate preparation of crystalline and fluid phases; instead, it involves the preparation of a coexistence state in a system from the first in a course of simulations. Here, one should start with a regular configuration because of the high density. The melting of a regular configuration takes a long time. The present author and coworkers used the following method to circumvent this problem [42]: at first, the system was divided into three parts—the crystal block, melt block, and interfacial region. An fcc crystal of $\phi = \phi_s$ was located in the crystal block, and a close-packed fcc crystal was present at the center of the melt block. We determined the width of the interfacial region (d_{int}) from the following relation:

$$\frac{\pi\sigma^3}{6}\frac{N}{(L_{melt} + d_{int} + L_{cryst})A} = \phi_{entire}, \tag{4}$$

where N is the total number of particles, L_{melt} and L_{cryst} are, respectively, the lengths (normal to the interface) of the melt and crystal blocks, A is the cross-sectional area of the system (parallel to the interface), and ϕ_{entire} is the volume fraction of the entire system, which fell around the center of the $\phi_f < \phi < \phi_s$ region. That is, d_{int} was calculated before determining the initial configuration of the particles. A snapshot is shown in Figure 1. The close-packed fcc configuration was placed in the melt block in order to obtain a high collision rate between HSs. The simulation was further modified in order to accelerate the randomization of the melt configuration. In Reference [42], the mass of crystal particles was set 1000 times heavier than that of melt particles. Equation (3) shows that heavier particles scarcely move, whereas lighter particles move well. Accordingly, the melt particles adopted a disordered configuration after several MD runs, as shown in Figure 2a. Figure 2 shows snapshots for an $N = 3787$ system. It took 5×10^5 total collisions to obtain the configuration shown in Figure 2a. Note that the 3D periodic boundary condition (PBC) was imposed on the configuration. Therefore, the right and left crystals form one block. After the melt block became molten, we made all the masses identical. A snapshot after equilibration of the entire system is shown in Figure 2b. Figure 2b shows a trajectory of the particles after 2×10^6 total collisions during the simulation of identical masses. This trajectory was constructed by connecting the particle positions observed for over 70 collisions per particle. The numerals indicated on the horizontal axis correspond to the labels of the crystal layers. The mass ratio can also be set to a value other than 1000, such as 100 or 10000. Optimization of the mass ratio would result in acceleration of the equilibration. When the mass ratio was moderate (such as that reported in [42]), the crystal particles showed slight movement. The crystal block will therefore move a little toward equilibrium in an equilibration process of the melt block.

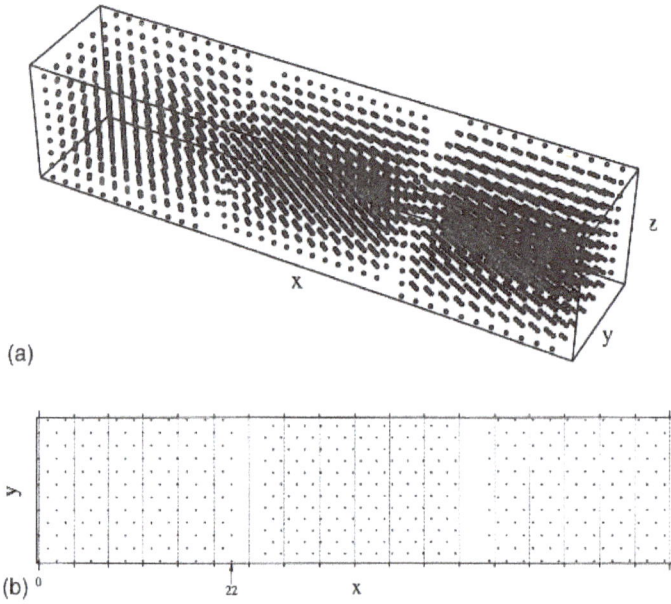

Figure 1. An initial configuration for the preparation of a hard-sphere crystal–fluid coexistence state (reproduced from Reference [42] with permission from the publisher): (**a**) a 3D snapshot; (**b**) a 2D projection.

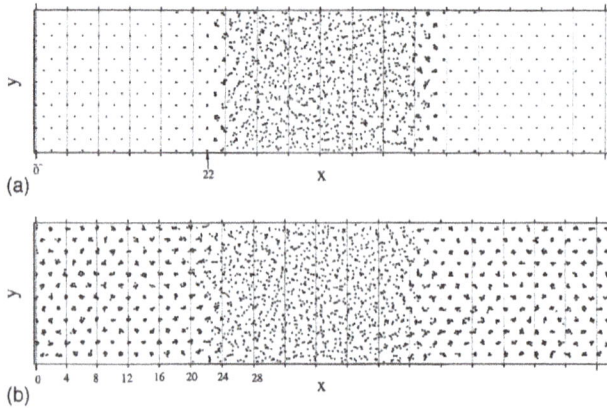

Figure 2. Configuration of a hard-sphere crystal–fluid coexistence state (reproduced from Reference [42] with permission from the publisher): (**a**) after the melt block became molten; (**b**) after equilibration of the entire system.

This method was developed by Davidchack and Laird [9]. They initially moved only the melt particles in the equilibration process of the melt block, and then moved only the crystal particles. These processes were repeated. Equilibration of the interface configuration is confirmed by observing the evolution of configuration in repeated processes—in particular, when the change in the switching of the motionless particles was observed. As mentioned in Section 1.1, the revision of the value of the HS phase transition condition was achieved by their well-equilibrated crystal–fluid coexistence state. This direct coexistence method has attracted more attention than the other methods for determining the phase equilibrium condition, as described in Reference [44]. Accurate knowledge of the equilibrium condition would help in determining the crystal growth condition. Distances from phase boundaries to the state point of the crystal growth condition in a phase diagram is a useful piece of information. Direct coexistence simulations also afford atomistic information about the interface, as described in the subsequent subsection.

3.2. Interface Properties

Researchers working on crystal growth classify the interface structure into two types: a sharp interface and a diffuse interface [45]. In the former (such as during crystal growth from vapors), essentially a layer-by-layer growth mode takes place. In the latter (such as for crystal growth from a melt), the crystal growth events occur simultaneously over a number of layers. The experimental observation of an interface between dense phases is difficult because of low accessibility to the interface through a dense phase. Atomistic information from computer simulations complements such an experimental observation.

The information on the diffuseness of an interface is essential to determine whether or not the multilayer mode can occur. Density profiles of the HS crystal/fluid interfaces along the interface normal are shown in Figure 3. The density profile shown in Figure 3b corresponds to the snapshot depicted in Figure 2b. The highlighted numeral "22" is an end position of the crystal block (Figure 1b). The end positions are also highlighted for (100) and (111) interface systems. These interfaces had widths several layers thick and compact interfaces, similar to those in the Lennard–Jones and soft-sphere systems [46–53]. While crystal growth researchers were surprised by the bonding picture (as described in Section 1.1), realization of crystal/fluid interfaces in soft-sphere systems (inverse 12th-power potential [46,50,51] and Weeks–Chandler–Andersen potential [48]) was a convincing proof for the present author that the HS interfaces can be constructed successfully by an MD simulation. If the width of the HS interface was infinitely broad (e.g., inversely proportional to the bond-breaking energy as calculated by Temkin's multi-layer model [54]; the numerical result shown in References [13,55–57] implies this property, and the analytical order parameter profile under continuum approximation [58–60] provides evidence), it would become difficult to grow crystals in a two-phase coexistence state via an interface. In other words, when the width of the interface is limited by the system size, the interface structure cannot be controlled through thermodynamic parameters only.

Layer-by-layer investigation of the interfacial region was performed. The trajectories of the particles (not shown in this review) were essentially the same as those reported by Davidchack and Laird [9], except that the system was relatively small. The number of particles used in the simulations reported in Reference [42] were 3888, 3788, and 4230 for (100), (110), and (111) interfaces, respectively. The sizes of the cross-sections parallel to the interface were $9.41\sigma \times 9.41\sigma$, $9.40\sigma \times 9.97\sigma$, and $9.79\sigma \times 9.60\sigma$, respectively. It should be remarked that, with respect to the intra-layer order, several layers were liquid crystalline. That is, despite the definite layered structure, the structure of several layers was far from crystalline with respect to the intra-layer order. Radial distribution function ($g^{(2D)}(r)$) was calculated for each layer. The 2D projections of the position vectors parallel to the interface were achieved for each layer and then a histogram of distances between all pairs of projection vectors was obtained. One can find liquid crystalline properties in these profiles. Results for the left interface (for the (100) interface simulation, both the right and left interfaces were almost symmetric) are shown in Figure 4. The second peak at $r \cong 2\sigma$ is indicative of the short-range order of a dense liquid. When two HSs are in contact with each other, the interparticle distance equals the HS diameter (i.e., $r = \sigma$), and when three HSs are in contact with each other on a line, two particles at both ends of the line give $r = 2\sigma$. Such configurations often appear for random arrangements of particles at a high density. The peak at $r \cong 2\sigma$ is confirmed in Figure 4a. In Figure 4b–d, a peak at $r \cong 1.6\sigma$ is shown, which reflects the lattice structure. The second-neighbor distance in an fcc (100) plane (a square lattice) is $\sqrt{2} \times l_{nn}$, where l_{nn} is the nearest distance that is slightly greater than σ, because the HS phase transition density is 0.5 in volume fraction. Along with the third peak at $r \sim 2.5\sigma$, a shoulder at $r \sim 2.2\sigma$ can also be seen in Figure 4d; however, the corresponding peak becomes very weak (as shown in Figure 4c). The third peak is attributed to the third-neighbor distance of the square lattice. While fine structures retain highly-ordered lattice, slight disorders such as those caused by thermal vibration broaden the peak, and sometimes it becomes difficult to distinguish shoulders from the main peak. Besides the smectic-A like structure (Figure 4a), a smectic-B like order has also been suggested.

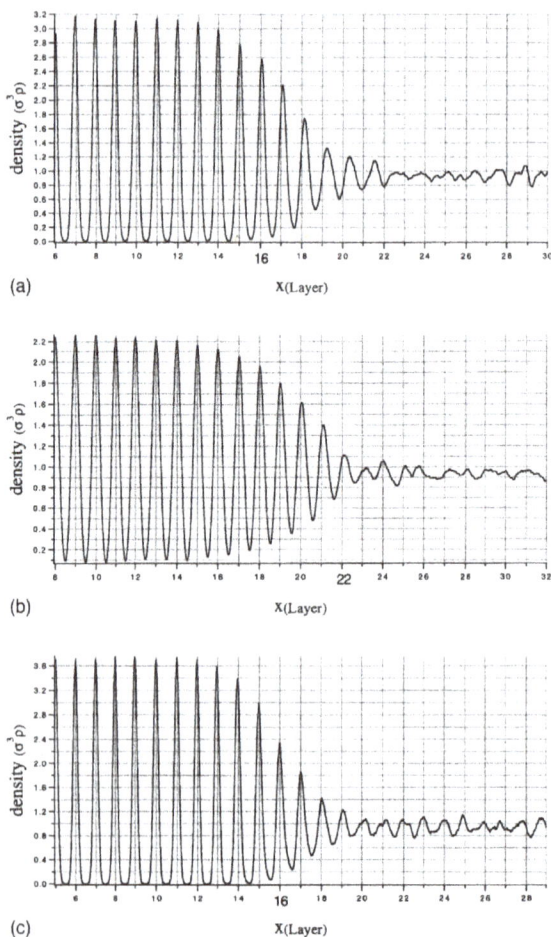

(a)

X(Layer)

(b)

X(Layer)

(c)

X(Layer)

Figure 3. Density profiles of the hard-sphere crystal/fluid interface (reproduced from Reference [42] with permission from the publisher): (**a**) (100); (**b**) (110); (**c**) (111) interfaces.

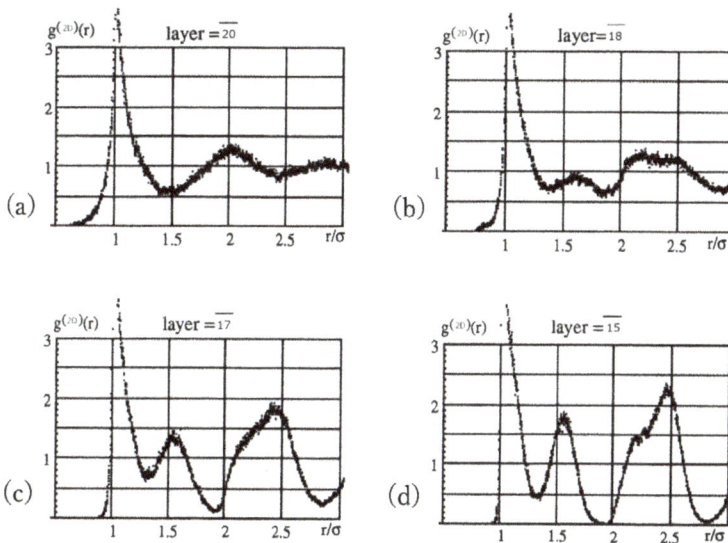

Figure 4. 2D radial distribution functions for several layers: (a) for 20th layer left from the layer indicated as "0" in Figure 3a; (b) for 18th; (c) for 17th; and (d) for 15th.

Drift of the interface was significant for both (110) and (111) interfaces (and thus, the right and left interfaces were asymmetric). This review does not discuss radial distributions for (110) and (111) interfaces because of poor statistics. In addition, diffusivity profiles were calculated from mean square displacements. Because of the small size of the system, statistics for mean square displacements could not be improved. Nevertheless, anisotropy in diffusivity was evident for the mean square displacements of particles—in particular, in the left 18th layer corresponding to Figure 4b. The diffusion coefficient in the parallel direction to the interface was greater than that in the normal direction.

The other crucial parameter is interfacial free energy. The nucleation rate is calculated based on the nucleation work employing the interfacial free energy (e.g., Reference [61]). The equilibrium shape of a crystallite is determined by the orientational dependence of interfacial free energy, also known as the Gibbs–Curie–Wulff rule. Davidchack and Laird developed an interface simulation to calculate the HS crystal/fluid interfacial free energy [62]. In 1986, Broughton and Gilmer computed the interfacial free energy for the Lennard–Jones system by introducing a cleavage potential to calculate the work of formation of the interface [63]. A cleavage wall, instead of the potential, was used to calculate the formation work of the interface. Values published in Reference [62] were $\gamma^*_{100} = 0.62 \pm 0.01$, $\gamma^*_{110} = 0.64 \pm 0.01$, and $\gamma^*_{111} = 0.58 \pm 0.01$ for {100}, {110}, and {111} interfaces, respectively. Here, the asterisk indicates the reduced unit $[k_B T/\sigma^2]$ (i.e., $\gamma^* \equiv \sigma^2 \gamma/k_B T$). These values were revised in 2005 [64], and recently revised again by Davidchack as $\gamma^*_{100} = 0.5820 \pm 0.019$, $\gamma^*_{110} = 0.5590 \pm 0.020$, and $\gamma^*_{111} = 0.5416 \pm 0.031$, and added $\gamma^*_{120} = 0.5669 \pm 0.020$ [65]. For comparison with values determined by other methods, see References [11,66].

4. Hard-Sphere System under Gravity

Earlier in Section 1, it has been mentioned that competition between the potential energy (potential well) and thermal energy gives rise to condensation. Thus, in systems comprising pure repulsive interaction, condensation does not occur. The HS system has only one type of characteristic energy, which is the thermal energy. However, if the system is kept in an external field, another characteristic energy can arise. It is expected that in a colloidal system in gravity (acceleration due to gravity g), a competition between the thermal energy and gravitational energy gives rise to phenomena like phase transitions. Hence, in such cases, it is convenient to introduce a dimensionless parameter $g^* = mg\sigma/k_B T$. This parameter is sometimes called the "gravitational number" and coincides with the Péclet number, which is defined as the ratio between the drift and diffusion. Parameter g^* is also expressed in terms of gravitational length, $l_g \equiv k_B T/mg$ (i.e., $g^* = \sigma/l_g$. If kinetics is the main concern, the concept of the Péclet number becomes more appropriate. Throughout this paper, we shall use g^*.

Colloidal suspensions crystallize by sedimentation [67]. This is a typical well-known method of colloidal crystallization, which is as intuitive as colloidal crystallization through the evaporation of dispersion media. Hence, it can be said that densification, in general, is the driving force of the Alder transition. The colloidal crystallization by sedimentation can in turn be understood as a phenomenon caused by the competition between $mg\sigma$ and $k_B T$. Because of phase transitions caused by competing factors, it is expected that various phenomena will also occur in the presence of gravity. An instance of such a phenomenon is the effect of gravity that reduces the stacking disorder in crystalline colloidal sediment, as reported by Zhu et al. [23]; this has been discussed at the end of Section 1.2.

4.1. Sedimentation

HS colloidal crystals obtained by settling under normal gravity are a mixture of fcc and hcp [68]. Contrarily, Zhu et al. reported that under microgravity, HS colloidal sediments were of rhcp structure [23]. This discovery cannot be understood intuitively, because the difference in the stacking sequence does not alter the particle number density. A large number of studies have investigated this phenomenon. It was shown by Kegel and Dhont that the rhcp structure, on aging, changes to the faulted-twinned fcc structure under *milligravity* [24]. The same trend was also obtained by Martelozzo et al. [69]. In contrast, Dolbnya et al. reported a slightly different result: after 1–1.5 years of gravitational annealing, the sediment acquired the fcc structure, including the absence of any stacking disorder [25].

MC simulation study of HSs in gravity were first carried out in 1994 [70]. However, the system sizes were very small, and accordingly, the number of stacking layers was also small. A main concern was the intra-plane order in gravity. Simulation for a large system was performed in 2002 [71]. An intra-plane structure was investigated with respect to the nucleation dynamics at the bottom. It was found that the effect of gravity obstructs the crystal growth from the point of view of nucleation dynamics. This seemingly contradicts the effect of gravity that reduces defects. It was in the mid-2000s that the present author and his coworkers performed computer simulations of HSs to investigate the stacking order in gravity [72,73]. MC simulations with constant volumes were conducted, and the strength of gravity was changed stepwise so as to avoid the trapping of systems in a metastable state such as glassy or polycrystalline states [73]. A series of snapshots is shown in Figure 5. The box length of a system containing $N = 1664$ particles was $L_x = L_y = 6.27\sigma$ and $L_z = 49.23\sigma$. Bottom ($z = 0$) and top ($z = L_z$) walls of this system were hard walls and the PBC was imposed in the x and y directions. After preparing a random configuration, g^* was increased up to 1.5 with increments of $\Delta g^* = 0.1$ and then decreased. The system was maintained at each g^* for 2×10^5 MC cycles. Here, an MC cycle (MCC) was defined in a way that an average of one MC particle move per particle was included in one MCC. The maximum displacement of MC moves was fixed at 0.06σ. It is confirmed from Figure 5 that defects in the crystal grown on the bottom wall dramatically vanished at $g^* \cong 0.9$. Hence, it can be said that we successfully demonstrated that defects are reduced under the effect of gravity. This g^* value corresponds to the condition in which the gravitational energy that moves a particle about one interparticle distance becomes comparable to the thermal energy, $k_B T$. In other words, the gravitational length is comparable to the interparticle spacing. However, this is a one-particle picture. This gives only a primitive explanation for the gravitational stabilization of the layered structure near the bottom wall. Detailed observation of defect disappearance would deepen the understanding of this mechanism.

A closer look at the evolution of 3D snapshots was undertaken previously [74]. Prior to elaborating on these investigations, it should be noted that the growth direction in Figure 5 is $\langle 001 \rangle$. Unique stacking sequence in this direction is a key idea of colloidal epitaxy [22]. A stacking fault is marked with a red line in Figure 6a–e. Contrary to the intuition that stacking disorder does not change the particle number density, a sink of the center of gravity is accompanied with the shrinkage of the stacking fault, as shown in Figure 6f. A magnification of the snapshot around the lower end of the stacking fault is shown in Figure 7. It should be remembered that a stacking fault is terminated by a Shockley partial dislocation. In addition, it should also be noted that the Shockley partial dislocation terminating an intrinsic stacking fault—such as that shown in Figure 7—accompanies a particle deficiency of 1/3 lattice line. This accounts for the sink of the center of gravity accompanied with the shrinkage of the stacking fault. It was observed that the buoyancy of a partial dislocation acts as a driving force for the disappearance of defects in gravity. In addition to the buoyancy, a stress field yields a driving force that causes the Shockley partial dislocation to move upward along a glide plane [75]. Further, the present author and his coworker calculated a cross term between the elastic field due to a Shockley partial dislocation and that caused by gravity [76]. This effect also acts as a driving force for the upward motion of the Shockley partial dislocation. Although numerical estimation has been carried out using an elastic constant for the HS crystal, this effect is not limited to colloidal crystals. However, the condition of coherent growth (which was reported in Reference [77]) was used.

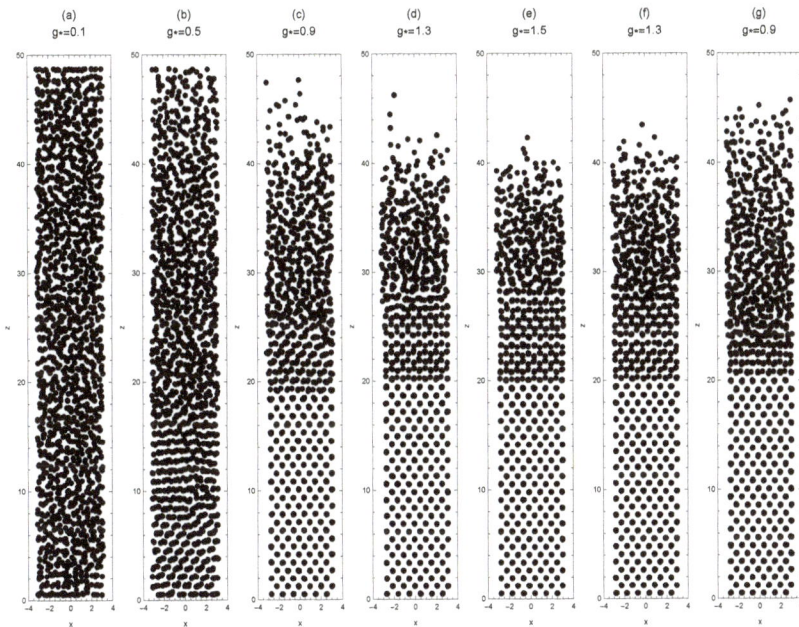

Figure 5. Snapshots of Monte Carlo simulation of $N = 1664$ hard spheres (HSs) (reproduced from Reference [73] with permission from the publisher). Values of g^* are indicated on the top.

The glide mechanism of a partial dislocation for the reduction of stacking disorder was revealed in Reference [74]. As mentioned in Section 1.2, the fcc–bcc interfacial free energy was determined by Pronk and Frenkel. Using this value for stacking fault energy (free energy), they estimated the time for stacking fault to be annealed out of a sediment [33]. It was assumed that the melting and recrystallization mechanism was involved here. The estimated time was a few months long, which was fairly consistent with Dolbnya et al.'s result of 1–1.5 years to obtain a perfect fcc stacking. To evaluate the time required for the glide mechanism, the activation barrier for the partial dislocation to move to the next lattice position is needed. While we do not have such a value at present, this mechanism is apparently smooth, because only particles along a dislocation line are involved. Therefore, this mechanism would account for defect reduction in a short period or that occurring immediately after the crystal growth. If a dislocation glide is mediated by a kink motion along the dislocation, the motion becomes smoother. It is possible that defect behaviors with intermediate time scales are related to other mechanisms, such as the climb-up of dislocation and jog-related phenomena.

Immediately after our study [73], a grand canonical MC simulation study of HSs in gravity was performed by Marechal and Dijkstra [78]. They focused on the dynamical aspect of crystallization. They also confirmed the occurrence of crystallization in two successive bottom layers reported by Biben [70]. This phenomenon was later confirmed experimentally [79]. Marechal et al. performed a Brownian dynamics simulation study of the stacking disorder from a dynamical point of view [80]. It was found that fast sedimentation introduces hcp stacking sequence in fcc stacking. Accordingly, a trade-off is suggested: while the gravity effect reduces the stacking disorder, it also accelerates the sedimentation simultaneously. This dilemma has been addressed in detail in Section 4.2. In an earlier study [73], we proposed the use of a centrifuge rotor to control the strength of a gravitational field. Besides examining the stepwise control of a gravitational field in simulations, different types of control are also possible through the variation of the centrifuge rotation velocity.

The results reported in References [73,74] also faced criticism. The driving force of the fcc $\langle 001 \rangle$ growth in these simulations was stress arising from the simulation box of PBC, which indeed was artificial. Our rebuttal was that stresses forcing the fcc $\langle 001 \rangle$ growth could be produced without the PBC. In reality, the fcc $\langle 001 \rangle$ growth has already been realized by utilizing a template in the colloidal epitaxy [22]. This motivated computer simulations on the colloidal epitaxy, as presented in the next subsection.

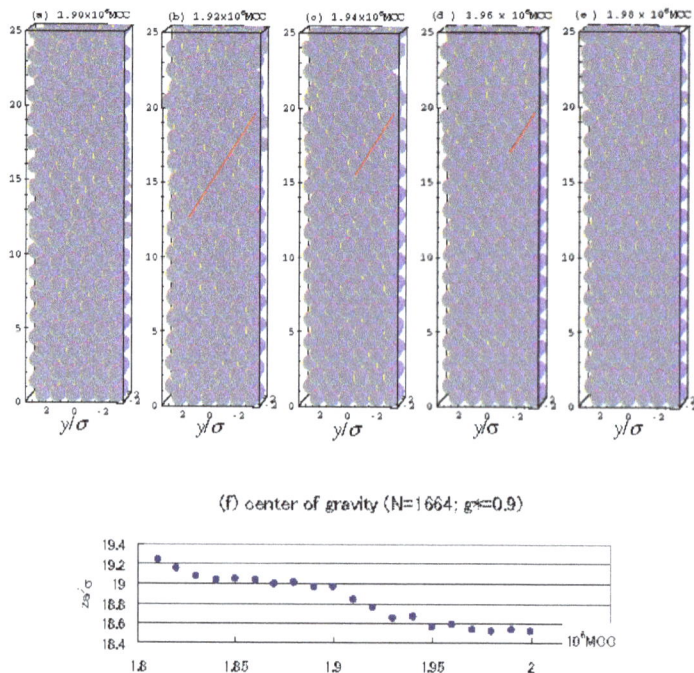

Figure 6. Evolution of 3D snapshots and the center of gravity during the disappearance of defects, as shown in Figure 5(c) (reproduced from Reference [74] with permission from the publisher). MCC: Monte Carlo cycle.

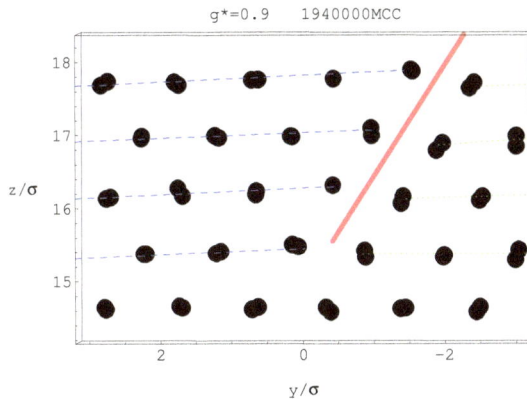

Figure 7. A magnified snapshot of Figure 6(c) (reproduced from Reference [74] with permission from the publisher).

4.2. Colloidal Epitaxy

In 1997, van Blaaderen et al. developed colloidal epitaxy [22]. As mentioned in Section 1.2, a key idea was the uniqueness of the stacking sequence in fcc $\langle 001 \rangle$. Interfacial free energy of the HS crystal against a flat hard wall is lowest for the fcc {111} face among low-indexed faces such as {110} and {100} [81–83]. Therefore, fcc {111} plane faces if one grows an HS crystal on a flat hard wall. This phenomenon has already been reported in an MD simulation study [84], and was also confirmed by a recent MC simulation study [85]. In reality, dense colloidal suspensions exhibit the same trend. Section 1.2 has already explained that it is difficult to avoid the stacking disorder in fcc {111} stacking by using the free energy difference.

In the preceding subsection, we discussed that fcc (001) stacking has many advantages, in addition to its uniqueness. Hence, it is possible to further develop this method. A shortcoming of colloidal epitaxy is that the obtained colloidal crystals are thin. Lin et al. successfully grew thick colloidal crystals with about 30 layers on a square grid template [86]. An advantage of the square grid is that it allows the lattice constant of the bottom layer to be adjusted via on-line matching of the template. Colloidal epitaxy on a square grid template has been reported (e.g., [87–89]). The present author employed a square grid as a template. In addition to the advantage just mentioned above, the computational cost to calculate interactions between particles and the pattern on the template is lower for a simple square grid than for an fcc (001) array. Crystallization of HSs on a patterned wall has already been investigated by MC simulations [90–92], and has also been recently investigated by MD simulations [93]. Enhancement of crystallization by various patterns has also been shown in the absence of a gravitational field. While gravity should be excluded to investigate the pure substrate effect, interplay between these two effects may yield some benefits. In addition, it is not easy to make the effect of gravity negligible in reality.

As a first stage, we merely replaced the flat bottom wall with a square patterned wall [94,95]. That is, the strength of gravity was increased in a stepwise manner. In Reference [95], the size of a horizontal system was doubled to that of Reference [94]. In simulations on a flat bottom wall (Reference [73]), the stacking sequence was changed to {111} stacking by doubling the size of the horizontal system. In contrast, in the simulation carried out on a square pattern, the stacking direction remained as $\langle 001 \rangle$. The pitch of the square grid was determined so that the lattice constant of the bottom crystal layer became slightly smaller than that for the simulation reported in Reference [73] at $g^* = 0.9$. In References [94,95], our interest was not limited to the detection of the threshold g^* value ($\sim$0.9), but was also extended to the phenomena occurring in a relatively wide range around the threshold. Accordingly, this review also examined the defect behavior under conditions when

g^* is slightly greater than the threshold value. The side-to-side distance between adjacent grooves was 0.338σ. The grid was made of grooves with width 0.707σ. Accordingly, the periodicity of the square grid was 1.478σ. The groove width was selected so that the diagonal length of a square in the cross-section of transverse and longitudinal grooves "coincides" to σ (for detail, see note 18 of Reference [96]). Snapshots of the $N = 26624$ system ($L_x = L_y = 25.09\sigma$ and $L_z = 200\sigma$) are shown in Figure 8. A two-fold—not three-fold—stacking sequence is an indication of fcc (001) stacking. Thickening of the crystal with increasing g^* is also seen. A triangular-shaped defect, as shown in Figure 8, is remarkable. This extended defect was later identified with a stacking fault tetrahedron [97].

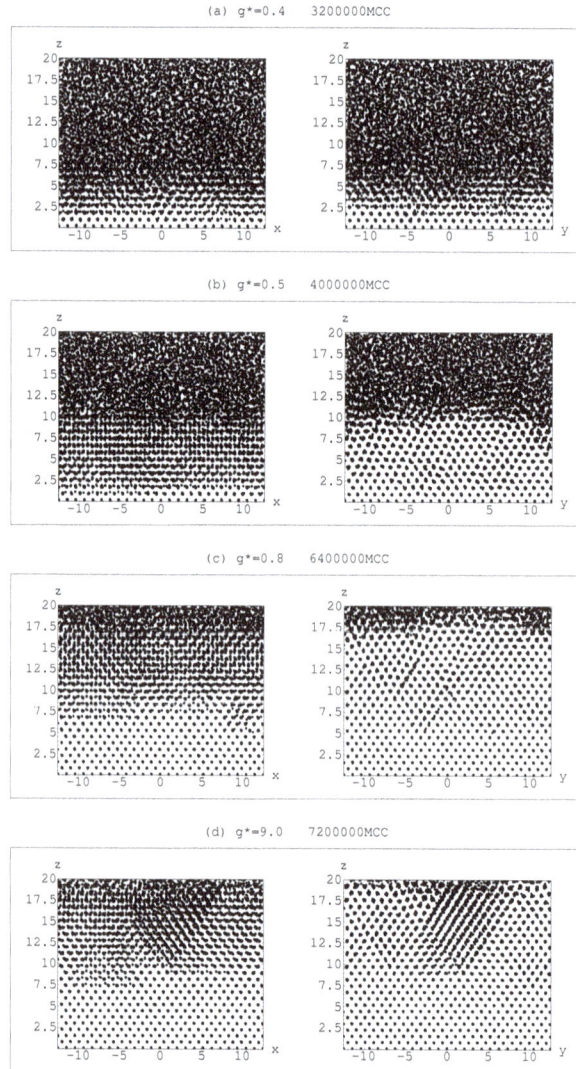

Figure 8. Snapshots (xz and yz projections) of a hard-sphere crystal grown on a square pattern with stepwise g^* increase (reproduced from Reference [95]).

After successfully replacing the stress that forced the growth direction to fcc ⟨001⟩, we examined the sudden switch-on of gravity [101] (some details are given in Section 4.3 because they relate more closely to this section). In the case of a flat bottom wall, this gravitation resulted in polycrystallization [72]. In this respect, it was surprising that in the case of a square grid template, we could successfully avoid the polycrytallization/amorphization without controlling the g^* value. This is another advantage of colloidal epitaxy, apart from the uniqueness of the stacking sequence. In other words, crystalline nucleation on a template and successive growth upward can overcome the crystallization inside. Of course, we cannot completely rule out the possibility that the crystal grown from a nucleation inside already becomes large so that it dominates the growth front of the crystal grown from the bottom when the growth front reaches the inside. Hence, it can be concluded that polycrystallization/amorphization can be avoided in colloidal epitaxy under a gravitational condition that the system polycrystallizes if the bottom wall is flat. Such a phenomenon has already been reported experimentally [98].

Here, it is important to note the contrasting wall effect. While the present and preceding subsections focus on the effects that promote the crystal growth in a wider sense, studies on walls that inhibit the crystallization on them are also interesting. Homogeneous nucleation inside a system was investigated using such walls by a Brownian dynamics simulation [99]. Combination of walls that promote and inhibit the crystallization may be used to create complicated structures.

4.3. Gravitational Annealing/Tempering

Annealing is a method of material processing. It is well known that the properties of GaN materials were successfully improved by a thermal annealing [100]; in this example, Mn-doped GaN were treated in N_2-ambient above 700 °C for 20 min to lower the resistivity. In contrast, in quenching, the temperature is decreased at a certain rate during the treatment of materials. After an annealing treatment, one can, for example, treat the materials at a temperature higher than the annealing temperature. Through such a tempering treatment, material properties can be modified. The aforementioned simulation under a constant gravitational condition corresponds to annealing. This subsection first briefly describes the gravitational annealing simulations, and then introduces the simulation results of gravitational tempering.

Let us look at some snapshots of simulations under constant gravitational conditions (Reference [101]). Snapshots of the $N = 26,624$ systems are shown. In this simulation, both the system size and the template were the same as those reported in Reference [95], except for $L_z = 1000\sigma$. The results of gravitational annealing are shown in Figures 9 and 10. Figure 9b shows that a crystal first formed at the bottom by sedimentation (Figure 9a) and then grew. A comparison of the xz projections of Figure 9b,c shows an improved crystallinity owing to gravitational annealing. In particular, region $z/\sigma \lesssim 11$ clearly showed better and improved crystallinity than that in the above region. However, one cannot have a simple conclusion on the defect reduction from yz projection. Region $11 \lesssim z/\sigma \lesssim 17$ shows increases in the number of defects. On the other hand, the defect around $(x/\sigma,y/\sigma,z/\sigma) \sim (5,7,12)$ extends in a three-dimensional region. This means that at least 3D analysis is necessary. One cannot judge on the crystallinity, merely based on the extent of defect structures. In addition, a tiny defect structure around $(x/\sigma,y/\sigma,z/\sigma) \sim (-1,0,2)$ should not be ignored. Hence, it can be said that improvement of crystallinity by gravitational annealing at $g^* = 1.6$ was successful but partial.

Let us look at the snapshots of gravitational annealing at $g^* = 1.4$ (Figure 10). Seemingly, Figure 10b shows that crystallinity of the crystalline layers formed on the bottom (Figure 10a) was already better than that for $g^* = 1.6$. Improved crystallinity can be more clearly seen in Figure 10c than in cases when $g^* = 1.6$. The extent of a less-defective crystalline region is evidently larger for the $g^* = 1.4$ case than for the $g^* = 1.6$ case. In addition, although a tiny defect structure remained for $g^* = 1.6$, no such defects were observed for $g^* = 1.4$. In contrast, the dots around $z/\sigma \sim 7.5$ resulting from the overlaid projected points widened. Based on these observations, a deviation of the particle position from its regular lattice point is suggested. Alleviation of gravitational tightness can amplify the lattice vibration of individual particles or the collective motion of particles. The latter may be indistinguishable from lattice defects such as the wandering of lattice lines. However, the overall crystallinity is higher for $g^* = 1.4$ than that for $g^* = 1.6$. Improvement in crystallinity through gravitational annealing has been shown in simulations, although the optimization of annealing under the gravity condition has not yet been completed.

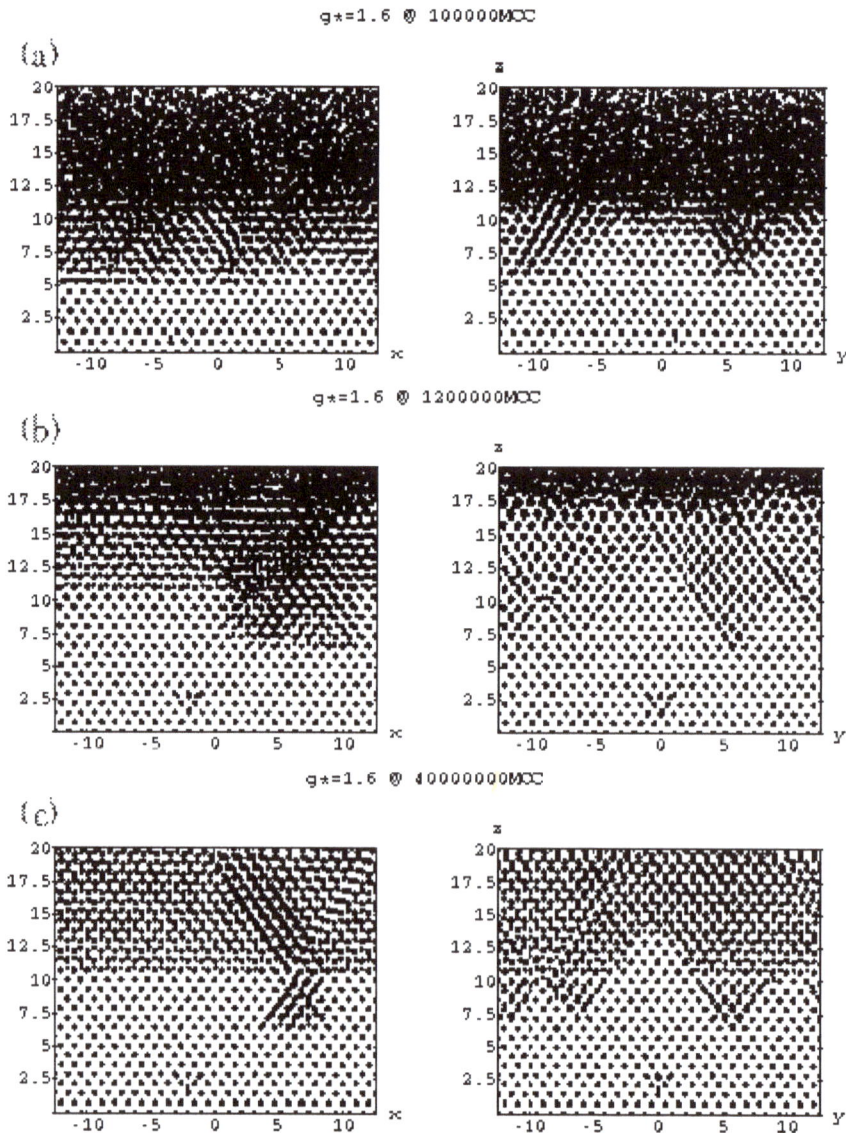

Figure 9. Snapshots of a hard-sphere crystal grown on a square pattern under constant gravitational condition at $g^* = 1.6$ (reproduced from Reference [96]): xz and yz projections at (**a**) 1.0×10^5th; (**b**) 1.2×10^6th; and (**c**) 4.0×10^7th MCCs.

g*=1.4 @ 10000MCC

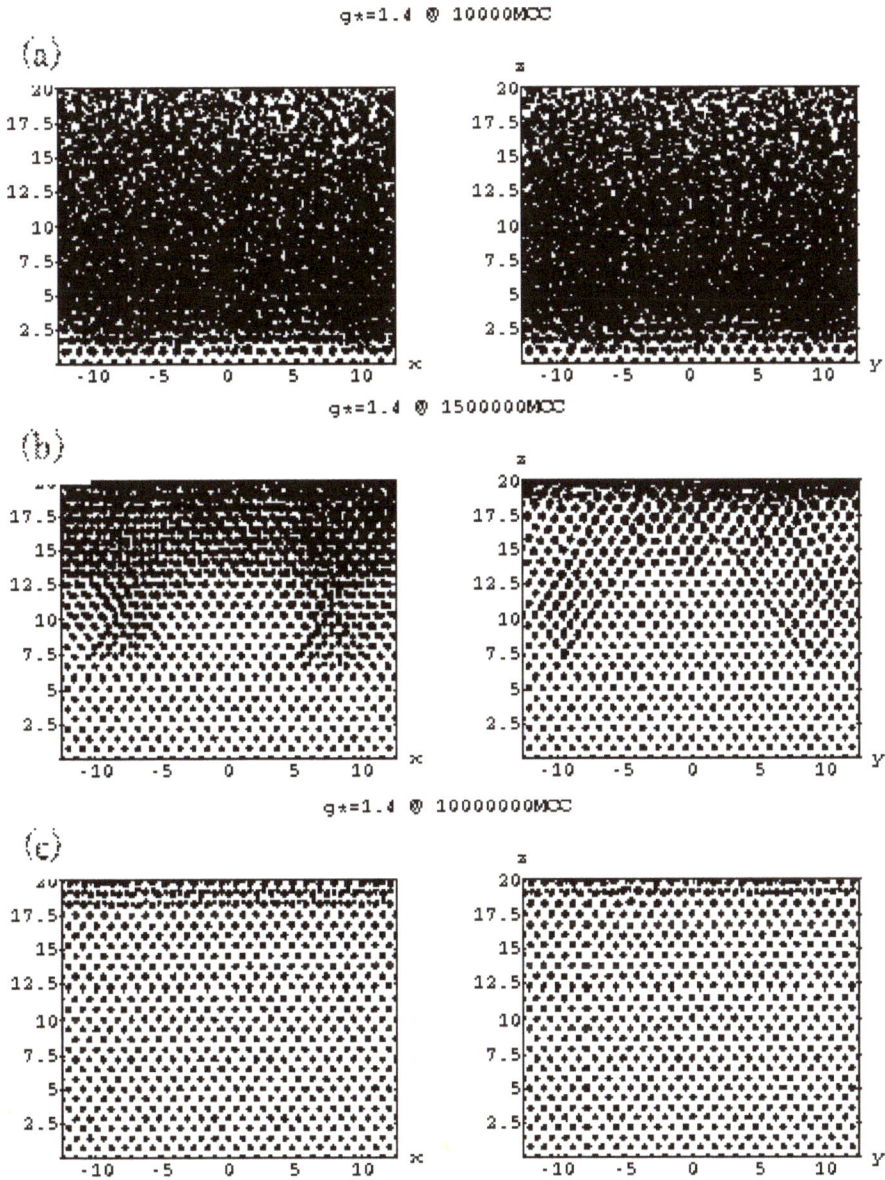

g*=1.4 @ 1500000MCC

g*=1.4 @ 10000000MCC

Figure 10. Snapshots of a hard-sphere crystal grown on a square pattern under a constant gravitational condition at g^* = 1.4 (reproduced from Reference [96]): xz and yz projections at (**a**) 1.0×10^4th, (**b**) 1.5×10^6th, and (**c**) 1.0×10^7th MCCs.

Inspired by these results, we proposed gravitational tempering to reduce defects in colloidal crystals [96]. After growing a colloidal crystal by sedimentation under a rather strong gravitational condition, treatment of the crystal at a moderate gravitational condition may improve the crystallinity. Alleviation of gravitational tightness is key for such a treatment. During crystal growth, defects that hardly move under a gravitational condition become mobile if the gravitation is weakened. However, weakening of the gravitational condition decreases the thermodynamic driving force of crystallization. In addition to the optimization of the tempering gravity, duration of the tempering program should also be optimized. We have successfully demonstrated MC simulations of gravitational tempering to reduce defects [102]. Sedimentation and gravitational tempering were tested at $g^* = 1.6$ and 1.4: g^* was kept at 1.6 for the first 2×10^7 MCCs, then decreased to 1.4 and maintained for 2×10^7 MCCs, and again 1.6 for the last 2×10^7 MCCs. The system size and the template were entirely the same as those reported in References [96,101]. Evolution of several parameters and snapshots of a simulation are shown in Figure 11. The result of a single simulation of five runs has been demonstrated here, and a dramatic defect disappearance can be seen (see the animation provided in the Supporting Information at http://pubs.acs.org/doi/suppl/10.1021/cg401884k). Evolution of the center of gravity z_G is plotted in Figure 11a in black. After a steep sink of z_G corresponding to sediment formation, it decreased gradually and became almost constant. Figure 11b is a snapshot of the crystal observed at the end of gravitational annealing after crystal growth by sedimentation at $g^* = 1.6$. Defect structure did not disappear by this gravitational annealing. Just after switching g^* to 1.4, z_G promptly increased. After some time, z_G suddenly sank slightly and then became almost constant. Snapshots just before and after this sink are shown in Figure 11c,d, respectively. An extended defect around $(x/\sigma, y/\sigma, z/\sigma) \sim (-10,3,15)$ disappeared. The slight sink of z_G was caused by this defect disappearance. Hence, it can be concluded that an extended defect that could not disappear through a gravitational annealing treatment did disappear. Defective layers are seen at the top region of yz projections in Figure 11c,d. Shrinking of this defective region was expected by treatment at $g^* = 1.6$ for the last 2×10^7 MCCs. Unfortunately, this defective region remained unchanged (snapshot is not shown). The numbers of particles of fcc-like, hcp-like, and other crystalline types in the observed region are plotted together in Figure 11. The kink positions in the evolution curves of these numbers coincide with those in z_G, except for at the 4×10^7th MCC. The triangular-shaped extended defect (a stacking fault tetrahedron) disappeared through conversion from hcp or other types of stacking to fcc ones. In Section 4.1, it was mentioned that a Shockley partial dislocation that terminates an intrinsic stacking fault accompanies the particle deficiency of 1/3 lattice line. The particle deficiency at the end of a stacking fault tetrahedron arises from the extended region over a triangle. The particles that underwent conversion between other types of crystals are also distributed over this triangle. Therefore, changes in the numbers and z_G were detectably large. In a previous study [102], we showed another result. Such changes were not detected for the shrinkage of only a few stacked planar defects.

In Section 4.1, we pointed out two aspects of the stacking disorder. It is generally observed that in crystal growth, crystals that grow rapidly contain many defects. A high degree of stacking disorder in rapid sedimentation is natural in this respect. An advantage of colloidal crystallization under a strong gravitational condition is enlargement of the grown crystal, rather than the rapid growth. It has been shown that even if the crystals contain a large number of defects, the defects in large crystals can be reduced by post-growth processing. Another advantage is the possibility of developing a technique of defect control. Hilhorst et al. successfully introduced a desired defect in a colloidal crystal in sedimentation by utilizing a template with a defective lattice structure [26]. For the functionalization of a colloidal crystal such as an optical waveguide, designed defects must be incorporated in the crystal. Such defect engineering can be performed by the sedimentation method using a template with regular lattice by varying the lattice spacing, as pointed out in Reference [97] and as explained as follows. It has been experimentally found that defects in colloidal crystals grown on a template are not triangular, but parallelepiped [103]. The defect structure of partial parallelepiped outer shape has also been reported in Reference [102]; results of two runs of five of the simulations have been reported, and in a run a triangular outer-shaped defect structure was seen as described above, and in the other, the outer shape of the defect structure was partial parallelepiped. As already mentioned, in order to match the template with the lattice spacing of the bottom layer at $g^* = 1.6$, the pitch of the square grid was set slightly smaller than the lattice spacing at $g^* = 0.9$ of Reference [73]. Under the stress provided by this pattern, stacking fault tetrahedra must dominate. By varying the lattice spacing of the template pattern, the dominant defect type may change. The effect of lattice spacing of the template pattern on the defect has already been argued by Jensen et al. [98]. As a recent development, the sedimentation method with colloidal epitaxy has been applied to control the structure of a grain boundary [104].

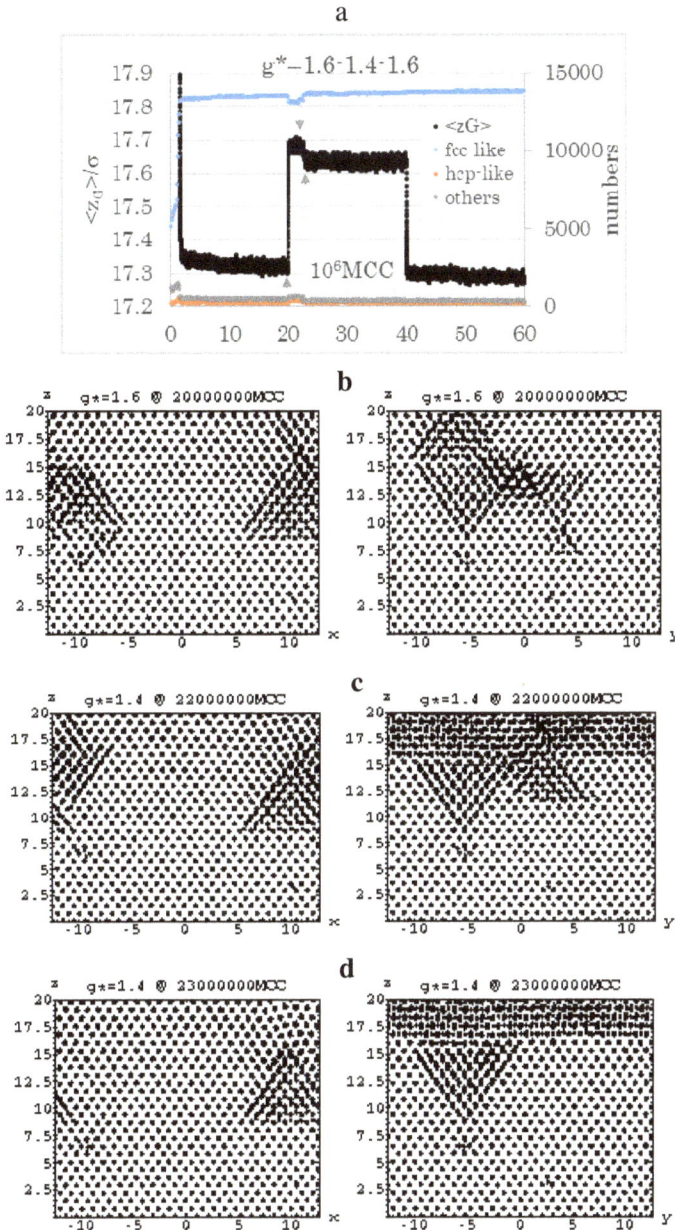

Figure 11. Result of the simulation of gravitational tempering (reproduced from Reference [102]): (**a**) the center of gravity and the number of the three types of crystalline particles; (**b**–**d**) snapshots.

5. Discussion

A main focus of the present review is the crystal–fluid coexistence in equilibrium, and the other is the crystal growth. The subsequent subsection first introduces an aspect of crystallization included in the equilibrium simulations, and then a feasibility study is presented on gravitational tempering.

5.1. Hard-Sphere Crystallization without External Field

The results of Reference [42] and successive development were discussed in Section 3. However, some of the results were not published because of poor statistics. Results with good statistics were published by other researchers (such as in Reference [9]) before the present author could improve the statistics. Besides the small size of the system, interface drift is another problem. The initial velocities are generally assigned using random numbers with zero average. Even if the averaged velocity over the entire system vanishes, the average over a particular part (e.g., in an interfacial region) cannot be taken as zero. Particular attention must be paid to prevent the interface drift. The present author investigated the crystallization and melting induced by a mass flow in the melt [105]. The driving force of HS crystallization is densification. Therefore, densification induced by a mass flow must cause HS crystallization.

This phenomenon can be understood based on the principle of mass conservation. In a conventional MD simulation, momentum and energy as well as mass are conserved. Treating the materials as a Navier–Stokes–Fourier fluid, the present author derived relations between thermodynamic quantities of two phases coexisting in a non-equilibrium steady-state [106]. Relations for the velocity of the interface, net mass-flow velocity, and thermodynamic variables such as temperature, pressure, and density derived from conservation laws are symmetric with respect to crystallization and melting. In contrast, entropy production changes its sign when we invert the flow velocity; the crystallization and melting are swapped with each other by this operation. This gives rise to a criterion that can be used for determination of the stability of the non-equilibrium steady-state interface. This formulation was extended to the isothermal case [107]. However, related studies have not been performed.

5.2. Processing of Colloidal Crystals Using Centrifugation

The utilization of a centrifuge rotor proposed in Reference [73] has been described in Section 4.1. This section first provides an introductory description on spontaneous sedimentation under normal gravitational condition. Subsequently, we discuss colloidal crystallization by centrifugation sedimentation. In other words, the feasibility of the experimental realization of gravitational tempering is discussed.

Typical values of g^* in normal gravity are described prior to determining them using a centrifuge rotor. Van Blaaderen et al. conducted a pioneering work on colloidal epitaxy and reported $g^* \sim 1$ [22]. Ackerson et al. carried out a light scattering experiment under the gravitational conditions $g^* \sim 0.3$ and 0.1 [108]. As mentioned in Section 4.1, Kegel and Dohnt conducted a milligravity experiment [24]. Gravity conditions as small as g^* on the order of 10^{-4} have been reported by matching the density of dispersion media to that of a dispersion phase. In a similar experimental study by Martelozzo et al., the gravitational conditions $g^* \sim 0.02$ and 0.004 were reported [69]. A real-space observation of the sedimentary growth of colloidal crystals was performed by Hoogenboom et al. under the gravitational conditions $g^* \sim 1$ and 4 [109]. In Dolbnya et al.'s experiment on 1–1.5 year long gravitational annealing, the gravitational conditions were $g^* \sim 0.03$ [25]. In a real-space observation by Royall et al., the gravitational conditions $g^* \sim 1$ and 2 were reported [110]. In an experimental study that confirmed the occurrence of crystallization in two successive bottom layers in gravity by Ramsteiner et al., the gravitational condition was $g^* \sim 7$ [79]. These values indicate that a wide range of the g^* value has already been covered. The main issue is in situ tuning of g^*. Varying g^* by changing the density of a dispersion medium and/or mass/diameter of dispersing particles cannot be applied in gravitational tempering because it cannot be carried out in situ. On the contrary, in situ observation is possible by methods such as laser scanning confocal microscopy [22,79,109–111].

A number of studies have already been conducted using a centrifuge rotor. Megens et al. investigated colloidal crystals by small-angle X-ray scattering [112]. They centrifuged colloidal suspensions at ∼400 G for more than 5 h to densify them. Here, G stands for normal gravity. In addition to spontaneous sedimentation, centrifugation experiments were also carried out by Ackerson et al. [108]. They used σ = 1000 nm and 760 nm particles, and reported that for smaller particles, the system failed to crystallize when the centrifugal force was greater than ∼6 G. Although this threshold value differs by two orders from Megens et al.'s centrifugation condition, it was not an error; in addition to the difference in other conditions, they focused on the kinetic aspect. Although the particle surface was not modified so as to exhibit the HS-like phase behavior, Suzuki et al. grew colloidal crystals by centrifugation at 82, 185, and 329 G [113]. Jensen et al. performed experiments on colloidal epitaxy at 9000 G [98]. These studies indicate that gravity condition can be controlled in centrifugation. In situ control of g^* is possible for gravitational tempering programs through centrifugation. A key issue for realizing gravitational tempering has been shown in Suzuki et al.'s gravitational annealing experiment [114]. In this study, we succeeded in removing striations from a silica colloidal crystal grown at 9 G by gravitational annealing at 50 G for 5 days. Photographs before and after gravitational annealing were been presented therein. However, these results were based on ex situ observations. In situ monitoring of disappearing defects is quite difficult during centrifugation. Even more difficult is establishing a setup of an optical system for confocal microscopy in a centrifuge rotor. A plan to rotate a conventional microscope together with a cuvette is under investigation.

6. Summary

We have reviewed here simulation studies of hard-sphere (HS) systems in the context of crystal growth. After a general introduction of models of colloidal particles, simulation methods are discussed. The review consists of two main parts: one on the crystal–fluid coexistence state in equilibrium, and the other the HS systems in gravity. The former part initially describes how the present author constructed the crystal–fluid coexistence state of hard spheres for the first time. Next, some interfacial properties derived from coexistence simulations are introduced. The latter part gives reviews of HS systems in gravity. Firstly, the simulation results of crystallization of hard spheres on a flat wall in gravity are presented. Then, the simulations on a template of a square pattern are described. Finally, the processing of colloidal crystals using gravity is discussed. Gravitational tempering is a new technique proposed by the present author and his coworkers. The review also discusses non-equilibrium phenomena included in equilibrium simulations. It then investigates how mass flow in melt induces crystallization and melting at the interfaces. Feasibility of gravitational tempering has also been argued in the review. The review shows that a wide range of gravitational conditions can even be observed in normal gravity. Gravitational condition can be varied in situ through centrifugation. Difficulty in in situ observation of the defect behavior has also been pointed out.

Acknowledgments: The author thanks the members of 2014 JAXA/ISAS WG "Studies Using Model Colloidal Aggregation and Crystallization under Controlled Gravity Field" for a fruitful discussion on the feasibility study described in Section 5.2. It is gratefully acknowledged that, in addition to the editors, the persons in charge of Editage company, to whom the author has ordered the English editing, understood him as a person with an autoimmune disease of Adult Still Disease (connective tissue disorder), having problems with joints and, in his present case, eyestrain (therefore, he has problems in typing and reading tiny characters).

Conflicts of Interest: The author declares no conflict of interest.

References

1.   Wood, W.W.; Jacobson, J.D. Preliminary Results from a Recalculation of the Monte Carlo Equation of State of Hard Spheres. *J. Chem. Phys.* **1957**, *27*, 1207–1208.

2. Alder, B.J.; Wainwright, T.E. Studies in Molecular Dynamics. II. Behavior of s Small Number of Elastic Spheres. *J. Chem. Phys.* **1957**, *27*, 1208–1209.

3. Alder, B.J.; Wainwright, T.E. Phase Transition for a Hard Sphere System. *J. Chem. Phys.* **1960**, *33*, 1439–1451.

4. Percus, J.K. (Ed.) *The Many Body Problem*; Interscience: New York, NY, USA, 1968; Chaps. XXII and XXVIII.

5. Kirkwood, J.G. Statistical Mechanics of Fluid Mixtures. *J. Chem. Phys.* **1935**, *3*, 300–313.

6. Onsager, L. The Effect of Shape on the Interaction of Colloidal Particles. *Ann. NY. Acad. Sci.* **1949**, *51*, 627–659.

7. Kimura, H. Statistical Theory of Liquid Crystalline Orderings in Hard Rod Fluids—Nematic, Choresteric, Smectic and Columnar Phases. In *Ordering in Macromolecular Systems*; Springer-Verlag: Berlin, Germany, 1994; pp.125–138.

8. Hoover, W.G.; Ree, F.H. Melting Transition and Communal Entropy for Hard Spheres. *J. Chem. Phys.* **1968**, *49*, 3609–3617.

9. Davidchack, R.L.; Laird, B.B. Simulation of the Hard-sphere Crystal-melt Interface. *J. Chem. Phys.* **1998**, *108*, 9452–9462.

10. Zykova-Timan, Z.; Horbach, J.; Binder, K. Monte Carlo Simulations of the Solid-liquid Transition in Hard Spheres and Colloid-polymer Mixtures. *J. Chem. Phys.* **2010**, *133*, 014705.

11. Fernández, L.A.; Martín-Mayor, V.; Seoane, B.; Verrocchio, P. Equilibrium Fluid-Solid Coexistence of Hard Spheres. *Phys. Rev. Lett.* **2012**, *1–8*, 165701.

12. Jackson, K.A. Mechanism of Growth. In *Liquid Metals and Solidification*; American Society of Metals: Cleveland, OH, USA, 1958; pp. 174–186.

13. Woodruff, D.P. *The Solid-Liquid Interface*; Cambridge University Press: London, UK, 1973.

14. Alfrey, T., Jr.; Braford, E.B.; Vanderhoff, J.W. Optical Properties of Uniform Particle-Size Latexes. *J. Opt. Soc. Am.* **1954**, *44*, 603–609.

15. Lux, V.W.; Klier, M.; Wesslau, H. Über Bragg-Reflexe mit Sichtbarem Licht an Monodispersen Kunststofflatices. I. *Ber. Bunsenges. Phys. Chem.* **1963**, *67*, 75–83.

16. Derjaguin, B; Landau, L. Theory of the Stability of Strongly Charged Lyophobic Sols and of the Adhesion of Strongly Charged Particles in Solutions of Electrolytes. *Acta Phys. Chim.* **1941**, *14*, 633–662.

17. Verway, E.J.W.; Overbeek, J.T.G. *Theory of the Stability of Lyophobic Colloids*; Elsevier: Amsterdam, The Netherlands, 1984.

18. Watachi, M.; Toda, M. An Evidence for the Existence of Kirkood-Alder Transition. *J. Phys. Soc. Jpn.* **1972**, *32*, 1147.

19. Anti, L.; Goodwin, J.W.; Hill, R.D.; Ottewill, R.H.; Owens, S.M.; Papworth, S. The Preparation of Poly(methyl methacrylate) Lattices in Non-aqueous Media. *Coll. Surf.* **1986**, *17*, 67–78.

20. Pusey, P.N.; van Megen, W. Phase Behavior of Concentrated Suspensions of Nearly Hard Colloidal Spheres. *Nature* **1986**, *320*, 340–342.

21. Underwood, S.M.; Taylor, J.R.; van Megen, W. Sterically Stabilized Colloidal Particles as Model Hard Spheres. *Langmuir* **1994**, *10*, 3550–3554.

22. van Blaaderen, A.; Ruel, R.; Wiltzius, P. Template-directed Colloidal Crystallization. *Nature* **1997**, *385*, 321–324.

23. Zhu, J.; Li, M.; Rogers, R.; Mayer, W.; Ottewill, R.T. STS-73 Space Shuttle Crew; Russel, W.B.; Chaikin, P.M. Crystallization of Hard-sphere Colloids in Microgravity. *Nature* **1997**, *387*, 883–885.

24. Kegel, W.K.; Dhont, K.G. "Aging" of the Structure of Hard Colloidal Spheres. *J. Chem. Phys.* **2000**, *112*, 3431–3436.

25. Dolbnya, I.P.; Petukhov, A.V.; Aarts, D.G.A.L.; Vroege, G.J.; Lekkerkerker, H.N.W. Coexistence of RHCP and FCC Phases in Hard-Sphere Colloidal Crystals. *Europhys. Lett.* **2005**, *72*, 962–968.

26. Hilhorst, J.; de Winter, D.A.M.; Wolters, J.R.; Post, J.A.; Petukov, A.V. Defect Engineering in Sedimentary Colloidal Photonic Crystals. *Langmuir* **2013**, *29*, 10011–10018.

27. Ohtaka, K. Energy Band of Photons and Low-energy Photon Diffraction. *Phys. Rev. B* **1979**, *19*, 5057–5067.

28. Yablonovitch, E. Inhibited Spontaneous Emission in Solid-State Physics and Electronics. *Phys. Rev. Lett.* **1987**, *58*, 2059–2062.

29. John, S. Strong Localization of Photons in Certain Disordered Dielectric Superlattices. *Phys. Rev. Lett.* **1987**, *58*, 2486–2489.

30. Frenkel, D.; Ladd, A.J.C. New Monte Carlo Method to Compute the Free Energy of Arbitrary Solids. Application to the fcc and hcp Phases of Hard Spheres. *J. Chem. Phys.* **1984**, *81*, 3188–3198.
31. Woodcock, L.V. Entropy Difference Between the Face-centered Cubic and Hexagonal Close-packed Crystal Structures. *Nature* **1997**, *385*, 141–143.
32. Bruce, A.D.; Wilding, N.B.; Ackerson, G.A. Free Energy of Crystalline Solids: A Lattice-Switch Monte Carlo Method. *Phys. Rev. Lett.* **1997**, *79*, 3002–3005.
33. Pronk, S.; Frenkel, D. Can Stacking Faults in Hard-sphere Crystals Anneal out Spontaneously? *J. Chem. Phys.* **1999**, *110*, 4589–4592.
34. Mau, S.-C.; Huse, D.A. Stacking Entropy of Hard-sphere Crystals. *Phys. Rev. E* **1999**, *59*, 4396–4401.
35. Elser, V. Phonon Contribution to the Entropy of Hard-sphere Crystals. *Phys. Rev. E* **2014**, *89*, 052404.
36. Alder, B.J.; Wainwright, T.E. Studies in Molecular Dynamics. I. General Method. *J. Chem. Phys.* **1959**. *31*, 459–466.
37. Rapaport, D.C. *The Art of Molecular Dynamics Simulation*, 2nd ed.; Cambridge University Press: Cambridge, UK, 1997.
38. Quentrec, B.; Brot, C. New Method for Searching for Neighbors in Molecular Dynamics Calculations. *J. Comp. Phys.* **1975**, *13*, 430–432.
39. Rapaport, D.C. The Event Scheduling Problem in Molecular Dynamic Simulation. *J. Comp. Phys.* **1980**, *34*, 184–201.
40. Isobe, M. Simple and Efficient Algorithm for Large Scale Molecular Dynamics Simulation in Hard Disk System. *Int. J. Mod. Phys. C* **1999**, *10*, 1281–2183.
41. Bannerman, M.N.; Sargant, R.; Lue, L. DynamO: A Free $O(N)$ General Event-Driven Molecular Dynamics Simulator. *J. Comp. Chem.* **2011**, *32*, 3329.
42. Mori, A.; Manabe, R.; Nishioka, K. Construction and Investigation of a Hard-sphere Crystal-melt Interface by a Molecular Dynamics Simulation. *Phys. Rev. E* **1995**, *51*, R3831–R3833.
43. Kyrlidis, A.; Brown, R. Density-functional Theory and Atomistic Simulation of the Hard-sphere Melt-solid Interface. *Phys. Rev. E* **1995**, *51*, 5832–5845.
44. Vega, C.; Sanz, E.; Abascal, J.L.F.; Noya, E.G. Determination of Phase Diagrams via Computer Simulation: Methodology and Applications to Water, Electrolytes and Proteins. *J. Phys. Condens. Matter* **2008**, *20*, 153101.
45. Cahn, J.W. Theory of Crystal Growth and Interface Motion in Crystalline Materials. *Acta Metal.* **1960**, *8*, 554–562.
46. Cape, J.N.; Woodcock, L.V. Molecular Dynamics Calculation of Phase Coexistence Properties: The Soft-Sphere Melting Transition. *Chem. Phys. Lett.* **1978**, *59*, 271–274.
47. Ladd, A.J.C.; Woodcock, L.V. Structure of the Lennard-Jones (100) Crystal-liquid Interface. *J. Phys. C Solid State Phys.* **1978**, *11*, 3565–3576.
48. Hiwatari, Y.; Stoll, E.; Schneider, T. Molecular-dynamics Investigation of Solid-liquid Coexistence. *J. Chem. Phys.* **1978**, *68*, 3401–3404.
49. Broughton, J.Q.; Abraham, F.F. A Comparison of the fcc (111) and (100) Crystal-Melt Interface by Molecular Dynamics Simulation. *Chem. Phys. Lett.* **1980**, *71*, 456–459.
50. Cape, J.N.; Woodcock, L.V. Soft-sphere Model for the Crystal-liquid Interface: A Molecular Dynamics Simulation of the Surface Stress. *J. Chem. Phys.* **1980**, *73*, 2420–2429.
51. Ueda, A.; Takada, J.; Hiwatari, Y. Molecular Dynamics Studies on Solid-Liquid Interface of Soft-Core Model. *J. Phys. Soc. Jpn.* **1981**, *59*, 307–314.
52. Broughton, J.Q.; Bonissent, A.; Abraham, F.F. The fcc (111) and (100) Crystal-Melt Interfaces: A Comparison by Molecular Dynamics Simulation. *J. Chem. Phys.* **1981**, *74*, 4029–4039.
53. Broughton, J.Q.; Gilmer, G.H. Molecular Dynamics of the Crystal-fluid Interface. V. Structure and Dynamics of Crystal-melt Systems. *J. Chem. Phys.* **1986**, *84*, 5749–5758.
54. Temkin, D.E. Molecular Roughness of the Crystal-Melt Boundary.   In *Crystallization Process*; Consultant Bureau: New York, NY, USA, 1996; pp. 15–23.
55. Chernov, A.A. *Modern Crystallography III*; Springer-Verlag: Berlin, Germany, 1984.
56. Vere, A.W. *Crystal Growth*; Plenum: New York, NY, USA, 1987.
57. Markov, I.V. *Crystal Growth for Beginners*; World Scientific: Singapore, 1995.
58. Homma, S.; Yoshida, U.; Nakano, H. Theory of Solid-Liquid Interface. *J. Phys. Soc. Jpn.* **1981**, *50*, 2175–2179.

59. Ishibashi, Y. On the Activation Energy of Solid-Melt Interface in the Temkin Model. *J. Phys. Soc. Jpn.* **1986**, *55*, 2099–2101.

60. Mori, A.; Maksmov, I.L. On the Temkin Model of Solid-Liquid Interface. *J. Cryst. Growth* **1999**, *200*, 297–304.

61. Mori, A.; Suzuki, Y. Grand Potential Formalism of Interfacial Thermodynamics for Critical Nucleus. *Natl. Sci.* **2013**, *5*, 631–639.

62. Davidchack, R.L.; Laird, B.B. Direct Calculation of the Hard-Sphere Crystal/Melt Interfacial Free Energy. *Phys. Rev. Lett.* **2000**, *85*, 4751–4754.

63. Broughton, J.Q.; Gilmer, G.H. Molecular Dynamics of the Crystal-fluid Interface. VI. Excess Surface Free Energies of Crystal-melt Systems. *J. Chem. Phys.* **1986**, *84*, 5759–5768.

64. Laird, B.B.; Davidchack, R.L. Direct Calculation of the Crystal-Melt Interfacial Free Energy via Molecular Dynamics Computer Simulation. *J. Phys. Chem. B* **2005**, *109*, 17802–17812.

65. Davidchack, R.L. Hard Sphere Revisited: Accurate Calculation of the Solid-liquid Interfacial Free Energy. *J. Chem. Phys.* **2010**, *133*, 23470.

66. Härtel, A.; Oettel, M.; Rozas, R.E.; Egelhaaf, S.U.; Horbach, J.; Löwen, H. Tension and Stiffness of the Hard Sphere Crystal-Fluid Interface. *Phys. Rev. Lett.* **2012**, *108*, 226101.

67. Davis, K.E.; Russel, W.B.; Glantschnig, W.J. Disorder-to-Order Transition in Settling Suspensions of Colloidal Silica: X-Ray Measurements. *Science* **1989**, *245*, 507–510.

68. Pusey, P.N.; van Megen, W.; Bartlett, P.; Ackerson, B.J.; Rarity, J.G.; Underwood, S.M. Structure of Crystals of Hard Colloidal Spheres. *Phys. Rev. Lett.* **1989**, *63*, 2753–2756.

69. Martelozzo, V.C.; Schofield, A. B; Poon, W.C.K.; Pusey P.N. Structural Aging of Crystals of Hard-sphere Colloids. *Phys. Rev. E* **2002**, *66*, 021408.

70. Biben, T.; Ohnesorge, R.; Löwen, H. Crystallization in Sedimentation Profiles of Hard Spheres. *Europhys. Lett.* **1994**, *28*, 665–670.

71. Volkov, I.; Cieplak, M.; Koplik, K.; Banavar, J.R. Molecular Dynamics Simulations of Crystallization of Hard Spheres. *Phys. Rev. E* **2002**, *66*, 061404.

72. Yanagiya, S.-I.; Mori, A.; Suzuki, Y.; Miyoshi, Y.; Kasuga, M.; Sawada, T.; Ito, K.; Inoue, T. Enhancement of Crystallization of Hard Spheres by Gravity: Monte Carlo Simulation. *Jpn. J. Appl. Phys.* **2005**, *44*, 5113–5116.

73. Mori, A.; Yanagiya, S.-I.; Suzuki, Y.; Sawada, T.; Ito, K. Monte Carlo Simulation of Crystal-fluid Coexistence States in the Hard-Sphere System under Gravity with Stepwise Control. *J. Chem. Phys.* **2006**, *124*, 174507.

74. Mori, A.; Suzuki, Y.; Yanagiya, S.-I.; Sawada, T.; Ito, K. Shrinking Stacking Fault through Glide of the Shockley Partial Dislocation in Hard-sphere Crystal under Gravity. *Mol. Phys.* **2007**, *105*, 1377–1383.

75. Mori, A.; Suzuki, Y.; Matsuo, S. Disappearance of a Stacking Fault in Hard-Sphere Crystals under Gravity. *Prog. Theor. Phys. Suppl.* **2009**, *178*, 33–40.

76. Mori, A.; Suzuki, Y. Interplay between Elastic Fields due to Gravity and a Partial Dislocation for a Hard-sphere Crystal Coherently Grown under Gravity: Driving Force for Defect Disappearance. *Mol. Phys.* **2010**, *108*, 1731–1738.

77. Mori, A.; Yanagiya, S.-I.; Suzuki, Y.; Sawada, T.; Ito, K. Crystal Structure of Hard Spheres under Gravity by Monte Carlo Simulation. *Sci. Technol. Adv. Mater.* **2006**, *7*, 296–302.

78. Marechal M.; Dijkstra, M. Crystallization of Colloidal Hard Spheres under Gravity. *Phys. Rev. E* **2007**, *75*, 061404.

79. Ramsteiner, I.B.; Jensen, K.E.; Weitz, D.A.; Spaepen, F. Experimental Observation of the Crystallization of Hard-sphere Colloidal Particles by Sedimentation onto Flat and Patterned Surfaces. *Phys. Rev. E* **2009**, *79*, 011403.

80. Marechal M.; Hermes, M.; Dijkstra, M. Stacking in Sediments of Colloidal Hard Spheres. *J. Chem. Phys.* **2011**, *135*, 034510.

81. Heni, M.; Löwen, H. Interfacial Free Energy of Hard-Sphere Fluids and Solids near a Hard Wall. *Phys. Rev. E* **1999**, *93*, 7057–7065.

82. Laird, B.B.; Davidchack, R.L. Wall-Induced Prefreezing in Hard Spheres: A Thermodynamic Perspective. *J. Phys. Chem. C* **2007**, *111*, 15952–15956.

83. Laird, B.B.; Davidchack, R.L. Calculation of the Interfacial Free Energy of a Fluid at a Static wall by Gibbs–Cahn Integration. *J. Chem. Phys.* **2010**, *132*, 204101.

84. Courtemanche, D.J.; van Swol, F. Wetting State of a Crystal-Fluid System of Hard Spheres. *Phys. Rev. Lett.* **1992**, *69*, 2078–2081.

85. Auer, S.; Frenkel, D. Line Tension Controls Wall-Induced Crystal Nucleation in Hard-Sphere Colloids. *Phys. Rev. Lett.* **2003**, *91*, 015703-1.

86. Lin, K.-H.; Crocker, J.C.; Prasad, V.; Schofield, A.; Weitz, D.A.; Lubensky, T.C.; Yodh, A.G. Entropically Driven Colloidal Crystallization on Patterned Surfaces. *Phys. Rev. Lett.* **2000**, *85*, 1770–1773.

87. Yi, D.K.; Seo, E.-M.; Kim, D.-Y. Surface-modulation-Controlled Three-dimensional Colloidal Crystals. *Appl. Phys. Lett.* **2002**, *80*, 225–227.

88. Zhang, J.; Alsayed, A.; Lin, K.H.; Sanyal, S.; Zhang, F.; Pao, W.-J.; Balagurusamy, V.S.K.; Heiney, P.A.; Yodh, A.G. Template-directed Convective Assembly of Three-dimensional Face-centered Cubic Colloidal Crystals. *Appl. Phys. Lett.* **2002**, *81*, 3176–3178.

89. Stenger, N.; Rehspringer, J.-L.; Hirlimann. C. Template-directed Self-organized Silica Beads on Square and Penrose-like Patterns. *J. Lumin.* **2006**, *121*, 278–281.

90. Heni, M.; Löwen, H. Surface Freezing on Patterned Substrates. *Phys. Rev. Lett.* **2000**, *85*, 3668–3671.

91. Heni, M.; Löwen, H. Precrystallization of Fluids Induced by Patterned Substrates. *J. Phys. Condens. Matter* **2001**, *13*, 4675–4696.

92. Cacciuto, A.; Frenkel, D. Simulation of Colloidal Crystallization on Finite Structured Templates. *Phys. Rev. E* **2005**, *72*, 041604.

93. Xu, W.-S.; Sun, Z.-Y.; An, L.-J. Heterogeneous Crystallization of Hard Spheres on Patterned Substrates. *J. Chem. Phys.* **2010**, *132*, 144506.

94. Mori, A. Disappearance of Stacking Fault in Colloidal Crystals under Gravity. *World J. Eng.* **2011**, *8*, 117–122.

95. Mori, A. Monte Carlo Simulation of Growth of Hard-sphere Crystals on a Square Pattern. *J. Cryst. Growth* **2011**, *318*, 66–71.

96. Mori, A.; Suzuki, Y.; Matsuo, S. Possibility of Gravitational Tempering in Colloidal Epitaxy to Obtain a Perfect Crystal. *Chem. Lett.* **2012**, *41*, 1069–1071.

97. Mori, A.; Suzuki, Y. Identification of Triangular Shaped Defect Often Appeared in Hard-sphere Crystals Grown on a Square Pattern under Gravity by Monte Carlo Simulations. *Physica B* **2014**, *452*, 58–65.

98. Jensen, K.E.; Pennachio, D.; Recht, D.; Weitz, D.A.; Spaepen, F. Rapid Growth of Large, Defect-free Colloidal Crystals. *Soft Matter* **2013**, *9*, 320–328.

99. Russo, J.; Maggs, A.C.; Bonn, D.; Tanaka, H. The Interplay of Sedimentation and Crystallization in Hard-sphere Suspensions. *Soft Matter* **2013**, *9*, 3769–3783.

100. Nakamura, S.; Pearton, S.; Fasol, G. *The Blue Laser Diode: The Complete Story*; Springer: Berlin, Germany, 2000.

101. Mori, A.; Suzuki, Y.; Matsuo, S. Monte Carlo Simulation of Defects in Hard-sphere Crystal Grown on a Square Pattern. *World J. Eng.* **2012**, *9*, 37–44.

102. Mori, A.; Suzuki, Y.; Sato, M. Gravitational Tempering in Colloidal Epitaxy To Reduce Defects Further. *Cryst. Growth Des.* **2014**, *14*, 2086–2088.

103. Schall, P.; Cohen, I.; Weitz, D.A.; Spaepen, F. Visualization of Dislocation Dynamics in Colloidal Crystals. *Science* **2004**, *305*, 1944–1948.

104. Maire, E.; Redston, E.; Gulda, M.P.; Weitz, D.A.; Spaepen, F. Imaging Grain Boundary Grooves in Hard-sphere Colloidal Bicrystals. *Phys. Rev. E* **2016**, *94*, 042604.

105. Mori, A. Effect of Mass Flow in Melt on the Motion of Crystal-Melt Interface of Hard Spheres: A Molecular Dynamics Study. *J. Phys. Soc. Jpn.* **1997**, *66*, 1579–1582.

106. Mori, A. Hydrothermodynamic Consideration on the Steady-state Motion of a Solid/liquid Interface. *J. Chem. Phys.* **1999**, *110*, 8679–8686.

107. Mori, A. Steady-State Interface Motion: Formulation Extended to the Isothermal Case. *J. Phys. Soc. Jpn.* **1999**, *68*, 876–880.

108. Ackerson, B.J.; Paulin, S.E.; Johnson, B. Crystallization by Settling in Suspensions of Hard Spheres. *Phys. Rev. E* **1999**, *59*, 6903–6913.

109. Hoogenboom, J.P.; Derks, D.; Vergeer, P.; van Blaaderena, A. Stacking Faults in Colloidal Crystals Grown by Sedimentation. *J. Chem. Phys.* **2002**, *117*, 11320–11328.

110. Royall, C.P.; Dzubiella, J.; Schmidt, M.; van Blaaderen, A. Nonequilibrium Sedimentation of Colloids on the Particle Scale. *Phys. Rev. Lett.* **2007**, *98*, 188304.

111. Hoogenboom, J.P.; Vergeer, P.; van Blaaderena, A. A Real-space Analysis of Colloidal Crystallization in a Gravitational Field at a Flat Bottom Wall. *J. Chem. Phys.* **2003**, *119*, 3371–3383.

112. Megens, M.; van Kats, C.M.; Bösecke, P.; Vos, W.L. Synchrotron Small-Angle X-ray Scattering of Colloids and Photonic Colloidal Crystals. *J. Appl. Crystallogr.* **1997**, *30*, 637–641.

113. Suzuki, Y.; Sawada, T.; Tamura, K. Colloidal Crystallization by a Centrifugation Method. *J. Cryst. Growth* **2011**, *318*, 780–783.

114. Suzuki, Y.; Endoh, J.; Mori, A.; Yabutani, T.; Tamura, K. Gravitational Annealing of Colloidal Crystals. *Defect Diffus. Forum* **2012**, *232–235*, 555–558.

© 2017 by the author. Licensee MDPI, Basel, Switzerland. This article is an open access article distributed under the terms and conditions of the Creative Commons Attribution (CC BY) license (http://creativecommons.org/licenses/by/4.0/).

crystals

MDPI

Article

In Silico Prediction of Growth and Dissolution Rates for Organic Molecular Crystals: A Multiscale Approach

Ekaterina Elts [1,*], Maximilian Greiner [2] and Heiko Briesen [1]

[1] Chair for Process System Engineering, Technical University of Munich, 85354 Freising, Germany; heiko.briesen@mytum.de
[2] Barry Callebaut Belgium N.V., 9280 Lebbeke-Wieze, Belgium; maximilian_greiner@barry-callebaut.com
*   Correspondence: ekaterinaelts@mytum.de; Tel.: +49-8161-71-3727

Academic Editor: Hiroki Nada
Received: 27 June 2017; Accepted: 18 September 2017; Published: 25 September 2017

Abstract: Solution crystallization and dissolution are of fundamental importance to science and industry alike and are key processes in the production of many pharmaceutical products, special chemicals, and so forth. The ability to predict crystal growth and dissolution rates from theory and simulation alone would be of a great benefit to science and industry but is greatly hindered by the molecular nature of the phenomenon. To study crystal growth or dissolution one needs a multiscale simulation approach, in which molecular-level behavior is used to parametrize methods capable of simulating up to the microscale and beyond, where the theoretical results would be industrially relevant and easily comparable to experimental results. Here, we review the recent progress made by our group in the elaboration of such multiscale approach for the prediction of growth and dissolution rates for organic crystals on the basis of molecular structure only and highlight the challenges and future directions of methodic development.

Keywords: molecular dynamics; kinetic Monte Carlo; continuum simulations; crystal growth; crystal dissolution; multiscale simulations

---

1. Introduction

Crystal growth and dissolution processes are an area of vital interest for pharmaceuticals, agrochemicals, organic electronics and other technologies. Owing to their significance to many different fields, those processes have been studied for over a century [1–3]. However, the prediction of crystal growth and dissolution kinetics for novel organic compounds still presents a major challenge.

The theories of crystal growth and dissolution are extensively discussed in the literature [2–5]. Both processes proceed by analogous mechanisms [6,7] and involve two main steps: (1) surface reaction and integration/disintegration of the surface species and (2) mass transfer of this species from/toward the bulk solution across the diffusion layer that surrounds the crystal [8]. Thus, the actual process of crystal growth and dissolution occurs at the molecular level, which also concerns crystal packing. Crystal structure databases, which provide crystal packing information for a huge variety of molecular structures, are to a large extent based on experimental X-ray diffraction analysis. The major drawback of this method is that it requires a flawless crystal of very high quality. Thus, the compound has to be synthesized and crystallized before information about the crystal structure can be obtained. Knowledge of the crystal packing is essential for predicting growth and dissolution properties. Consequently, to predict growth and dissolution properties from nothing but the molecular structure, simulation techniques to predict the crystal structure must be available. However, not only the nanoscale aspects are significant for understanding the crystal growth and dissolution processes. There are many important phenomena associated with crystal growth that occur on a mesoscale comprising hundreds of nanometer to tens of microns and occurring over long time scales (microseconds and longer) [9], like, e.g., the evolution of crystal surface structures due to the formation of terraces, which range in size from 0.1 to 1 microns, or step bunches, which can be as large as 100 microns [9]. Even larger length scales and longer time scales are needed to incorporate concentration effects and calculate face displacement velocities. An *in silico* prediction of macroscopic growth and dissolution properties thus necessarily requires a multiscale modeling approach that integrates all of these aspects in one unified simulation framework.

The standard simulation technique for modeling molecular-level behavior to reveal how the molecular details influence the growth and dissolution kinetics is molecular dynamics (MD). In this method, Newton's equations of motion are solved numerically for all atoms to track the time evolution of the systems and to derive the kinetic and thermodynamic properties of interest [10]. However, MD can, at most, only probe behavior on the nanometer and nanosecond scale. Thus, most investigations of growth and dissolution processes using MD are reported for relatively small and simple molecules like, e.g., urea [11–13] and glycine [14–17], or even for simple model systems, such as hard spheres and Lennard–Jones particles [18–21], whereas most organic molecules, especially those used as active pharmaceutical ingredients (APIs), form more complex crystal structures, and it is extremely challenging to capture their crystal growth using fully-atomistic simulations [22]. Replacing atomistic details with lower resolution, coarse-grained (CG) beads, in which groups of co-localized atoms are treated as a single interaction site, allows one to overcome the complexity of molecules and the long time scale associated with the crystallization [22,23]. However, the interpretation of time is problematic in CG models [24]. The time scale needs to be calibrated by directly comparing with experimental data or dynamics from atomic simulations for the system at hand [24]. Thus, mainly the usefulness of CG models to obtain relative characteristics, like, e.g., to estimate the role of additives on the crystal growth of different API molecules was demonstrated so far [22,23]. Enhanced-sampling methods such as umbrella sampling [25], metadynamics [26] and forward flux sampling [27] have emerged as useful tools for understanding the mechanisms involved in crystallization. A basic idea, common to these rare event sampling methods, is that a biased potential is added to the system either to drive it along a predefined reaction coordinate or to prevent it from repeating already explored trajectories [23]. However, the efficiency of these methods strongly depends on the accuracy of the choice of reaction coordinates and may be inefficient in the case of a large number of degrees of freedom of the molecule under consideration. Moreover, upon introduction of an extra term into the system Hamiltonian, the actual dynamics of the system is to some extent hampered [28]. Recently, Salvalaglio et al. [13] demonstrated that using well-tempered metadynamics, applied within MD, one can quantitatively estimate the ratio between growth rates and thus predict the crystal habits and their dependence on additive concentration and supersaturation. However, the authors stressed that the approach does not allow computing absolute growth rates. The Reuter group also established a method for a quick prediction of approximate dissolution rates at low undersaturation based on the combination of hyperdynamics and metadynamics approaches [29–31]. This method relies on the classic rotating spiral model of Burton, Cabrera and Frank (BCF) [32], which assumes that dissolution (growth) proceeds via rotating spirals of step edges at screw dislocations and that dissolution (incorporation) of molecular units takes place primarily at kink defects along these step edges.

Kinetic Monte Carlo (kMC) is the method of choice in mesoscale modeling of dissolution or growth. In a kMC simulation, the growth or dissolution of a crystal is approximated by involving rare and independent state transition events, like, e.g., transition of molecule from solution to a kink or step site at the crystal surface. At each step in the simulation, the next event is determined on the basis of the probability proportional to the rate for that event. The time of the next event is determined by the overall rate for the microscopic surface processes and a suitably-defined random number. If a set of relevant states is defined and the transition rate constants are known, then the time evolution of the system can be modeled. Thus, the definition of a minimal set of distinct states and the estimation of corresponding transition rates are the main challenges for kMC simulations. Many kMC studies consider states on the basis of their nearest-neighbor coordination [33,34] or next-nearest-neighbor coordination [11,35]. Alternatively, the problem of state definition can be solved by identifying the most significant factors defining site reactivity with the help of electronic structure calculations and MD simulations for selected sites on the crystal surface [36]. Different approaches based on MD [11,20,33–35], accelerated MD [31,37], ab initio MD [38] or even DFT techniques [39] have been reported to determine rate constants, though the last three imply significant computational effort making them less attractive for systems with a high number of potential transitions, as well as for systems consisting of complex molecules with a high internal degree of freedom [40]. Significant steps toward the multiscale modeling of crystallization have been presented in studies conducted by Piana and co-workers [11,33,35], who first combined MD and kMC approaches to investigate the growth of a urea crystal from solution. Their simulations successfully predicted the different crystal morphologies of urea in solutions of methanol and water. However, as this only required relative rate constants, the model is still restricted in its applicability. Process time is an important factor in predicting crystallization processes, which demands that kinetic information including absolute rate constants is correctly accounted for. Further, in their studies, Piana et al. benefited from two special properties of urea: firstly, urea dissolution and growth has been shown to be fast in experiment and simulation, and secondly, urea is a small molecule, which has no facile torsional degrees of freedom [33]. For most substances, these simplifications are not necessarily true. Moreover, the concentration effects, relevant at the macroscopic scale, are not considered in their study, hindering the direct comparison with the experimental growth rates.

On the macroscopic scale, continuum methods are gainfully applied [41–47], handling physics expressed by continuum partial differential equations. These simulations are important for understanding the crystal growth and dissolution processes in their complexity accounting for advection and diffusion processes. They, however, do not shed light on how the molecular details influence the growth and dissolution kinetics. Molecular dynamics simulations are at most used only to predict physical parameters such as diffusivity or solubility, which are then employed to compute scale continuum transport models [48]. In most cases, the macroscopic simulation relied on experimental data; thus, the derivation of predictions for novel compounds still presents a major obstacle.

Thus, the protocol for *in silico* prediction of crystal growth/dissolution rates on the basis of molecular structure should comprise and join all of the corresponding steps from the prediction of molecular packing and crystal shape to continuum simulation of growth and dissolution processes. Figure 1 presents a multiscale protocol, elaborated based on our previously published studies and findings [40,49–52].

In this paper, we review all of the aspects relevant to establishing of such a protocol and highlight the challenges arising at each individual step. The first two steps, prediction of (1) crystal structure from molecular structure and (2) crystal shape, are necessary to provide the information for all further simulations and, thus, to initialize the multiscale protocol. These steps were completely omitted in our previous studies; however, in Section 2, we give a short survey of the actual state of knowledge and techniques in this field. The third step involves MD simulations to take into account molecular-level processes and to obtain process rate constants for kMC simulations, as was initially proposed by Piana et al. [11,33,35]. In comparison to Piana et al., we consider the three-dimensional crystal model in MD simulations, enabling dissolution to be seen on MD time scales even for highly hydrophobic and poorly water-soluble pharmaceutical ingredients. This, as well as a choice of force field for MD simulations and the necessity and ways to hold constant solution concentration in MD simulations are considered in Section 3 of this paper. Our approach to properly transfer the microscopic information from MD to kMC simulations and all procedures needed for that (state identification, detection of only rare, uncorrelated transition events in MD simulations, etc.) are described in Section 4. The fourth step of the multiscale protocol represents kMC simulations to calculate crystal face displacement velocities based on the MD rates and thus at the same constant solution concentration. This step is described in Section 5 of the paper. The fifth and last step involves continuum simulations to describe concentration-dependent crystal growth/dissolution on the macroscopic scale, where the results can be directly compared with the experimental ones. The scale continuum transport model, as well as our approach to transfer the mesoscopic information from kMC simulations to the macroscopic level are described in Section 6 of the paper. To demonstrate our approach, the dissolution rates for aspirin were predicted and validated by comparison with experimental assessment of aspirin dissolution using a Jamin-type interferometer [53], as described in Section 7 of the manuscript. Aspirin was chosen as a model substance as it is a well-studied compound, where the literature provides a broad array of information like crystal structure [54], polymorphs [55,56], morphology [57], critical nucleus size [58] needed to initialize the multiscale simulation protocol, as well as experimental data on crystal dissolution [59,60] for comparison with the simulation results and validation of the different steps of our protocol. Thus, all of the simulation results presented in the paper are for molecular aspirin crystals. However, our approach introduces the flexibility to handle different organic model compounds and not restricted to aspirin. A summary of our findings and some concluding remarks, as well as the information about future directions and challenges can be found in Section 8.

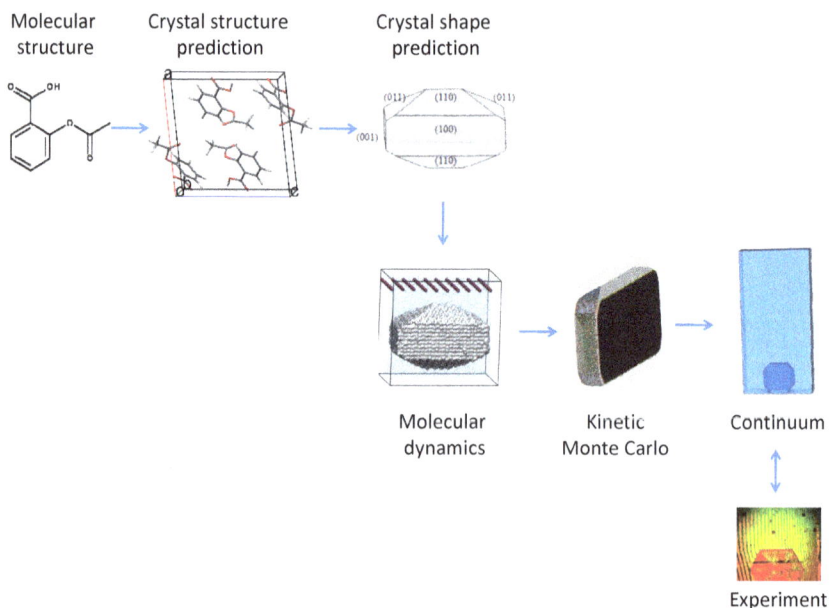

Figure 1. Protocol for *in silico* prediction of crystal growth/dissolution rates on the basis of the molecular structure only (case scenario: aspirin).

2. Crystal Structure and Shape Prediction

The prediction of crystal structure at the atomic level is one of the most fundamental challenges in condensed matter science [61]. Crystal structure prediction (CSP) approaches have been evolving rapidly in the past few decades and have now grown into an overwhelmingly vast, diversified and active field of research [62,63]. The ever increasing computer power allowed making many successful predictions of organic crystal structures [63]. The best starting point to keep track of the existing range of approaches is constituted by the accounts of the Blind Tests of CSP organized by the Cambridge Crystallographic Data Centre every few years [64]. These contests ask participants to predict the crystal structures of organic molecules starting from only the 2D molecular structure, which are then compared with the experimentally obtained, but not publicly released data to reveal the particularly effective techniques. Despite the excessive ramification of approaches developed to date, the whole CSP process can generally be subdivided into three distinct steps, each with their own methodologies and challenges: (1) conformational exploration of the molecule(s), (2) generation of candidate packing arrangements and (3) (re-)ranking of candidate structures using some form of fitness function [64]. Successful crystal engineering relies on the knowledge of molecular conformation, i.e., the overall shape of molecular building-blocks, as well as the relative arrangement of functional groups within a molecule that can participate in structure-directing interactions [65]. Molecular shape can be easily predicted for rigid molecules, but becomes more challenging as molecular flexibility is increased and the molecules of interest have a choice of conformers when self-assembling into a crystal [65]. Different conformers may lead to very different crystal packing arrangements, ultimately influencing the properties of the crystal [65]. Therefore, conformational flexibility can be seen as a principal obstacle to crystal engineering. Commercially available and industrially employed CSP approaches like the Polymorph Predictor within Accelrys' Materials Studio rely on a rigid molecule approximation and only concentrate on the evaluation of different packing arrangements. This is clearly insufficient for flexible APIs, which in the crystal frequently adopt geometries that are significantly different from the stable gas-phase conformer. This is particularly relevant for the engineering of pharmaceutical materials, where solid form properties of active pharmaceutical ingredients may be manipulated, either by selection between polymorphs, or the design of multi-component crystals, such as salts or co-crystals. Thus, there is also still no general solution to the main CSP challenge, i.e., the existence of crystal polymorphism [62,63]. Each polymorph may differ from others in physicochemical properties, such as density, solubility, bioavailability, mechanical strength, dissolution rate, to name but a few [23]. Knowledge of different polymorphs is essential to the pharmaceutical industry for choosing the crystalline form with the most efficient therapeutic performance and to exclude the crystallization of any unwanted form.

To see which surfaces are relevant to study growth and dissolution processes, not only molecular packing, but also crystal shape should be predicted *in silico*. They both are needed to initialize the multiscale growth and dissolution protocol. Crystals reveal a large variety of shapes, depending on their chemical composition and structure, as well as on the growth conditions, such as supersaturation, temperature, solvent and even impurities. For novel compounds, such external parameters will not be necessarily known, but one still needs to determine the facets that dominate the morphology from the known crystal structure for the next steps of the proposed multiscale protocol. Under conditions of extremely slow growth, the shape of a crystal is determined by thermodynamics: the crystal tends to grow to a shape of a polyhedron having minimum surface energy [66]. Such an equilibrium shape is thus obtained, according to the Gibbs thermodynamic principle, by minimizing the total surface free energy associated with the crystal-medium interface. The procedure is based on the Gibbs–Wulff theorem, also known as Wulff construction [67], and provides the equilibrium crystal shape from separately computed surface free energies of all low-index crystal facets. Critical for this shape are thereby not the absolute surface free energies, but only their relative ratios. Less stringent (and efficient) approaches may and have therefore been employed to their computation. One means to this end is to neglect (or crudely approximate) vibrational free energy contributions to the surface free energies and to use static surface energies obtained for $T = 0\,$K optimized surface structures. This reduces the computational burden to such an extent that in fact first-principles electronic structure methods like DFT may directly be employed to produce these numbers. This approach is for instance commonly followed to determine nanoparticle shapes in heterogeneous catalysis [68–70]. It has also been used for growth applications from solutions, where also any solvent influences have been neglected [71,72]. This neglect extends over both the solvent influence on the surface vibrational properties and the electrostatic effects on the surface energies due to the solvent dielectric properties. However, these various levels of approximation are generally not deeply investigated, and further efforts should be made to conduct systematic analysis of the level of theory required to reliably predict the crystal shape. Moreover, one can notice that the Gibbs condition is not normally realized in practice as crystal growth in crystallizers is significantly removed from the equilibrium state [73]. Under most conditions, the shape of a crystal is determined by kinetics rather than thermodynamics [73]. Thus, it certainly might be a limitation for the whole protocol, if crystal structure and shape predictions completely fail. Other methods are also available to deduce morphology of a crystalline material from its internal crystal structure, like, e.g., the Bravais–Friedel–Donnay–Harker (BFDH) method [74–76], which uses the crystal lattice and symmetry to generate a list of possible growth faces and their relative growth rates, or the attachment energy model [77], which takes into account the energetics of crystal interactions in addition to the crystal geometry. These methods in general provide good predictions of vapor-grown crystals or crystals grown in systems in which the solvent does not interact strongly with the solute. However, in many instances, the simulated crystal shapes differ from the experimental ones because of the kinetic effects due to supersaturation, solvent and impurities dominating the crystal growth process [78]. The more is known about crystal growth conditions, the better the crystal shape can be predicted. For that, a number of sophisticated methods exists that account for the effects of such external factors [79–83]. However, for the next step of the protocol, e.g., MD simulations, it is essential that the dominant faces and edges are presented. Their actual size and relations to the surrounding surfaces are less important, as shown in the next section.

Currently, no simple general solution to predict crystal packing and crystal shape data for compounds with conformational flexibility exists; neither are we concerned with the elaboration of crystal structure and shape prediction methods. Instead, we relied on the experimental data for aspirin [57,84] in our work [40,52]. Aspirin was modeled in its protonated state using unit cell parameters of the polymorphic form I [84]. The aspirin nanocrystal for MD simulations was built according to the experimental morphology, typical for aspirin crystallized from ethanol-water solutions [57].

3. MD Simulations

MD simulations are widely used to study crystal growth and dissolution and are of the utmost importance in unraveling the microscopic details of these processes. However, simulations are presently affected by several shortcomings, which hinder a reliable comparison with experimental growth and dissolution rates and limit growth/dissolution studies to systems and conditions often far from those investigated experimentally [28]. These weaknesses can be classified into two main categories: (1) limitations related to the accuracy of the computational model used to represent the system and (2) shortcomings due to the finite-size effects. In this section, we describe our attempts to resolve these issues and to obtain reliable results in MD simulations: from the choice of the parameter set to describe intermolecular and intramolecular interactions and simulations of nanocrystal in its experimentally determined shape to avoiding finite-size effects during crystal growth and dissolution simulations.

3.1. Choosing the Force Field

To successfully simulate crystal growth and dissolution, i.e., to obtain the results corresponding to the experimental ones quantitatively or at least qualitatively, the molecular interactions in both crystal and solution environments have to be accurately described. Thus, the choice of force field with a suitable combination of a mathematical formula and associated parameters that are used to describe molecular energy plays a pivotal role. In the literature, a broad variety of force field parameter sets can be found. The difficulty in choosing a suitable force field mainly arises from the strict focus of most common force fields on particular properties. Moreover, some of these are designed to be compound specific; thus, their applicability to novel compounds is limited. For screening purposes those force fields are favored for which software packages and online resources are provided to facilitate the generation of force field parameter files for common simulation packages. Thus, the most general approach to find the parameters for novel compounds like, e.g., small-molecule drugs is to use common biomolecular force fields designed for protein interactions and ligand-docking simulations. Here, small-molecule drugs of varying configuration and containing various functional groups are included in the force fields parameter sets. Examples for such force fields are, besides others, the CHARMM [85] (Chemistry at HARvard Molecular Mechanics) force field, the AMBER [86] (Assisted Model Building and Energy Refinement) force field or, in an extended form, the Generalized AMBER Force Field [87] (GAFF), and the OPLS [88] (Optimized Potential for Liquid Simulations) force field. These force fields are designed to adequately describe protein interactions, where the screening of multiple small-molecule compounds is an important field of interest. For that, force field generating tools have been developed, e.g., the SWISS PARAM server [89] for the CHARMM-compatible force field based on the Merck Molecular Force Field (MMFF) [90], the ACPYPE (AnteChamber PYthon Parser interfacE) [91] wrapper script around the ANTECHAMBER software [92] for the GAFF or the PRODRG server [93] for the GROMOS [94] (GROningen MOlecular Simulation package) force field. The latter, however, has been shown to produce sub-optimal results [95], especially when it comes to calculating the charges. Additionally, force fields can be obtained directly from software packages in commercial software, such as Maestro from the Schrödinger software package creating OPLS force field parameters.

Choosing an appropriate force field remains a challenge and is crucial to the results obtained. For example, studying the crystal growth of glycine from an aqueous solution, Banerjee and Briesen [14] monitored dissolution, although growth was expected for the applied supersaturated conditions. Cheong and Boon [16] compared different force fields and charges for the simulation of glycine crystal growth and found the heat of solution to be an important criterion when choosing the force field for crystal growth simulation. Heat of solution is the change in the enthalpy when 1 mol of a substance is dissolved in a solvent, thus an important quantity for crystallization, as it incorporates both the crystal and the solution phase and quantifies the tendency for crystallization and dissolution. A positive value indicates that the crystal dissolution is an endothermic process. The crystalline state is then favored over the dissolved state, and one can also expect that crystal growth would be obtainable with a suitable supersaturation.

In addition to the heat of solution, we considered several other criteria to make a proper choice in force field selection for crystal growth and dissolution simulations [49]. Structural, thermodynamic and interface-specific parameters were evaluated for several API ingredients (aspirin, ibuprofen, paracetamol). Apart from the importance of the heat of solution for the evaluation of the solid-to-solution phase transitions, it was concluded that the interaction energies might give valuable information about the choice of the force field whenever a dominant interaction within the crystal was present, such as hydrogen bond pairs between carboxyl groups in the case of aspirin and ibuprofen. Moreover, the lattice parameters of the unit cell after its relaxation with the corresponding force field (further referred to as a *relaxed* unit cell) have been shown to provide reliable information, which is important for the stability of the crystal structure. The distance threshold and the orientation tolerance parameters, characterizing maximal deviations from the reference crystal molecular positions and orientations, have also proven to be appropriate indicators of a correct representation of the crystal-water interfaces by the force fields [49]. The consideration of all of these aspects led us to the choice of the Merck molecular force field for the simulation of aspirin crystal dissolution in our further works [40,50].

3.2. Superiority of Three-Dimensional over Two-Dimensional Dissolution Simulations

Most investigations of crystallization processes using MD consider the distinct faces of the crystal in contact with the solvent [11,12,14–16,33,35,96–98]. However, time scales accessible in regular MD simulation are often not sufficient to resolve dissolution from the perfectly flat interfaces. Recently published experimental and MD simulation data on the dissolution of paracetamol highlighted the significance of the so-called "corner and edge effect" [99], indicating that in particular corners and edges between facets serve as the initial sites for dissolution and may be accessed by MD.

The crystal representation used in the paracetamol study [99], however, was in no way related to the true experimental morphology, and due to its limited size, the employed crystal was not stable over the whole simulation time. This prevents a direct comparison to experiment or use as a predictive-quality protocol. We overcame these limitations with MD simulations of an entire aspirin nanocrystal in its experimentally determined shape revealing the (100), (001), (011) and (110) faces, cf. Figure 2a [40,50]. It was cut from a pre-equilibrated supercell (310 K and ambient pressure of 1 bar) using the VMD [100] program. The size of the nanocrystal was around $10 \times 18 \times 14$ nm^3 [40] and thus exceeded the size of the critical diameter for stable nuclei [58], enabling one to properly sample dissolution while the crystal bulk remained in the same configuration as obtained from the relaxed unit cell. The system consisted of a total number of 9407 aspirin molecules surrounded with 177589 TIP3P water molecules [40]. The dissolution was simulated over 280 ns using version 4.6 of the GROMACS package and took about ten days on 1024 cores of SuperMUC System of Leibniz Supercomputing Centre [40]. Pronounced differences were observed in the face-specific dissolution behavior. A dissolution mechanism via receding edges was found for the (001) plane, which is in good agreement with experimental results [40,50]. However, while the proposed dissolution mechanism for the (100) plane is terrace sinking on a rough surface, no pronounced dissolution of the perfectly flat face was seen. The most obvious reason why there is almost no dissolution of the (100) aspirin crystal face in the MD simulations is that simulation times were too short. While this is necessarily true, another reason for the stability of the (100) plane might be the strong deviation of the initial surface structure used in the simulations from experimentally observed ones. Danesh et al. [60] have shown that the (100) faces are very rough, which is in good agreement with etching patterns obtained by Wen et al. [101] for pure water as an etching medium. The roughness of the (100) crystal face is expected to be of major importance for crystal dissolution. Unfortunately, the feasible simulation cell size does not allow for the construction of proper, rough surfaces for MD simulations.

Thus, one can notice that the principal limitation to the range of applicability of the given multiscale approach is determined by the rate of the molecular crystallization and dissolution steps, as well as size limitations in MD. In order to calculate a reliable rate for a transition event, this event should occur at least a few times during the MD simulation [35]. A too small number of events of a certain type would lead to zero or almost zero rates. As the MD rates are input rates for the kMC algorithm, they have a strong influence on the end results primarily affecting a given slow growing/dissolving surface, as well as the crystal habit, determined by the relative growth/dissolution rates. The problem is that the typical times currently accessible to a single MD simulation are 10^{-7} s, which might prevent the observation of some of the slowest steps. This also prevents the observation of the formation of new faces and edges, which are not present in the experimental morphology, but could be expected to evolve during the kMC simulations, like, e.g., the development of a new (010) face due to easy detachment of molecules with a low number of neighbors from the tip formed by the (110) faces. However, as the basic assumption of the given multiscale approach is that individual molecular steps are independent of each other, it is possible to combine the results of multiple independent MD simulations to calculate the transition rates. Thus, to obtain all rates for kMC simulations, three different starting configurations were considered for aspirin nanocrystal, cf. Figure 2: (a) crystal in its experimental shape, (b) crystal obtained by cutting the tip formed by the (110) faces and, thus, with the (010) face exposed and (c) block structure revealing only the (100), (010) and (001) faces to obtain rates for (010)/(100) and (010)/(001) edges [40]. Thereby, we also were able to sample dissolution events for the slowest (100) face. We assumed that it is the orientation of the hydrogen bonds that plays an important role for stabilization of the (100) face. As the hydrogen bonds are directed perpendicular to the (100) plane, the molecules have strong interactions with the underlying layer of aspirin molecules. Thus, to increase the sampling of dissolution events, alternative terminations were applied to this slowly dissolving face in the starting configuration (b), cf. Figure 2b: from the two opposite (100) faces, one face was constructed such that the surface layer had hydrogen-bonding partners, whereas on the opposite side, the surface layer was cut such that no hydrogen-bonding partners were present.

Additionally, another approach was applied to improve the sampling of dissolution events for the (100) faces in this structure: voids were introduced to reduce the number of nearest neighbors for the molecules on the surface, thus reducing their stability in the crystal lattice, cf. Figure 2b. While the second approach led to no significant differences in the dissolution properties on the (100) face, cutting the crystal so that the hydrogen-bonding partners were removed allowed us to register much more transition events [40].

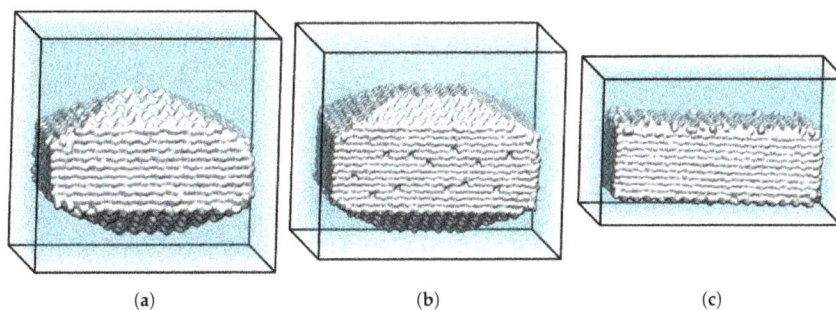

(a)   (b)   (c)

Figure 2. Representations of aspirin crystal morphology in MD simulations: (**a**) Starting configuration of the nanocrystal in its experimental shape, (**b**) starting configuration with the (010) face and (**c**) block structure revealing only the (100), (010), and (001) faces.

3.3. Constant Chemical Potential

With the presently available computational resources, classical MD calculations can typically study systems of size up to 10^5–10^6 atoms. However, such size limitations are particularly dramatic in the simulation of phase transformations, such as crystal growth from solution or dissolution of crystals [13,102–105]. While for a macroscopic system, the solutions' chemical potential does not change in the time scale accessible by MD simulation, the standard MD simulations cannot guarantee this. During the growth or dissolution, the concentration of the solute changes, resulting in a change of chemical potential eventually affecting the process itself. In MD, this should be handled by finite-size corrections. Recently, some works appeared, which discuss finite-size effects and methods to expand the information from MD simulations of phase transformations towards the limit of a macroscopic system [13,102–105]. However, most of them consider only nucleation processes, where the correction term is introduced based on the classical nucleation theory and modified liquid drop model for describing nucleation in a finite closed model [102], sharing many approximations and shortcomings. Thus, the direct application of this method to the crystal growth and dissolution problem is not trivial. Grand-canonical simulations could in principle eliminate finite-size effects by imposing a constant chemical potential in the solution phase. This is achieved by means of trial insertion and deletion of molecules. However, low acceptance rates render such approaches computationally infeasible [106]. Furthermore, insertions of particles near the growing crystal or insertion of fractional particles may lead to unphysical effects, which may ultimately obscure the true growth mechanisms and rates [106]. An approach to allow for simulations under constant chemical potential was proposed by Perego et al. [105]. Their CμMD method allows to maintain a region containing the growing/dissolving crystal and its immediate surrounding at constant solution concentration, while the remainder of the simulation box acts as a molecular reservoir. This is achieved by implementing an external force, which regulates the passage of molecules from and to the control region. Consequently the time scale accessible for this method is limited by the amount of solute molecules in the molecular reservoir. The method was successfully applied to study planar crystal-solution interfaces. However, significant changes are needed to extend the method to other geometries [105]. Moreover, the parametrization of the CμMD method is not trivial and needs a preliminary tuning and testing to provide an effective decoupling between the reservoir region and the growing crystal [105].

A simple approach to simulate crystal dissolution at constant undersaturation of the surrounding medium at comparably low computational cost was devised and demonstrated by our group [50]. In our MD simulations of aspirin nanocrystal dissolution [40,50], we proposed to use virtual atoms (sticky dummy atoms), which have a strong interaction potential with dissolved aspirin molecules, whereas interactions with water are excluded. To ensure that the virtual atoms do not interfere with the crystal bulk and thus have no impact on the crystal dissolution behavior, a plane of 100 sticky dummy atoms is introduced into the water slab at a distance of at least two cutoffs from the crystal (see Figure 3a). This still allows for free diffusion of aspirin molecules in the solvent. As was shown in our work [50], almost exactly half of the molecules, which have been in the liquid once, have returned to the crystal, where they could adsorb or reintegrate into the crystal lattice, which agrees nicely with the random walk expected for free diffusion. The other half reach within the cutoff radius of sticky dummy atoms and get trapped immediately and irreversibly, cf. Figure 3b, where the final configuration after 280 ns of simulation is shown. Thus, the number of aspirin molecules diffusing freely in solution is kept at a low value over the whole simulation time, and continuous dissolution of the crystal at almost zero concentration conditions can be monitored.

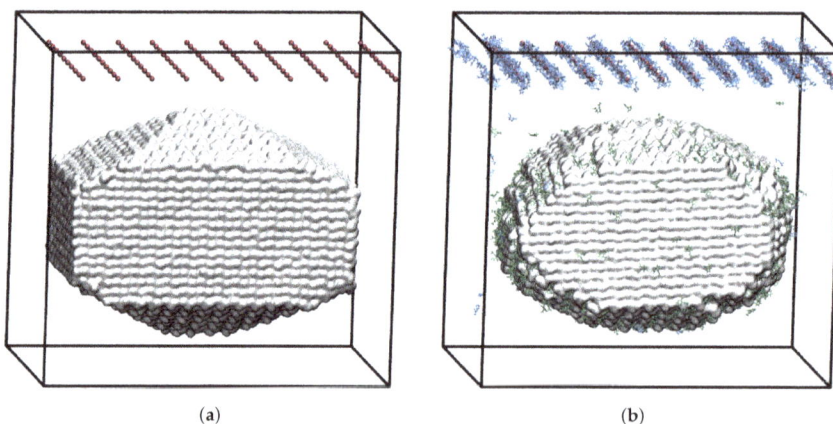

(a)   (b)

Figure 3. Sticky dummy atoms (red points above the nanocrystal) in action: (**a**) initial and (**b**) final snapshots from MD simulation of aspirin nanocrystal in its experimental form with crystal-like molecules plotted as white surface, liquid-like molecules in blue and adsorbed molecules in green. Water molecules are not shown for clarity.

4. Linking Nanoscale and Microscale

The principal objective of MD simulations in the multiscale strategy by Piana et al. [33] is to obtain state transition rates for kMC simulations. The main task for a quantification of these rate constants from the obtained MD trajectories is thereby the unambiguous definition and identification of distinct molecular states. To characterize the crystal growth and dissolution process, order parameters need to be established that discriminate between "crystal-like" and "solution-like" states. For aspirin dissolution, structural order parameters in the form of the orientation and number of neighbors proved to be sufficient [51]. Crystal-like molecules are thereby defined as those having their orientations within a certain threshold value from the orientations of the molecules from the relaxed unit cell used to construct the crystal (called reference orientations) and at least one neighbor among the crystal molecules, as described in detail in our papers [40,51]. However, the identification of an appropriate set of order parameters can be far from trivial in many other cases [107], and a significant number of approaches to differentiate one state from the other can be found in the literature [12,20,33,96–98,108–117].

Crystalline molecules can then be further subdivided in different substates according to their local coordination environment, e.g., on the basis of the number of neighboring crystalline molecules. The state transition rate constants for the kMC simulations can be calculated from MD simulations by just counting the corresponding transition events, as well as the possibilities for these transitions to occur [40]:

$$k_{A \to B} = \frac{1}{\Delta t \cdot n_{\Delta t}} \frac{\sum_{n_{\Delta t}} n_{A \to B}}{\sum_{n_{\Delta t}} n_A}, \tag{1}$$

where $n_{A \to B}$ is the number of transitions from state A to state B and n_A is the number of molecules in state A calculated at each time interval Δt, while $n_{\Delta t}$ is the number of time intervals during the whole simulation.

The central idea behind a kMC simulation is a coarse-graining of the time evolution to discrete rare events and focusing on the corresponding Markovian state-to-state dynamics [118]. Thus, only rare, uncorrelated events are of interest. The analysis of processes occurring at the solid/solution interface during crystal growth and dissolution simulations requires an effective way to detect only rare, uncorrelated transitions from one state to another. Because of the oscillatory behavior of molecules, this is not a trivial problem. The oscillatory behavior of molecules (especially of surface molecules interacting with the solution) leads to spurious recrossings at the boundary between liquid and solid state, which severely complicate the reliable identification of significant transitions that are relevant over longer time scales [51]. The analysis problem arising due to such fast non-Markovian dynamics becomes especially acute for systems consisting of complex molecules with a high degree of conformational flexibility [51]. Due to strong fluctuations of the molecular position and orientation, numerous transitions between crystal-like substates with different number of neighbors, as well as from the crystal-like to the solution-like state can be registered. Consideration and calculation of all of these transition events would lead to strong overestimation of transition rates. The number of such fast, non-Markovian transitions can be reduced by analyzing the data averaged over some time interval, as done by Piana et al. [11,33,35]. However, the resulting transition rates are very sensitive to the choice of the time interval [20,51]. Reily and Briesen paid more attention to the choice of sampling interval and tried to avoid the counting of recrossing events. Namely, the time interval was chosen on the basis of the velocity autocorrelation function (VAF) of solute particles near the equilibrated interface, as well as the VAF of solute particles in the bulk of the crystal slab from the same simulation. Only transitions where the particles remained in the new state for two consecutive time intervals were counted [20]. To improve the estimation of molecular states during crystal growth and dissolution MD simulations, we proposed a new approach [51], based on Kalman filtering [119], making it possible to focus on rare, uncorrelated transition events, i.e., effective dynamics of the Markov chain. The idea is to consider the fluctuations of orientation and position of each crystal molecule as noisy measurements of the "true" (corresponding to the Markovian molecular state) molecular orientation and position and to estimate these "true" values using the Kalman filter algorithm. For the application of a Kalman filter [119], information on the measurement noise variance and the process noise variance is needed. Often, these parameters are just tuned to obtain good filter performance. To avoid this level of arbitrariness, we introduced a scheme to define all filter parameters and thus to provide a way for robust and reliable molecular state definition. According to this scheme, filter parameters as well as tolerance parameters are determined from short preliminary MD simulation, in which the crystal structure still stays stable. The details on the method and its application to processing of MD simulations data can be found in [51]. To analyze nanocrystal MD simulations and calculate rates for individual transition events, like incorporation or dissolution of a solute molecule into/from a particular crystal surface or edge, a scheme to classify all crystal molecules into different edge and surface categories was introduced. There, the neighbor-based approach combined with the geometry-based resolution algorithm is used to identify whether a molecule under consideration belongs to a certain surface or edge from a predefined set of surfaces and edges. Thereby, related faces or edges may be considered to be of the same type (like, e.g., the opposite ($\bar{1}$00) and (100) faces further referred to as (100) for MD and kMC simulations), as demonstrated in Figure 4.

Figure 4. Result of the application of state classification scheme for an aspirin nanocrystal at the beginning of MD simulation. The molecules located at flat faces are gray for the (100) face, blue for the (001) face, light blue for the (011) face and light gray for the (110) face. Molecules located on edges formed by the intersection of single-indexed faces are indicated in light green, and purple is used to show molecules on the edges formed by the intersection of double-indexed faces. Orange indicates molecules on the edges formed by single- and double-indexed faces. Water and sticky dummy atoms are not shown. Reprinted with permission from [40]. Copyright 2016 American Chemical Society.

For each type (e.g., particular edge or surface), rates dependent on the neighbor count can be calculated. Usually, the data points acquired from MD simulations are not distributed over the whole range of the number of neighbors and, thus, need to be interpolated to be used as an input for kMC simulations. Thereby the event count for each specific number of neighbors can be used as the weight for fitting. As the rates for different numbers of neighbors can differ by several orders of magnitude, a logarithmic scale is used. In [40], for each type j, the logarithmic values of the dissolution rates $y_i = \ln(k_{ij})$ over the corresponding number of neighbors x_i were plotted and approximated with the power law function $y = a \cdot x^b + c$, where a, b and c are fitting parameters. This function was chosen for fitting, as it is simple and gave a reasonable fit for all types. In this case, the dependence of rates for each type j from the number of neighbors can be rewritten as $k_j = \exp\left(a \cdot x^b + c\right)$. Comparing this with the Arrhenius equation $k = A \exp\left(-\frac{E_a}{RT}\right)$, one finds for the activation energy: $E_a(x) = -aRTx^b$ in the case of $A = \exp(c)$. This expression gives us the dependence of activation energy from the number of neighbors x, as well as from their spatial arrangement, varying for different edge and face types, and thus, represented by the use of type-specific coefficients a, b, c [40].

Altogether, for successfully linking of MD and kMC simulations, the whole range of analysis procedures should be performed on MD data, cf. Figure 5.

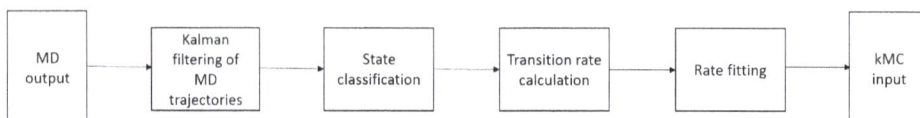

Figure 5. Linking MD and kMC approaches.

5. Kinetic Monte Carlo Simulations

A basic n-fold kMC algorithm [120,121] can be employed for the simulation of crystal growth and dissolution. In contrast to MD simulations, each molecule can be represented just by a point on a grid with the arrangement of the lattice sites according to the crystal structure. Moreover, only key growth and dissolution processes are considered for the system. These two simplifications make kMC computationally much less expensive, thus enabling the simulation of larger size and longer time scales compared to MD. For example, the aspirin crystal considered in our kMC studies [40] consisted of about 8.48 million molecules and dissolution was simulated for 7.5 ms, which took only about two days on a single processor. Each grid site in kMC simulations carries the information of whether it is occupied by a crystal molecule or vacant. For each occupied site, it can also be defined how many neighbors it has and to which crystal face or edge it belongs, i.e., its state, based on the applied state identification scheme. For the crystal dissolution simulations in [40], each kMC process corresponded to the removal of one of the crystal sites. All possible kMC processes were grouped into types to distinguish between states of the sites. The dissolution processes from a particular face or edge were considered as belonging to a particular type, cf. Figure 6. The processes within a certain type had different rates depending on the number of neighbors of the specific crystal site. These rates were calculated from MD simulations, as described above.

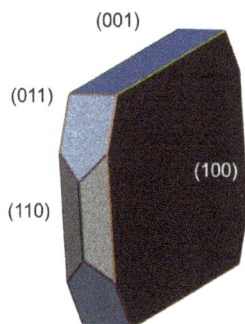

Figure 6. Representation of aspirin crystal in kMC simulations. The molecules located at flat faces are gray for the (100) face, blue for the (001) face, light blue for the (011) face and light gray for the (110) face. Molecules located on edges formed by the intersection of single-indexed faces are indicated in light green, and purple is used to show molecules on the edges formed by the intersection of double-indexed faces. Orange indicates molecules on the edges formed by single- and double-indexed faces.

On each kMC step, first, one of the available types (e.g., dissolution from a particular face or edge) was chosen in proportion to the transition probability obtained from the MD simulations. Then, a particular process within the selected type, i.e., one of the sites for dissolution belonging to the selected face or edge, was chosen using an effective linear selection method [122]. The occupancy of the selected site was then changed to vacant and for all neighbors of the selected site, the state identification scheme was applied to update their states. After updating the system time, the next kMC step could be performed until the maximum number of steps was reached.

In comparison to MD simulations, crystal face displacements could be directly observed in kMC simulations (see Figure 7), and the corresponding rates were calculated. The distance found between the face center calculated on each kMC step and the initial plane was stored during kMC simulations. Displacement rates were obtained from the linear regression of this distance over time. As the Monte Carlo technique is of a stochastic nature, a number of simulations is required to obtain statistically meaningful results. The typical way to determine the required number of replicates is to perform a convergence study by computing the standard deviations. For aspirin crystal dissolution, 25 independent simulations were used for calculation of average face displacement rates, and the relative standard deviations obtained were about 1–2% [40].

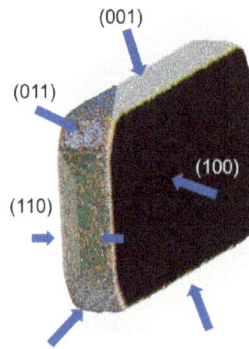

Figure 7. Face displacements from kMC simulations.

The evolution of the whole system is simulated based on the transition rates from MD simulations. Thus, the kMC face displacement rates are obtained at constant concentration (zero concentration in the case of aspirin dissolution simulations) and are only conditionally comparable to the experiment, where the concentration decreases during the crystal growth or rises during the crystal dissolution. To compensate for this, the continuum simulations based on the face displacement rates from kMC simulations can be performed, as described in the next section.

6. Coupling Molecular and Continuum Domains

According to Mullin et al. [4], the kinetics of crystal growth or dissolution of a face j of the crystal can be described by the empirical equation:

$$\mathbf{G}_j(t) = \mathbf{k}_j(T) \left(\frac{c(t)}{c_{sat}} - 1 \right)^{g_j} \tag{2}$$

where $\mathbf{G}_j$ is the face displacement velocity, $\mathbf{k}_j(T)$ is a rate constant for a specific temperature T for the respective face, $c(t)$ and c_{sat} are the bulk and saturation concentrations, correspondingly, and g_j is the order of the growth process. This equation does not take advection into account, which will be relevant when considering flow conditions, as planned in our future work (see also the Conclusions). Only processes in a stagnant fluid are considered here.

The order of the growth process in the Equation (2) may differ from 1, as this equation accounts not only for integration/disintegration, but also for the diffusion process. However, one can consider diffusion and integration/disintegration as two separate and independent processes. Taking the concentration on the crystal surface $c_{surf,j}(t)$ instead of the bulk concentration $c(t)$ in Equation (2) and setting g_j to 1 would allow one to split integration/disintegration from diffusion process:

$$\mathbf{G}_j(t) = \mathbf{k}_j(T) \left(\frac{c_{surf,j}(t)}{c_{sat}} - 1 \right) \tag{3}$$

For simplicity, we will further consider the case of an aspirin crystal dissolving in a chamber filled with water. With a surface concentration of zero at $t = 0$, the values obtained from combined MD and kMC simulations (cf. Figure 7) can be considered as initial values for the face displacement velocities, as they were obtained at constant zero concentration, provided by the above dummy atoms mechanism in MD simulations:

$$\mathbf{v}_{kMC,j} = \mathbf{G}_j(t = 0) = -\mathbf{k}_j(T) \tag{4}$$

Thus, we obtain the following expression for time-varying face displacement velocities:

$$G_j(t) = \mathbf{v}_{\mathrm{kMC},j}\left(1 - \frac{c_{surf,j}(t)}{c_{\mathrm{sat}}}\right) \tag{5}$$

The process of diffusion can be separately covered by the classical diffusion equation known as Fick's second law:

$$\frac{\partial}{\partial t}c(\mathbf{x},t) - D\nabla^2 c(\mathbf{x},t) = 0 \tag{6}$$

where D is the diffusion coefficient, the dependent variable c is the concentration in the bulk, the independent variables are the position $\mathbf{x}$ and time t.

Finding a unique solution requires an initial condition for the concentration in the bulk, as well as boundary conditions on the walls of the chamber and on the crystal surface. For crystal dissolution, the initial bulk concentration can be considered as zero. On the walls of the chamber, the flux in and out of the domain is zero. This is represented by the Neumann boundary condition:

$$D\frac{\partial c}{\partial \mathbf{x}}\mathbf{n}\bigg|_{\mathrm{walls}} = 0 \tag{7}$$

where $\mathbf{n}$ is the face normal of the walls of the chamber. The boundary condition on the crystal surface is face-specific and can be also represented by the Neumann boundary condition describing the flux over the respective crystal face:

$$-D\frac{\partial c}{\partial \mathbf{x}}\mathbf{n}\bigg|_j = G_j(t)\frac{\rho}{M} = \mathbf{v}_{\mathrm{kMC},j}\left(1 - \frac{c_{surf,j}(t)}{c_{\mathrm{sat}}}\right)\frac{\rho}{M} \tag{8}$$

where $\mathbf{v}_{\mathrm{kMC},j}$ is the rate constant for the face j directly obtained from kMC simulation (cf. Figure 7), $c_{surf,j}$ and c_{sat} are the surface and saturation concentrations, ρ is the density and M is the molecular mass of the crystal substance. The diffusion coefficient in Equation (6) as well as the saturation concentration can be easily calculated using MD simulations [40,52]; thus, the equation can be solved, and the face displacement velocities can be evaluated.

7. Continuum Simulations and Results

In continuum simulations partial differential equations are solved numerically, e.g., with the finite-element method. Comparing to molecular dynamics and kinetic Monte Carlo simulations, time and size scales can be extended by a large amount, and the simulations can cover typical experimental scales. Thus, a comparison of the respective results yields a good indicator of the simulation accuracy. In [52], a solution of the model Equation (6) with the corresponding boundary conditions (Equations (7) and (8)) was obtained for an aspirin crystal and compared with experimental results obtained using a Jamin-type interferometer [53] with the experimental setup, i.e., the size of the measurement chamber and aspirin crystal, chosen in agreement with the simulation settings. The simulations of the dissolution of an aspirin crystal in the measurement chamber of the interferometric device were performed using COMSOL multiphysics Version 5.0 [123]. Thereby, two different variants were considered: the values for diffusion coefficient D and solubility c_{sat} in model Equations (6)–(8) were either predicted *in silico* (Variant 1) or set to experimental values known from the literature [52,124] (Variant 2). In the first case the diffusion coefficient was calculated from the averaged mean square displacement of 20 individual MD simulations. The solubility of aspirin in water was calculated with the conductor-like screening model for realistic solvation (COSMO-RS [125]) using commercially available software COSMOtherm (Version C30 Release 15.01., COSMOlogic, Leverkusen, Germany), though it can be also calculated from MD simulations using, e.g., the direct coexistence method or chemical potential route [126], which however could be computationally prohibitive. In continuum simulations, the concentration was evaluated as the average over the whole domain and normalized with the saturation concentration. It is further referred to as relative saturation.

The simulation results yielded a relative saturation of 83% after 24 h of dissolution in the case of predicted parameters (Variant 1) and 67% in the case of experimental model parameters (Variant 2). The relative saturation measured using the Jamin-type interferometer [53] for three individual aspirin crystals with the same crystal dimensions as in simulations was in the range of 55–66% [52]. Thus, the simulation results in Variant 2 only slightly overestimated the experimental data. Overestimation of the final average concentration in the case of predicted model parameters was expected, as the predicted diffusion coefficient was about 20% higher than the known experimental value [52].

The displacement velocity of the (001) face over time was calculated from the total flux of aspirin over the selected crystal surface in simulations:

$$\mathbf{v}_{001} = MA_{001}^{-1}\rho^{-1} \int_{A_{001}} -D\frac{\partial c}{\partial \mathbf{x}}\mathbf{n}\,\mathrm{d}A, \tag{9}$$

using A_{001} as the surface area of the (001) face.

Experimentally, the face displacement velocity was obtained from tracking the surface position over time. For that, the crystal was divided into three segments as described in [52], and experimental results, presented in Figure 8, are given as averaged data points over these segments. Results for both simulations and the experiment exhibit comparably fast dissolution velocities at the beginning, whereas face displacement levels out toward the end. The simulation data obtained with experimental parameters (Variant 2) underestimate the face displacement velocity in the initial hours, while the face displacement velocity almost exactly matches the experimental findings in Variant 1. However, from Figure 8, it can be also seen that the process is diffusion-controlled. A sensitivity analysis performed for all parameters describing the dissolution process: face rate constants obtained from kMC simulations, diffusion coefficient and solubility also revealed a strong diffusion control, and thus, an expressed insensitivity of the simulation results to the actual intrinsic kMC disintegration rates [52].

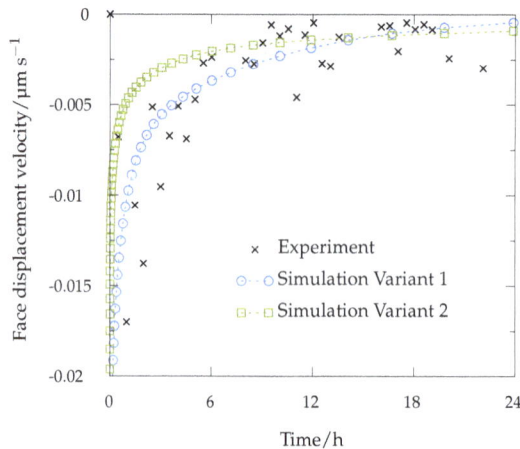

Figure 8. Face displacement velocity for the (001) face of aspirin crystal as obtained in the experiment and calculated from continuum simulations based on kMC face rate constants and (Variant 1) the predicted *in silico* values for the diffusion coefficient and solubility or (Variant 2) experimental values for the diffusion coefficient and solubility.

8. Conclusions and Outlook

Here, we described the multiscale modeling approach for *in silico* prediction of crystal growth/dissolution rates on the basis of molecular structure only, highlighted the difficulties in the realization of all single steps and illustrated our recent attempts to overcome these on the example of aspirin dissolution. Through a suitable combination of MD, kMC and continuum simulations, we were able to predict the aspirin dissolution rates at tractable computational cost and in good agreement with experimental results [52]. However, the dissolution of aspirin was found to be diffusion-controlled in both the simulation and experiment employing a stagnant fluid. Due to this, the obtained continuum simulation results are not very sensitive to face displacement rates obtained from kMC simulations, but rely mainly on the determined diffusion coefficient. Thus, the validation of the actual multiscale-integration approach is still limited. Additionally, the conditions of a stagnant fluid can generally be considered inadequate for actual oral administration of API crystals, and many reports [127–129] suggest that inclusion of hydrodynamic (fluid flow) factors is critical in describing any dissolution processes in the human gastrointestinal tract. The consideration of flow conditions, firstly, would make the study more realistic and, secondly, would allow for a much deeper understanding of the proposed multiscale-integration approach by overcoming the diffusion limitation. Instead of only solving a distributed mass transfer equation, momentum transfer will have to be solved in this case in a coupled manner via a CFD approach. In future work, we will study the dissolution behavior under flow conditions both simulation based and experimentally to validate the scale integration concepts for disintegration-controlled conditions.

The goal to achieve such an assessment on the basis of the molecular structure only has also not yet been met at the other end of the scale. In current aspirin simulations, not only the molecular structure, but also the crystal packing and the crystal shape were used as input. Such information is readily available for model APIs like aspirin. However, this is generally not the case for multicomponent systems, like, e.g., hydrates, salts and co-crystals, which are presently one of the primary strategies pursued to improve the solubility of poorly-soluble drugs [130,131]. The consideration of such systems will also definitely increase the complexity of the state transition description for the MD-kMC integration. In terms of future work, new sophisticated neighbor and orientation definition schemes will be elaborated to expand the scope of the multiscale approach toward multicomponent systems. The state identification scheme mentioned in Section 4 and described in detail in our previous work [40] supports correct assignment of molecules to different face and edge categories only if these are defined by low Miller indices of one and zero. Some complex compounds may, however, require accounting for higher-indexed faces. For that, a new algorithm is currently under development in our group. The proper realization of all protocol steps will enable simplifying the design of crystallizer operations, as well as support decision making in early stages of *in silico* drug development as it will consider release properties for an API on the basis of the molecular structure only.

Acknowledgments: This work has been supported by the Deutsche Forschungsgemeinschaft (DFG) through Grants BR 2035/4-2 and BR 2035/8-2 and has made use of the computer resources provided by the Leibniz Supercomputing Centre under Grant pr58la. The ScalaLife Competence Center (http://www.scalalife.eu/) is also acknowledged for the assistance with computing resources and the GROMACS code.

Author Contributions: All authors jointly contributed to devising the presented methodology; M.G. performed the simulations; E.E. elaborated and developed analysis tools to link MD and kMC techniques; M.G. and E.E. developed kMC code; H.B. initiated and supervised the investigations; E.E. wrote the paper.

Conflicts of Interest: The authors declare no conflict of interest.

References

1. Weeks, J.; Gilmer, G.H. *Advances in Chemical Physics*; Chapter Dynamics of Crystal Growth; John Wiley and Sons, Inc.: Hoboken, NJ, USA, 2007; Volume 40, pp. 157–228.
2. Dhanaraj, G.; Byrappa, K.; Prasad, V.; Dudley, M. (Eds.) *Springer Handbook of Crystal Growth*; Springer: Heidelberg, Germany, 2010.
3. Nishinaga, T. (Ed.) *Handbook of Crystal Growth. Fundamentals: Thermodynamics and Kinetics*; Elsevier: Amsterdam, The Netherland, 2014.
4. Mullin, J. *Crystallization: Third Edition*; Butterworth-Heinemann: London, UK, 1997.

5.   Hurle, D. (Ed.)   *Handbook of Crystal Growth*; Fundamentals. (a): Thermodynamics and Kinetics; (b): Transport and Stability; North Holland Elsevier Science Publishers: Amsterdam, The Netherlands, 1993; Volume 1.

6.   Dove, P.; Han, N. Kinetics of Mineral Dissolution and Growth as Reciprocal Microscopic Surface Processes Across Chemical Driving Force.   Perspectives on Inorganic, Organic and Biological Crystal Growth: From Fundamentals to Applications Directions. *Am. Inst. Phys. Conf. Ser.* **2007**, *916*, 215–234.

7.   Clark, J.N.; Ihli, J.; Schenk, A.S.; Kim, Y.Y.; Kulak, A.N.; Campbell, J.M.; Nisbet, G.; Meldrum, F.C.; Robinson, I.K.   Three-Dimensional Imaging of Dislocation Propagation During Crystal Growth and Dissolution. *Nat. Mater.* **2015**, *14*, 780–784.

8.   Veesler, S.; Puel, F. *Handbook of Crystal Growth. Fundamentals: Thermodynamics and Kinetics*; Part A, Chapter Crystallization of Pharmaceutical Crystals; Elsevier: Amsterdam, The Netherland, 2014; Volume I, pp. 915–951.

9.   Derby, J.J.; Chelikowsky, J.R.; Sinno, T.; Dai, B.; Kwon, Y.I.; Lun, L.; Pandy, A.; Yeckel, A. Large-Scale Numerical Modeling of Melt and Solution Crystal Growth. *AIP Conf. Proc.* **2007**, *916*, 139–158.

10.   Allen, M.P.; Tildesley, D.J. *Computer Simulation of Liquids*; Oxford University Press Inc.: New York, NY, USA, 1989.

11.   Piana, S.; Gale, J.D. Understanding the Barriers to Crystal Growth: Dynamical Simulation of the Dissolution and Growth of Urea from Aqueous Solution. *J. Am. Chem. Soc.* **2005**, *127*, 1975–1982.

12.   Salvalaglio, M.; Vetter, T.; Giberti, F.; Mazzotti, M.; Parrinello, M. Uncovering Molecular Details of Urea Crystal Growth in the Presence of Additives. *J. Am. Chem. Soc.* **2012**, *134*, 17221–17233.

13.   Salvalaglio, M.; Vetter, T.; Mazzotti, M.; Parrinello, M. Controlling and Predicting Crystal Shapes: The Case of Urea. *Angew. Chem. Int. Ed.* **2013**, *52*, 13369–13372.

14.   Banerjee, S.; Briesen, H. Molecular dynamics simulations of glycine crystal-solution interface. *J. Chem. Phys.* **2009**, *131*, 184705.

15.   Gnanasambandam, S.; Rajagopalan, R.   Growth Morphology of α-Glycine Crystals in Solution Environments: An Extended Interface Structure Analysis. *CrystEngComm* **2010**, *12*, 1740–1749.

16.   Cheong, D.W.; Boon, Y.D. Comparative Study of Force Fields for Molecular Dynamics Simulations of α-Glycine Crystal Growth from Solution. *Cryst. Growth Des.* **2010**, *10*, 5146–5158.

17.   Parks, C.; Koswara, A.; Tung, H.H.; Nere, N.K.; Bordawekar, S.; Nagy, Z.K.; Ramkrishna, D. Nanocrystal Dissolution Kinetics and Solubility Increase Prediction from Molecular Dynamics: The Case of α-, β-, and γ-Glycine. *Mol. Pharm.* **2017**, *14*, 1023–1032.

18.   Volkov, I.; Cieplak, M.; Koplik, J.; Banavar, J.R. Molecular Dynamics Simulations of Crystallization of Hard Spheres. *Phys. Rev. E* **2002**, *66*, 061401.

19.   Lemarchand, C.A. Molecular dynamics simulations of a hard sphere crystal and reaction-like mechanism for homogeneous melting. *J. Chem. Phys.* **2012**, *136*, 234505.

20.   Reilly, A.; Briesen, H. Modeling crystal growth from solution with molecular dynamics simulations: Approaches to transition rate constants. *J. Chem. Phys.* **2012**, *136*, 034704.

21.   Gruhn, T.; Monson, P.A. Molecular dynamics simulations of hard sphere solidification at constant pressure. *Phys. Rev. E* **2001**, *64*, 061703.

22.   Mandal, T.; Huang, W.; Mecca, J.M.; Getchell, A.; Porter, W.W.; Larson, R.G. A framework for multi-scale simulation of crystal growth in the presence of polymers. *Soft Matter* **2017**, *13*, 1904–1913.

23.   Mandal, T.; Marson, R.L.; Larson, R.G. Coarse-grained modeling of crystal growth and polymorphism of a model pharmaceutical molecule. *Soft Matter* **2016**, *12*, 8246–8255.

24.   Ingólfsson, H.I.; Lopez, C.A.; Uusitalo, J.J.; de Jong, D.H.; Gopal, S.M.; Periole, X.; Marrink, S.J. The power of coarse graining in biomolecular simulations. *Wiley Interdiscip. Rev. Comput. Mol. Sci.* **2014**, *4*, 225–248.

25.   Kästner, J. Umbrella sampling. *Wiley Interdiscip. Rev. Comput. Mol. Sci.* **2011**, *1*, 932–942.

26.   Quigley, D.; Rodger, P. A metadynamics-based approach to sampling crystallisation events. *Mol. Simul.* **2009**, *35*, 613–623.

27.   Allen, R.J.; Valeriani, C.; ten Wolde, P.R.   Forward flux sampling for rare event simulations.   *J. Phys. Condens. Matter* **2009**, *21*, 463102.

28.   Sosso, G.C.; Chen, J.; Cox, S.J.; Fitzner, M.; Pedevilla, P.; Zen, A.; Michaelides, A. Crystal Nucleation in Liquids: Open Questions and Future Challenges in Molecular Dynamics Simulations. *Chem. Rev.* **2016**, *116*, 7078–7116.

29. Dogan, B.; Schneider, J.; Reuter, K. In Silico Prediction of Dissolution Rates of Pharmaceutical Ingredients. *Chem. Phys. Lett.* **2016**, *662*, 52–55.

30. Schneider, J.; Zheng, C.; Reuter, K. Thermodynamics of Surface Defects at the Aspirin/Water Interface. *J. Chem. Phys.* **2014**, *141*, 124702.

31. Schneider, J.; Reuter, K. Efficient Calculation of Microscopic Dissolution Rate Constants: The Aspirin-Water Interface. *J. Phys. Chem. Lett.* **2014**, *5*, 3859–3862.

32. Burton, W.K.; Cabrera, N.; Frank, F.C. The Growth of Crystals and the Equilibrium Structure of their Surfaces. *Philos. Trans. R. Soc.* **1951**, *243*, 299–358.

33. Piana, S.; Reyhani, M.; Gale, J.D. Simulating micrometre-scale crystal growth from solution. *Nature* **2005**, *483*, 70–73.

34. Reilly, A.; Briesen, H. A detailed kinetic Monte Carlo study of growth from solution using MD-derived rate constants. *J. Cryst. Growth* **2012**, *354*, 34–43.

35. Piana, S.; Gale, J.D. Three-dimensional kinetic Monte Carlo simulation of crystal growth from solution. *J. Cryst. Growth* **2006**, *294*, 46–52.

36. Kurganskaya, I.; Luttge, A. Kinetic Monte Carlo Simulations of Silicate Dissolution: Model Complexity and Parametrization. *J. Phys. Chem. C* **2013**, *117*, 24894–24906.

37. Stack, A.G.; Raiteri, P.; Gale, J.D. Accurate Rates of the Complex Mechanisms for Growth and Dissolution of Minerals Using a Combination of Rare-Event Theories. *J. Am. Chem. Soc.* **2012**, *134*, 11–14.

38. Chen, J.C.; Reischl, B.; Spijker, P.; Holmberg, N.; Laasonen, K.; Foster, A.S. Ab initio Kinetic Monte Carlo simulations of dissolution at the NaCl-water interface. *Phys. Chem. Chem. Phys.* **2014**, *16*, 22545–22554.

39. Dkhissi, A.; Estéve, A.; Mastail, C.; Olivier, S.; Mazaleyrat, G.; Jeloaica, L.; Rouhani, M.D. Multiscale Modeling of the Atomic Layer Deposition of HfO_2 Thin Film Grown on Silicon: How to Deal with a Kinetic Monte Carlo Procedure. *J. Chem. Theory Comput.* **2008**, *4*, 1915–1927.

40. Elts, E.; Greiner, M.; Briesen, H. Predicting Dissolution Kinetics for Active Pharmaceutical Ingredients on the Basis of Their Molecular Structures. *Cryst. Growth Des.* **2016**, *16*, 4154–4164.

41. Boetker, J.P.; Rantanen, J.; Rades, T.; Müllertz, A.; Østergaard, J.; Jensen, H. A New Approach to Dissolution Testing by UV Imaging and Finite Element Simulations. *Pharm. Res.* **2013**, *30*, 1328–1337.

42. Haddish-Berhane, N.; Nyquist, C.; Haghighi, K.; Corvalan, C.; Keshavarzian, A.; Campanella, O.; Rickus, J.; Farhadi, A. A multi-scale stochastic drug release model for polymer-coated targeted drug delivery systems. *J. Control. Release* **2006**, *110*, 314–322.

43. Bai, G.; Armenante, P.M. Hydrodynamic, mass transfer, and dissolution effects induced by tablet location during dissolution testing. *J. Pharm. Sci.* **2009**, *98*, 1511–1531.

44. Lamberti, G.; Galdi, I.; Barba, A.A. Controlled release from hydrogel-based solid matrices. A model accounting for water up-take, swelling and erosion. *Int. J. Pharm.* **2011**, *407*, 78–86.

45. D'Arcy, D.M.; Healy, A.M.; Corrigan, O.I. Towards determining appropriate hydrodynamic conditions for in vitro in vivo correlations using computational fluid dynamics. *Eur. J. Pharm. Sci.* **2009**, *37*, 291–299.

46. Boetker, J.; Raijada, D.; Aho, J.; Khorasani, M.; Søgaard, S.V.; Arnfast, L.; Bohr, A.; Edinger, M.; Water, J.J.; Rantanen, J. In silico product design of pharmaceuticals. *Asian J. Pharm. Sci.* **2016**, *11*, 492–499.

47. Kindgen, S.; Rach, R.; Nawroth, T.; Abrahamsson, B.; Langguth, P. A Novel Disintegration Tester for Solid Dosage Forms Enabling Adjustable Hydrodynamics. *J. Pharm. Sci.* **2016**, *105*, 2402–2409.

48. Haddish-Berhane, N.; Rickus, J.L.; Haghighi, K. The role of multiscale computational approaches for rational design of conventional and nanoparticle oral drug delivery systems. *Int. J. Nanomed.* **2007**, *2*, 315–331.

49. Greiner, M.; Elts, E.; Schneider, J.; Reuter, K.; Briesen, H. Dissolution study of active pharmaceutical ingredients using molecular dynamics simulations with classical force fields. *J. Cryst. Growth* **2014**, *405*, 122–130.

50. Greiner, M.; Elts, E.; Briesen, H. Insights into Pharmaceutical Nanocrystal Dissolution: A Molecular Dynamics Simulation Study on Aspirin. *Mol. Pharm.* **2014**, *11*, 3009–3016.

51. Elts, E.; Greiner, M.; Briesen, H. Data Filtering for Effective Analysis of Crystal-Solution Interface Molecular Dynamics Simulations. *J. Chem. Theory Comput.* **2014**, *10*, 1686–1697.

52. Greiner, M.; Choscz, C.; Eder, C.; Elts, E.; Briesen, H. Multiscale modeling of aspirin dissolution: From molecular resolution to experimental scales of time and size. *CrystEngComm* **2016**, *18*, 5302–5312.

53.   Eder, C.; Choscz, C.; Müller, V.; Briesen, H. Jamin-interferometer-setup for the determination of concentration and temperature dependent face-specific crystal growth rates from a single experiment. *J. Cryst. Growth* **2015**, *426*, 255–264.

54.   Sundaralingam, M.; Jensen, L.H. Refinement of the structure of salicylic acid. *Acta Crystallogr.* **1965**, *18*, 1053–1058.

55.   Payne, R.S.; Rowe, R.C.; Roberts, R.J.; Charlton, M.H.; Docherty, R. Potential polymorphs of aspirin. *J. Comput. Chem.* **1999**, *20*, 262–273.

56.   Bond, A.; Boese, R.; Desiraju, G. On the Polymorphism of Aspirin: Crystalline Aspirin as Intergrowths of Two "Polymorphic" Domains. *Angew. Chem. Int. Ed.* **2007**, *46*, 618–622.

57.   Hammond, R.B.; Pencheva, K.; Ramachandran, V.; Roberts, K.J. Application of Grid-Based Molecular Methods for Modeling Solvent-Dependent Crystal Growth Morphology: Aspirin Crystallized from Aqueous Ethanolic Solution. *Cryst. Growth Des.* **2007**, *7*, 1571–1574.

58.   Hammond, R.B.; Pencheva, K.; Roberts, K.J. A Structural-Kinetic Approach to Model Face-Specific Solution/Crystal Surface Energy Associated with the Crystallization of Acetyl Salicylic Acid from Supersaturated Aqueous/Ethanol Solution. *Cryst. Growth Des.* **2006**, *6*, 1324–1334.

59.   Kim, Y.; Matsumoto, M.; Machida, K. Specific Surface Energies and Dissolution Behavior of Aspirin Crystal. *Chem. Pharm. Bull.* **1985**, *33*, 4125–4131.

60.   Danesh, A.; Connell, S.D.; Davies, M.C.; Roberts, C.J.; Tendler, S.J.B.; Williams, P.M.; Wilkins, M.J. An in situ dissolution study of aspirin crystal planes (100) and (001) by atomic force microscopy. *Pharm. Res.* **2001**, *18*, 299–303.

61.   Woodley, S.M.; Catlow, R. Crystal structure prediction from first principles. *Nat. Mater.* **2008**, *7*, 937–946.

62.   Day, G. *Computational Pharmaceutical Solid State Chemistry*; Chapter Advances in Crystal Structure Prediction and Applications to Pharmaceutical Materials; John Wiley & Sons, Inc.: Hoboken, NJ, USA, 2016; pp. 87–115; ISBN 978-1-11-870068-6.

63.   Price, S. Predicting crystal structures of organic compounds. *Chem. Soc. Rev.* **2014**, *43*, 2098–2111.

64.   Reilly, A.M.; Cooper, R.I.; Adjiman, C.S.; Bhattacharya, S.; Boese, A.D.; Brandenburg, J.G.; Bygrave, P.J.; Bylsma, R.; Campbell, J.E.; Car, R.; et al. Report on the Sixth Blind Test of Organic Crystal-Structure Prediction Methods. *Acta Crystallogr. Sect. B* **2016**, *72*, 439–459.

65.   Thompson, H.P.; Day, G.M. Which conformations make stable crystal structures? Mapping crystalline molecular geometries to the conformational energy landscape. *Chem. Sci.* **2014**, *5*, 3173–3182.

66.   Nývlt, J.; Ulrich, J. *Addmixtures in Crystallization*; VCH: Weinheim, Germany, 1995.

67.   Wulff, G. Velocity of growth and dissolution of crystal faces. *Z. Kristallogr.* **1901**, *34*, 449–530.

68.   Rogal, J.; Reuter, K.; Scheffler, M. Thermodynamic stability of PdO surfaces. *Phys. Rev. B* **2004**, *69*, 075421.

69.   Ouyang, R.; Liu, J.X.; Li, W.X. Atomistic Theory of Ostwald Ripening and Disintegration of Supported Metal Particles under Reaction Conditions. *J. Am. Chem. Soc.* **2013**, *135*, 1760–1771.

70.   Garcia-Mota, M.; Rieger, M.; Reuter, K. Ab Initio Prediction of the Equilibrium Shape of Supported Ag Nanoparticles on α-Al$_2$O$_3$(0001). *J. Catal.* **2015**, *321*, 1–6.

71.   Lovette, M.A.; Robben Browning, A.; Griffin, D.W.; Sizemore, J.P.; Snyder, R.C.; Doherty, M.F. Crystal Shape Engineering. *Ind. Eng. Chem. Res.* **2008**, *47*, 9812–9833.

72.   Kuvadia, Z.B.; Doherty, M.F. Spiral growth model for faceted crystals of non-centrosymmetric organic molecules grown from solution. *Cryst. Growth Des.* **2011**, *11*, 2780–2802.

73.   Mayerson, A.S. (Ed.) *Handbook of industrial Crystallization*; Butterworth-Heinemann: Newton, MA, USA, 2002.

74.   Bravais, A. *Etudes Crystallographiques*; Academie des Sciences: Paris, France, 1913.

75.   Friedel, G. Etudes sur la loi de Bravais. *Bull. Soc. Fr. Miner.* **1907**, *30*, 326–455.

76.   Donnay, J.D.H.; Harker, D. A new law of crystal morphology extending the law of Bravais. *Am. Miner.* **1937**, *22*, 446–467.

77.   Hartman, P.; Bennema, P. The attachment energy as a habit controlling factor. I. Theoretical considerations. *J. Cryst. Growth* **1980**, *49*, 145–156.

78.   Rai, B. (Ed.) *Molecular Modeling for the Design of Novel Performance Chemicals and Materials*; CRC Press: Boca Raton, FL, USA, 2012.

79.   Li, J.; Doherty, M.F. Steady State Morphologies of Paracetamol Crystal from Different Solvents. *Cryst. Growth Des.* **2017**, *17*, 659–670.

80. Tilbury, C.J.; Green, D.A.; Marshall, W.J.; Doherty, M.F. Predicting the Effect of Solvent on the Crystal Habit of Small Organic Molecules. *Cryst. Growth Des.* **2016**, *16*, 2590–2604.

81. Shim, H.M.; Koo, K.K. Prediction of Growth Habit of β-Cyclotetramethylene-tetranitramine Crystals by the First-Principles Models. *Cryst. Growth Des.* **2015**, *15*, 3983–3991.

82. Zhang, C.; Ji, C.; Li, H.; Zhou, Y.; Xu, J.; Xu, R.; Li, J.; Luo, Y. Occupancy Model for Predicting the Crystal Morphologies Influenced by Solvents and Temperature, and Its Application to Nitroamine Explosives. *Cryst. Growth. Des.* **2013**, *13*, 282–290.

83. Yang, X.; Qian, G.; Zhang, X.; Duan, X.; Zhou, X. Effects of Solvent and Impurities on Crystal Morphology of Zinc Lactate Trihydrate. *Chin. J. Chem. Eng.* **2014**, *22*, 221–226.

84. Wilson, C.C. Interesting proton behaviour in molecular structures. Variable temperature neutron diffraction and ab initio study of acetylsalicylic acid: Characterising librational motions and comparing protons in different hydrogen bonding potentials. *New J. Chem.* **2002**, *26*, 1733–1739.

85. MacKerell, A.D.; Bashford, D.; Bellott, M.; Dunbrack, R.; Evanseck, J.D.; Field, M.J.; Fischer, S.; Gao, J.; Guo, H.; Ha, S.; et al. All-atom empirical potential for molecular modeling and dynamics studies of proteins. *J. Phys. Chem. B* **1998**, *102*, 3586–3616.

86. Hornak, V.; Abel, R.; Okur, A.; Strockbine, B.; Roitberg, A.; Simmerling, C. Comparison of Multiple Amber Force Fields and Development of Improved Protein Backbone Parameters. *Proteins Struct. Funct. Bioinf.* **2006**, *65*, 712–725.

87. Wang, J.M.; Wolf, R.M.; Caldwell, J.W.; Kollman, P.A.; Case, D.A. Development and testing of a general AMBER force field. *J. Comput. Chem.* **2004**, *25*, 1157–1174.

88. Jorgensen, W.L.; Maxwell, D.S.; Tirado-Rives, J. Development and testing of the OPLS all-atom force field on conformational energetics and properties of organic liquids. *J. Am. Chem. Soc.* **1996**, *118*, 11225–11236.

89. Zoete, V.; Cuendet, M.A.; Grosdidier, A.; Michielin, O. SwissParam, a Fast Force Field Generation Tool For Small Organic Molecules. *J. Comput. Chem.* **2011**, *32*, 2359–2368.

90. Halgren, T.A. Merck molecular force field. I. Basis, form, scope, parameterization, and performance of MMFF94. *J. Comput. Chem.* **1996**, *17*, 490–519.

91. Da Silva Sousa, A.; Alan, W.; Vranken, W.F. ACPYPE—AnteChamber PYthon Parser interfacE. *BMC Res. Notes* **2012**, *5*, 367.

92. Wang, J.; Wang, W.; Kollman, P.A.; Case, D.A. Automatic atom type and bond type perception in molecular mechanical calculations. *J. Mol. Graph. Model.* **2006**, *25*, 247–260.

93. Schüttelkopf, A.W.; van Aalten, D.M.F. PRODRG: A tool for high-throughput crystallography of protein-ligand complexes. *Acta Crystallogr.* **2004**, *60*, 1355–1363.

94. Oostenbrink, C.; Villa, A.; Mark, A.E.; van Gunsteren, W.F. A biomolecular force field based on the free enthalpy of hydration and solvation: The GROMOS force-field parameter sets 53A5 and 53A6. *J. Comput. Chem.* **2004**, *25*, 1656–1676.

95. Lemkul, J.A.; Allen, W.J.; Bevan, D.R. Practical Considerations for Building GROMOS-Compatible Small-Molecule Topologies. *J. Chem. Inf. Model.* **2010**, *50*, 2221–2235.

96. Hawtin, R.; Quigley, D.; Rodger, P. Gas hydrate nucleation and cage formation at a water/methane interface. *Phys. Chem. Chem. Phys.* **2008**, *10*, 4853–4864.

97. Liang, S.; Kusalik, P.G. Explorations of gas hydrate crystal growth by molecular simulations. *Chem. Phys. Lett.* **2010**, *494*, 123–133.

98. Jacobson, L.C.; Matsumoto, M.; Molinero, V. Order parameters for the multistep crystallization of clathrate hydrates. *J. Chem. Phys.* **2011**, *135*, 074501.

99. Gao, Y.; Olsen, K.W. Molecular Dynamics of Drug Crystal Dissolution: Simulation of Acetaminophen Form I in Water. *Mol. Pharm.* **2013**, *10*, 905–917.

100. Humphrey, W.; Dalke, A.; Schulten, K. VMD: Visual molecular dynamics. *J. Mol. Graph.* **1996**, *14*, 33–38.

101. Wen, H.; Li, T.; Morris, K.R.; Park, K. Dissolution Study on Aspirin and α-Glycine Crystals. *J. Phys. Chem. B* **2004**, *108*, 11219–11227.

102. Wedelkind, J.; Reguera, D.; Strey, R. Finite-size effects in simulations of nucleation. *J. Chem. Phys.* **2006**, *125*, 214505.

103. Salvalaglio, M.; Mazzotti, M.; Parrinello, M. Urea homogeneous nucleation mechanism is solvent dependent. *Faraday Discuss.* **2015**, *179*, 291–307.

104. Grossier, R.; Veesler, S. Reaching One Single and Stable Critical Cluster through Finite-Sized Systems. *Cryst. Growth Des.* **2009**, *9*, 1917–1922.

105. Perego, C.; Salvalaglio, M.; Parrinello, M. Molecular dynamics simulations of solutions at constant chemical potential. *J. Chem. Phys.* **2015**, *142*, 144113.

106. Zimmermann, N.E.R.; Vorselaars, B.; Quigley, D.; Peters, B. Nucleation of NaCl from Aqueous Solution: Critical Sizes, Ion-Attachment Kinetics, and Rates. *J. Am. Chem. Soc.* **2015**, *137*, 13352–13361.

107. Anwar, J.; Zahn, D. Uncovering Molecular Processes in Crystal Nucleation and Growth by Using Molecular Simulation. *Angew. Chem. Int. Ed.* **2011**, *50*, 1996–2013.

108. Steinhardt, P.J.; Nelson, D.R.; Ronchetti, M. Bond-orientational order in liquids and glasses. *Phys. Rev. B* **1983**, *28*, 784–805.

109. Lechner, W.; Dellago, C. Accurate determination of crystal structures based on averaged local bond order parameters. *J. Chem. Phys.* **2008**, *129*, 114707.

110. Radhakrishnan, R.; Trout, B.L. Nucleation of Hexagonal Ice (I_h) in Liquid Water. *J. Am. Chem. Soc.* **2003**, *125*, 7743–7747.

111. Brukhno, A.; Anwar, J.; Davidchack, R.; Handel, R. Challenges in molecular simulation of homogeneous ice nucleation. *J. Phys. Condens. Matter* **2008**, *20*, 494243.

112. Leyssale, J.M.; Delhommelle, J.; Millot, C. Reorganization and Growth of Metastable α-N_2 Critical Nuclei into Stable β-N_2 Crystals. *J. Am. Chem. Soc.* **2004**, *126*, 12286–12287.

113. Mettes, J.A.; Keith, J.B.; McClurg, R.B. Molecular crystal global phase diagrams. I. Method of construction. *Acta Crystallogr. Sect. A Found. Crystallogr.* **2004**, *60*, 621–636.

114. Zahn, D. Atomistic Mechanisms of Phase Separation and Formation of Solid Solutions: Model Studies of NaCl, NaCl-NaF, and Na($Cl_{1-x}Br_x$) Crystallization from Melt. *J. Phys. Chem. B* **2007**, *111*, 5249–5253.

115. Xu, S.; Bartell, L. Analysis of Orientational Order in Molecular Clusters. A Molecular Dynamics Study. *J. Phys. Chem.* **1993**, *97*, 13544–13549.

116. Kinney, K.E.; Xu, S.; Bartell, L.S. Molecular Dynamics Study of the Freezing of Clusters of Chalcogen Hexafluorides. *J. Phys. Chem.* **1996**, *100*, 6935–6941.

117. Santiso, E.E.; Trout, B.L. A general set of order parameters for molecular crystals. *J. Chem. Phys.* **2011**, *134*, 064109.

118. Reuter, K. First-Principles Kinetic Monte Carlo Simulations for Heterogeneous Catalysis: Concepts, Status, and Frontiers. In *Modeling and Simulation of Heterogeneous Catalytic Reactions: From the Molecular Process to the Technical System*; Deutschmann, O., Ed.; Wiley-VCH: Weinheim, Germany, 2011; pp. 71–111.

119. Kalman, R.E. A New Approach to Linear Filtering and Prediction Problems. *J. Basic Eng.* **1960**, *82*, 35–45.

120. Bortz, A.B.; Kalos, M.H.; Lebowitz, J.L. A new algorithm for Monte Carlo simulation of Ising spin systems. *J. Comput. Phys.* **1975**, *17*, 10–18.

121. Chatterjee, A.; Vlachos, D. An overview of spatial microscopic and accelerated kinetic Monte Carlo methods. *J. Comput. Aided Mater. Des.* **2007**, *14*, 253–308.

122. Burghaus, U.; Stephan, J.; Vattuone, L.; Rogowska, J. *A Practical Guide to Kinetic Monte Carlo Simulations and Classical Molecular Dynamics Simulations*; Nova Science Publishers, Inc.: New York, NY, USA, 2006.

123. Comsol AB. *COMSOL Multiphysics User'S Guide*; Comsol AB: Stockholm, Sweden, 2014.

124. Edwards, L.J. The Dissolution and Diffusion of Aspirin In Aqueous Media. *Trans. Faraday Soc.* **1951**, *47*, 1191–1210.

125. Klamt, A. Conductor-like Screening Model for Real Solvent: A New Approach to the Quantitative Calculation of Solvation Phenomena. *J. Phys. Chem. A* **1995**, *99*, 2224–2235.

126. Espinosa, J.R.; Young, J.M.; Jiang, H.; Gupta, D.; Vega, C.; Sanz, E.; Debenedetti, P.G.; Panagiotopoulos, A.Z. On the calculation of solubilities via direct coexistence simulations: Investigation of NaCl aqueous solutions and Lennard-Jones binary mixtures. *J. Chem. Phys.* **2016**, *145*, 154111.

127. Sugano, K. Theoretical comparison of hydrodynamic diffusion layer models used for dissolution simulation in drug discovery and development. *Int. J. Pharm.* **2008**, *363*, 73–77.

128. Chakrabarti, S.; Southard, M. Control of Poorly Soluble Drug Dissolution in Conditions Simulating the Gastrointestinal Tract Flow. 1. Effect of Tablet Geometry in Buffered Medium. *J. Pharm. Sci.* **1996**, *85*, 313–319.

129. Chakrabarti, S.; Southard, M.Z. Control of Poorly Soluble Drug Dissolution in Conditions Simulating the Gastrointestinal Tract Flow. 2. Cocompression of Drugs with Buffers. *J. Pharm. Sci.* **1997**, *86*, 465–469.

130. Sikarra, D.; Shukla, V.; Kharia, A.; Chatterjee, D. Techniques for solubility enhancement of poorly soluble drugs: An overview. *J. Med. Pharm. Allied Sci.* **2012**, *1*, 1–22.

131. Tiwle, R.; Ajazuddin; Giri, T.; Tripathi, D.; Jain, V.; Alexander, A. An exhaustive review on solubility enhancement for hydrophobic compounds by possible applications of novel techniques. *Trends Appl. Sci. Res.* **2012**, *7*, 596–619.

© 2017 by the authors. Licensee MDPI, Basel, Switzerland. This article is an open access article distributed under the terms and conditions of the Creative Commons Attribution (CC BY) license (http://creativecommons.org/licenses/by/4.0/).

crystals

MDPI

Article

Molecular Dynamics Analysis of Synergistic Effects of Ions and Winter Flounder Antifreeze Protein Adjacent to Ice-Solution Surfaces

Tatsuya Yasui [1], Tadashi Kaijima [1], Ken Nishio [2] and Yoshimichi Hagiwara [3,*]

[1] Division of Mechanophysics, Graduate School of Science and Technology, Kyoto Institute of Technology, Matsugasaki, Sakyo-ku, Kyoto 606-8585, Japan; kit.tatsu3334@gmail.com (T.Y.); tadashi.kaijima@gmail.com (T.K.)

[2] School of Science and Technology, Kyoto Institute of Technology, Matsugasaki, Sakyo-ku, Kyoto 606-8585, Japan; t.l.anri.05.0118@docomo.ne.jp

[3] Faculty of Mechanical Engineering, Kyoto Institute of Technology, Matsugasaki, Sakyo-ku, Kyoto 606-8585, Japan

* Correspondence: yoshi@kit.ac.jp; Tel.: +81-75-724-7324; Fax: +81-75-724-7300

Received: 19 June 2018; Accepted: 19 July 2018; Published: 22 July 2018

Abstract: The control of freezing saline water at the micrometer level has become very important in cryosurgery and cryopreservation of stem cells and foods. Adding antifreeze protein to saline water is a promising method for controlling the freezing because the protein produces a gap between the melting point and the freezing point. Furthermore, a synergistic effect of the solutes occurs in which the freezing point depression of a mixed solution is more noticeable than the sum of two freezing point depressions of single-solute solutions. However, the mechanism of this effect has not yet been clarified. Thus, we have carried out a molecular dynamics simulation on aqueous solutions of winter flounder antifreeze protein and sodium chloride or calcium chloride with an ice layer. The results show that the cations inhibit the hydrogen bond among water molecules not only in the salt solutions but also in the mixed solutions. This inhibition depends on the local number of ions and the valence of cations. The space for water molecules to form the hydrogen bonds becomes small in the case of the mixed solution of the protein and calcium chloride. These findings are consistent with the synergistic effect. In addition, it is found that the diffusion of ions near positively-charged residues is attenuated. This attenuation causes an increase in the possibility of water molecules staying near or inside the hydration shells of the ions. Furthermore, the first hydration shells of the cations become weak in the vicinity of the arginine, lysine and glutamic-acid residues. These factors can be considered to be possible mechanisms of the synergistic effect.

Keywords: ice surface; winter flounder antifreeze protein; sodium ions; calcium ions; synergistic effects; molecular dynamics simulation

---

1. Introduction

The control of freezing water at micrometer levels has become very important in recent years in various fields: (1) cryopreservation of food and food ingredients [1,2]; (2) cryopreservation of blood, stem cells and organs to be transplanted [3]; (3) cryosurgery [4]; and (4) ice slurry for cooling or energy storage [5].

Several methods have been developed for this control of freezing. Quick freezing is a promising method to avoid the rapid growth of ice crystals and the destruction of preserved materials. However, this type of freezing requires a lot of electrical energy. Another promising method is the use of an additive, which functions to lower the freezing point while retaining the melting point. The growth of ice can be arrested by controlling the preservation temperature in a gap between these two points. Antifreeze proteins (AFPs) are appropriate additives because they produce a wide gap between the two points despite their low concentration. Furthermore, they do not increase osmotic pressure significantly and are non-toxic.

Among various AFPs discovered, HPLC6, the major fraction of winter flounder AFP, has been widely studied. HPLC6 consists of 37 amino-acid residues, and forms a helical structure [6]. Although the majority of these amino-acid residues are alanine, four threonine residues are positioned at nearly identical distances on one line parallel to the helical axis. Bi-pyramidal ice crystals, covered with (20$\bar{2}$1) pyramidal faces, are usually observed in the HPLC solution when the temperature is between the two points in a quasi-equilibrium condition. The distance between the oxygen atoms on these pyramidal faces is nearly identical to the distance between the threonine residues. Thus, it had been hypothesized that the hydrogen atoms of the threonine residues were bonded permanently to the oxygen atoms on the pyramidal faces in the ice crystal and that the water molecules were prevented from bonding to the ice surface by the Gibbs–Thomson (or Kelvin) effect [7,8]. On the other hand, Haymet et al. [9] showed by using mutants of HPLC6 that the hydrophobic interaction between the methyl group in the threonine residues and the ice surfaces plays a more important role in the ice growth inhibition mechanism than the hydrogen bond. Furthermore, Baardsnes et al. [10] suggested, from their experimental results with another mutant, that three alanine residues and an adjacent threonine residue in the central part of the HPLC6 form a surface binding to the ice. Based on these results, Davies et al. [11] and Jorov et al. [12] discussed theoretically the key role of the alanine-rich surfaces of HPLC6 in the ice-growth inhibition mechanism.

To elucidate the ice growth inhibition mechanism of HPLC6 in more detail, molecular dynamics simulations were carried out [13–16]. The simulation results showed the following scenarios: (1) there is no significant gain of hydrogen bonds of HPLC6 to the water molecules at the ice surface [13]; (2) the specific side of HPLC6, including not only the threonine residues but also the alanine residues, is oriented toward the pyramidal faces [14]; and (3) HPLC6 binds stably to the pyramidal faces when the hydrophobic residues bind to the faces [15,16]. Taken together, these results show that the van der Waals' interactions play an important role in the binding of AFPs onto the ice surface.

Despite these results, the antifreeze mechanism due to HPLC6 in the winter flounder has not yet been fully understood. In particular, the effect of other solutes on the ice/AFP interaction has not yet been discussed in detail. Liquid, not only in winter flounder but also in food, cells, and organs, contains ions. Evans et al. [17] observed that the freezing point depression for the mixed aqueous solution of HPLC6 and sodium chloride is significantly more than the sum of the freezing point depression of the HPLC6 solution and that of salt solution (The freezing point depression of salt solutions occurs through a colligative mechanism, while this depression of AFP solutions and mixed solutions occurs through a non-colligative mechanism). This difference in the freezing point depression is due to the synergistic effect of the salt and the AFP, and is likely to be a result of the hydration shells surrounding the dissolved ions [17]. Kristiansen et al. [18] observed a similar effect for the mixed solutions of insect AFP and various salts including calcium chloride. Hagiwara and Aomatsu [19] obtained results of the depression of interface temperature in the unidirectional freezing of mixed solutions of HPLC6 and sodium chloride, which was consistent with the synergistic effects. They also discovered a decrease in the ion concentration with time at the interface of a mixed solution of HPLC6 and sodium permanganate. Hayakari and Hagiwara [20] carried out a molecular dynamics analysis on a mixed solution of HPLC6 and sodium chloride near the pyramidal face of an ice layer. They obtained the conclusion that the translational motion and freely-rotational motion of HPLC6 were attenuated by the ions as a result of the hydration of the ions. They also concluded that the approach of HPLC6 towards the ice surface with the settlement of rotation of HPLC6, caused by the hydration of ions, enhanced the interaction between a part of the protein including the threonine residues and the water molecules on the ice surface. This interaction is thought to be a possible mechanism of the synergistic effect. However, more research works, in particular molecular dynamics analysis for the mixture of AFP and other ions, are necessary for elucidating the synergistic effect.

In the present study, we carry out molecular dynamics simulations for the dilute aqueous solution of HPLC6, sodium ions, calcium ions and chloride ions near a thin ice layer covered with secondary prism or pyramidal faces, in order to analyze the interaction between the ice and AFP and the effect of ions on the interaction. The secondary prism face is the face of the ice showing the most growth [21]. Thus, the effects of solutes on the ice growth are expected to be the most noticeable. On the other hand, the pyramidal face is the face of the ice showing the least growth. Thus, the amino acid residues of HPLC6, which are bound on this face, are almost stationary. We selected the pyramidal faces only when we examined whether amino acid residues of HPLC6 influence the diffusion of the cations. It is preferable to attenuate the motion and rotation of HPLC6 for measuring the displacement and diffusion coefficient of cations.

2. Computational Procedures

We have been revising our in-house simulation program for many years [20,22–26]. In the present study, we have revised our program further.

2.1. Assumptions

In the main computation, we adopted the Canonical ensemble. Thus, the number of molecules, the volume of the computational domain, and the temperature were assumed to be unchanged. On the other hand, in the preliminary computation, we changed the temperature and energy to obtain the initial condition. Based on our previous study [20], we assumed that the impact of the change in the ensemble would be negligible in the period of 300 ps.

2.2. Governing Equations

The Newton–Euler equations for the translational and rotational motions of the molecules were solved at each time step, and were integrated with respect to time by using the Gear algorithm [27]. We used a 5-value Gear algorithm to the Newton equation for the translational motion, in which the time derivatives up to the fifth order were considered. We adopted a 4-value Gear algorithm to the Euler equation for the rotation. All the computations were carried out with a time step of 0.5 fs. Periodic boundary conditions were also imposed on the computational domain.

2.3. Temperature Scaling

We kept the temperature constant in a specific region for the specific period mentioned below. For this procedure, the statistical temperature, T, was defined by the total energy of the translational motion for all the molecules, K_T, and that of the rotational motion for the molecules, K_R. The temperature is written as follows:

$$T = \frac{1}{2}\left(\frac{2}{3k_BN}K_T + \frac{2}{3k_BN}K_R\right) = \frac{1}{3k_BN}\left(\frac{1}{2}\sum_{i=1}^{N}mv_i^2 + \frac{1}{2}\sum_{i=1}^{N}\left(I_{px_i}\omega_{px_i}^2 + I_{py_i}\omega_{py_i}^2 + I_{pz_i}\omega_{pz_i}^2\right)\right), \quad (1)$$

where k_B is the Boltzmann constant, N is the total number of molecules, m is the mass of molecules, and I_p is the inertia moment based on its principal axis. Temperature scaling was carried out locally for certain periods in which the translational velocity, v, and the angular velocity, ω, of each molecule were changed by the following equation:

$$v_i^{(new)} = v_i^{(old)}\sqrt{\frac{T_{pd}}{T}}, \quad \omega_i^{(new)} = \omega_i^{(old)}\sqrt{\frac{T_{pd}}{T}}, \quad (2)$$

where T_{pd} is the predetermined temperature.

2.4. Production of Mixtures

2.4.1. In the Case of Ice Layers with Prism Faces

We produced the mixture by using the following three-stage procedure in a similar way to that adopted in [26].

Stage 1: First, a small ice cube with *Ih* structure was formed. This procedure is the same as that described in our previous study [22]. A total of 360 water molecules were in a cubic domain of approximately 2.25 nm. In this domain, the center of mass of each molecule was allocated at the lattice of the *Ih* structure. The orientation of hydrogen atoms was randomly determined. Secondly, 24 ice cubes were connected to each other in order to form a large rectangular prism of ice. The dimensions of the new ice crystal were approximately 6.736 nm × 4.667 nm × 8.800 nm. The *x-*, *y-* and *z*-axes were located along the median, shortest, and longest edges of the rectangular prism, respectively.

Stage 2: A region of one-third in the *x*-direction of the rectangular prism was specified as the ice layer. The outside of the ice layer was defined as the liquid region. The computation started. The temperature for the liquid region was set at 370 K at the initial instant in order to ensure liquidization. It was decreased in steps to 250 K in a period of 350 ps (See Table 1). The restraint of the molecules in the ice was carried out at 10 K in order to reduce the impact of melting. Then, the temperature of the ice layer was increased in steps to 250 K in the period (See Table 1). The secondary prism faces were realized at the interface.

Table 1. Predetermined temperatures for producing the mixture of water an ice layer with prism faces.

Periods [ps]	0–50	50–100	100–150	150–200	200–250	250–300	300–350	350–
T_{pd} for ice [K]	10	50	100	150	200	250	260	250
T_{pd} for water [K]	370	370	340	310	290	280	260	250

Stage 3: In the cases with the HPLC6 model, some water molecules were removed from a cocoon-shaped region in the computation result at 350 ps. The longest axes of these regions, and thus the axes of the model, were parallel to the z-axis (See Figure 1). The model was introduced into the region so that the four threonine residues of the model faced the interface at this instant.

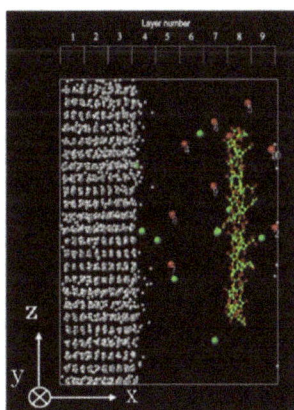

Figure 1. The computational domain with an ice layer with prism faces. White dots show oxygen atoms of water molecules in the bulk ice region (water molecules in the liquid region are not shown). Red dots and green dots show sodium ions and chloride ions, respectively. Many dots connected show the HPLC6 model. Nine layers are defined in the computational domain.

In addition, in the cases containing the ions, a number of water molecules from the computation result at 350 ps, which were selected arbitrarily, were replaced with sodium ions, calcium ions or chloride ions. Based on our previous study [20], we assumed that the impact of the introduction of solutes would be negligible in the period of 300 ps, except for the interaction between the solutes. This exception is due to the diffusion or slow motion of ions. To simulate the *NVT* ensemble after 350 ps, we adopted the method developed by Nosé [28]. The computations continued until 2 ns for all the cases. This period is not too short compared with the period of 4 ns in the case of the simulation for an ice layer and water with a HPLC6 model in Ref. [15].

2.4.2. In the Cases of Ice Layers with Pyramidal Faces

We produced the mixture by using the following five-stage procedure in the same way as that developed in [20].

Stage 1: The same ice cube as that described in Section 2.4.1 was used. 60 ice cubes were connected to each other in order to form another large rectangular prism of ice.

Stage 2: Water molecules inside two regions of the triangular prisms and two regions of the pentagonal prisms were removed in order to obtain a new rectangular prism ice crystal. The two smallest surfaces of this rectangular prism were covered with the pyramidal face of $(20\bar{2}1)$. The dimensions of the new ice crystal were 6.74 nm × 7.00 nm × 8.80 nm. The *x*-, *y*-, and *z*-axes were located along the shortest, median, and longest edges of the rectangular prism, respectively.

Stage 3: A region of approximately 32% in the *y*-direction of the rectangular prism was specified as the ice layer. One third of the layer in the middle of the *y*-direction was considered as the core of the ice layer. The outside of the ice layer was defined as the liquid region.

Stage 4: The computation started. The temperature for the liquid region was set at 370 K at the initial instant in order to ensure liquidization. It was decreased in steps to 280 K in a period of 100 ps (See Table 2). The restraint of the molecules in the ice was carried out at 0 K for the first 75 ps in order to reduce the impact of the melting. Then, the temperature for the ice layer was increased in steps for the first 95 ps and then maintained at 200 K in the core region of the ice layer and at 250 K in the surface region of the layer (See Table 2).

Stage 5: The method for introducing the protein model into the liquid water region is the same as that explained in Section 2.4.1. Consequently, the α-helical axis of HPLC6 aligned with the $\langle 10\bar{1}2 \rangle$ vector of the pyramidal face. Due to the periodic boundary conditions, the distance between the adjacent protein molecules is 6.74 nm in the *x*-direction. This distance is wider than that of approximately 5 nm, which is the footprint of the protein/ice-surface interaction measured by Grandum et al. [29]. They measured the grooves and ridges on the ice surface with a scanning tunnel microscope. Thus, the dimension of our computational domain is sufficiently large.

A number of water molecules from the computation result at 100 ps, which were selected arbitrarily, were replaced with sodium ions, calcium ions or chloride ions in the cases containing the ions.

The method developed by Nosé was used to simulate the *NVT* ensemble after 100 ps. The computations continued in the period of 1 ns.

Table 2. Predetermined temperatures for producing the mixture of water an ice layer with pyramidal faces.

Periods [ps]	0–50	50–60	60–65	65–70	70–75	75–80	80–85	85–90	90–95	95–100
T_{pd} for ice [K]	0	0	0	0	0	50	100	150	200	250 (200)
T_{pd} for water [K]	370	350	330	310	290	280	280	280	280	280

2.5. Potential Functions of Water Molecules

The TIP4P/Ice potential function proposed by Abascal et al. [30] was used for the potential function of the interaction between two water molecules. This potential was developed by the re-parameterization of the TIP4P potential [31], and has been used for ice. The TIP4P potential consists of the Coulomb potential and the Lennard–Jones potential. The numerical constants for TIP4P/Ice are almost the same as those for TIP4P except for the potential well depth of Lennard–Jones potential and the electric charges of the Coulomb potential. These parameters for TIP4P/Ice are approximately 36% and 13% higher than those of TIP4P, respectively. The Ewald summation (See Frenkel and Smit [32]) was used for the Coulomb potential in order to reduce the electrical force of distant molecules. The parameters for the Ewald summation; i.e., the width of the Gaussian distribution, the maximum value of the vector for the location of the image cell in the physical space, and the maximum value of the vector for the location of the image cell in the Fourier space—were the same as those adopted in our previous study [24].

2.6. Model of HPLC6

Table 3 shows the primary structure of HPLC6. We adopted a rigid, helical model of HPLC6, which was similar to that used by Nada and Furukawa [15]. The location of atoms was decided based on the result obtained by Sicheri and Yang [8]. The rigidity can be considered as an overestimation of the hydrogen bond between two adjacent sites in the axial direction of the model. Flexibility can be considered by adopting an algorithm, such as SHAKE developed by Ryckaert et al. [33]. However, we did not include any algorithm for the flexibility because the time scale (or frequency) of deformation of HPLC6 due to the interaction between the model and water molecules is not clear. In addition, the fluctuation of dihedral angles of the central amino-acid residues of HPLC6 was within 9 degrees, according to the molecular dynamics simulation with AMBER by Cheng and Merz [34]. Therefore, we think that the rigid model is not too unrealistic. If we obtain results, which clearly show the validity of flexible HPLC6 models from the viewpoint of the length scale and time scale, we will show our results by using a new, flexible model in a future paper.

The OPLS parameters developed by Jorgensen et al. [35] were adopted as the potential parameters of each site of the model. The sum of the electric charges assigned to the groups in the amino-acid residues were −1. The cutoff radius was set at 1.5 nm. This is to avoid the unrealistic interaction through the ice layer between a water molecule in solution near one side of the ice layer and a water molecule in solution near the other side of the ice layer. The center of the radius was set at the indicator point of each group of sites. The parameters for the Ewald summation for the water molecules, mentioned in Section 2.5, were used also for the Coulomb potential of the sites of amino-acid residues of HPLC6 in order to reduce the electrical force between the sites and distant water molecules.

Table 3. Winter flounder antifreeze protein.

	Primary Structure	Molecular Weight
Winter flounder AFP	DTASDAAAAAALTAANAKAAAELTAANAAAAAAATAR	3243 Da

D: Aspartic acid, T: Threonine, S: Serine, A: Alanine, L: Leucine, N: Asparagine, K: Lysine, E: Glutamic acid, R: Arginine.

2.7. Model of Ions

The models developed by Chowdhuri and Chandra [36] were used for the sodium and chloride ions, while the model proposed by Chialvo [37] was used for the calcium ions.

2.8. Computational Conditions

The computational conditions in the present study were compared with those in the literature [13–16,20] in Table 4. It is found from this table that the dimension of the liquid region and the concentration of solutes are reasonable.

Table 4. Comparison of computational conditions.

	Dalal [13]	Wierzbicki [14]	Nada [15]	Hayakari [20]	Todde [16]		Present Study	
Ice Face	Pyramidal	Pyramidal	Pyramidal	Pyramidal	Prism	Pyramidal	Prism	Pyramidal
Domain size [nm³]	8.4 × 4.0 × 7.8	5.6 × 12.0 × 7.6	7.0 × 5.7 × 7.0	6.7 × 7.0 × 8.8	6.1 × 21.7 × 5.9	18.1 × 6.7 × 7.5	6.7 × 4.7 × 8.8	6.7 × 7.0 × 8.8
Ice thickness [nm]	1.4	3.8	1.4	2.3	11	9	2.3	2.3
Number of H₂O molecules liquid	5484	11,292	6520	8570	12,288	14,724	5537 [1]	8541 [2]
Number of H₂O molecules solid	2841	5313	2160	4185	12,288	14076	2880	4185
Number of AFP molecules	1	2	1	1	1	1	1	1
AFP conc. [mg/mL]	32	31	26	19	13	11	29	19
Number of cations	-	-	-	24	-	-	10	15
Ion conc. [mol/L]	-	-	-	0.15	-	-	0.10	0.10
Temperature [K]	190	165	268	260	228	228	250	250
Ensemble	NVT	NVT	NPT	NV	NPT	NPT	NVT	NVT
Potential functions	TIP3P	TIP3P	6site	TIP4P/Ice	TIP4P	TIP4P	TIP4P/Ice	TIP4P/Ice

[1] 5527 in the case of CaCl₂; [2] 8526 in the case of CaCl₂.

3. Statistical Quantities

We discuss the following statistical quantities. These are the same as those defined in our previous studies [20,24–26] except for the tetrahedricity parameter.

3.1. Radial Distribution Function (RDF)

The RDF between an atom of a water molecule or an ion, and the oxygen atoms in a spherical shell is defined as follows (See, for example, Allen and Tildesley [38]):

$$g(r) = \frac{V}{4\pi r^2 \Delta r N^2} \left\langle \sum_i \sum_{j \neq i} \delta(r_i - r_j) \right\rangle, \tag{3}$$

where $r = |r_i - r_j|$ is the distance between two atoms (or between a site and an atom), V is the volume of the computation domain, Δr is the thickness of the shell, N is the number of oxygen atoms, and δ is the delta function. Δr was equal to 0.01 nm. Please note that the total number of water molecules in the spherical domain ($0 < r < \infty$) was not identical because the water molecules inside the ice layer were not included in the summation in this equation.

3.2. Hydrogen Bond Correlation Function (HBCF)

In our previous study [25], we used the hydrogen-bond correlation function (hereafter abbreviated to HBCF) to analyze the hydrogen bond among water molecules. The function is defined as follows:

$$C_{HB}(t) = \frac{\left\langle \sum_j^N n_{HBj}(t_0 + t) n_{HBj}(t_0) \right\rangle}{\left\langle \sum_j^N n_{HBj}(t_0) n_{HBj}(t_0) \right\rangle}, \tag{4}$$

where $n_{HBj}(t_0)$ is the number of water molecules bonded with a molecular j at a reference time t_0, and $n_{HBj}(t_0 + t)$ is the number of water molecules that have maintained the hydrogen bond with a molecular j for the period of t. The cluster structure was identified by only the distance of water molecules from a specific water molecule. In the present study, we focus on water molecules in the first hydration shell of a specific ion.

3.3. Diffusion Coefficient

The diffusion coefficient can be obtained from the following equation:

$$D_S = \lim_{t \to \infty} \frac{1}{6t} \left\langle |r_j(t_1 + t) - r_j(t_1)|^2 \right\rangle,$$ (5)

where t_1 is defined as an instant when the mean square displacement obeys the power law with time [39]. The mean square displacement, $D(t)$, is defined by the following equation:

$$D(t) = \frac{1}{N} \left\langle \sum_j |r_j(t_0 + t) - r_j(t_0)|^2 \right\rangle,$$ (6)

where r_j is the position vector of a molecular j, t is time, t_0 is a reference time, N is the number of molecules and the angle bracket shows a time mean value. The mean squared deviation (MSD) was obtained from the position vectors for the periods of 10 ps at an interval of 0.5 ps. The moving-averaged values of the MSD were obtained over the period of 50 ps.

3.4. Tetrahedricity Parameter

We adopted the tetrahedricity parameter [40] as an index to estimate water molecular structure in the liquid phase. The tetrahedricity parameter M_T is expressed by the following equation:

$$M_T = \frac{\sum_{i > j} (l_i - l_j)^2}{15 \langle l^2 \rangle}$$ (7)

where l_i are the lengths of the six edges of a tetrahedron. If the nearest four water molecules to a water molecule form a tetrahedron, M_T becomes nearly zero. On the other hand, if the distances between two of four water molecules differ from each other, M_T becomes nearly a unity. Thus, the tetrahedricity parameter shows the degree of the water structure.

4. Results and Discussion

We will discuss the results for the system of solutions with a thin ice layer covered with secondary prism faces in Sections 4.1–4.5. We will then discuss the results for the system of solutions with a thin ice layer covered with pyramidal faces in Section 4.6. The results obtained were thermodynamically non-equilibrium because the diffusion of ions and the interaction of solutes are too slow to capture their developed state.

4.1. Time Changes in the Ion Positions

Figure 2 shows time changes in the ion position in the *x*-direction. The origin of the ordinate corresponds to the center of the computational domain. The positions close to the top and bottom edges in the graphs indicate that the corresponding ions are located close to the ice surfaces. Figure 2a,b show the time changes in the cases without HPLC6, while Figure 2c,d show the time changes in the cases with HPLC6. It is found from these figures that the changes in the ion positions were within the range of 0.5 nm for the period of 1500 ps except for one or two ions in each case. The time changes are due to the diffusion of ions or the interaction between the ions and HPLC6.

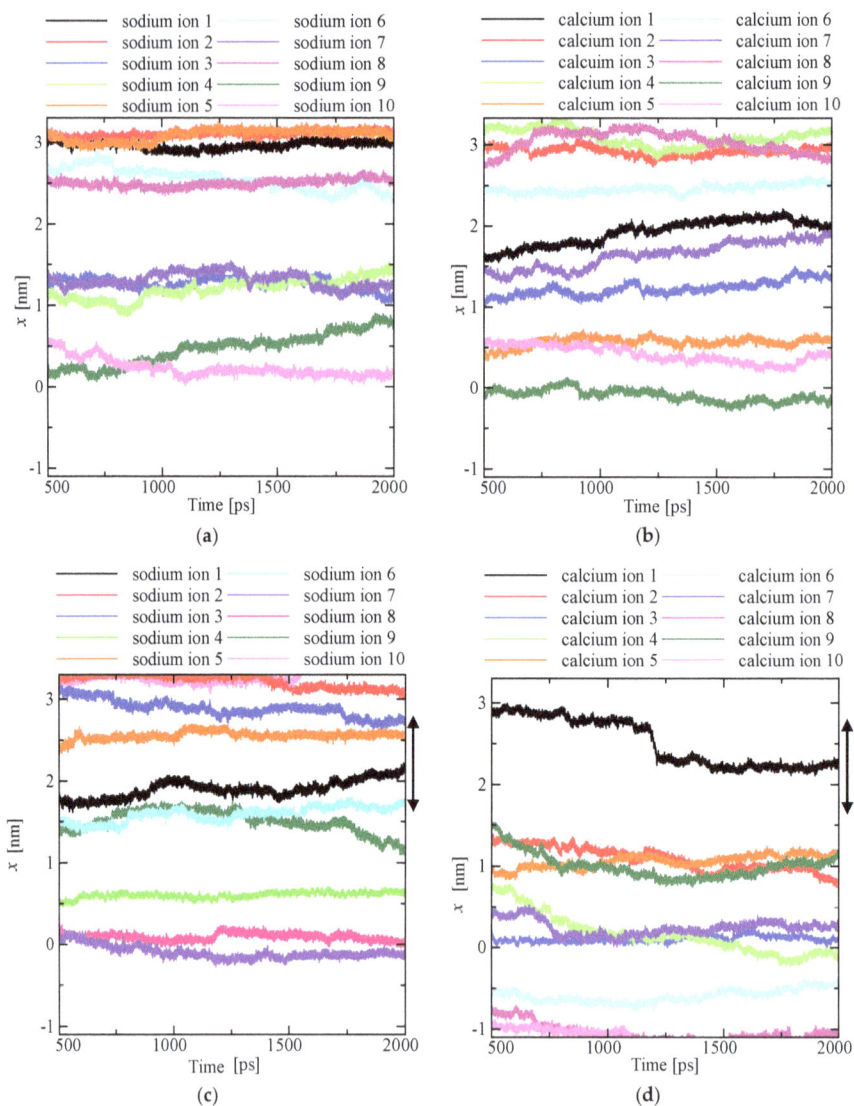

Figure 2. Time changes in the positions of ions; (**a**) Sodium ions in the NaCl solution, (**b**) Calcium ions in the CaCl$_2$ solution, (**c**) Sodium ions in the mixed solution, (**d**) Calcium ions in the mixed solution. The ice layer was located in the range of $-3.366 < x < -1.123$ [nm] during the period. The arrows in (**c**) and (**d**) show the range of location of HPLC6 during the period.

4.2. Radial Distribution Function (RDF)

Figure 3 shows typical examples of the RDF for water molecules around several ions in the mixed solutions. The peak values differ by margins of approximately 10% for the sodium ions and 45% for the calcium ions for 1500–1550 ps. On the other hand, the peak values differ by margins of approximately 4% for the sodium ions and 2% for the calcium ions for 1950–2000 ps. These changes in the differences are due to the changes in the RDFs for water molecules around the sodium ion 10, and the calcium ions 8 and 10. It is found from Figure 2c,d that these three ions gradually approached the ice surfaces by 1500 ps and that they gradually left from the locations adjacent to the ice surfaces by 2000 ps. Thus, the hydration shells of cations were not affected by the presence of HPLC6.

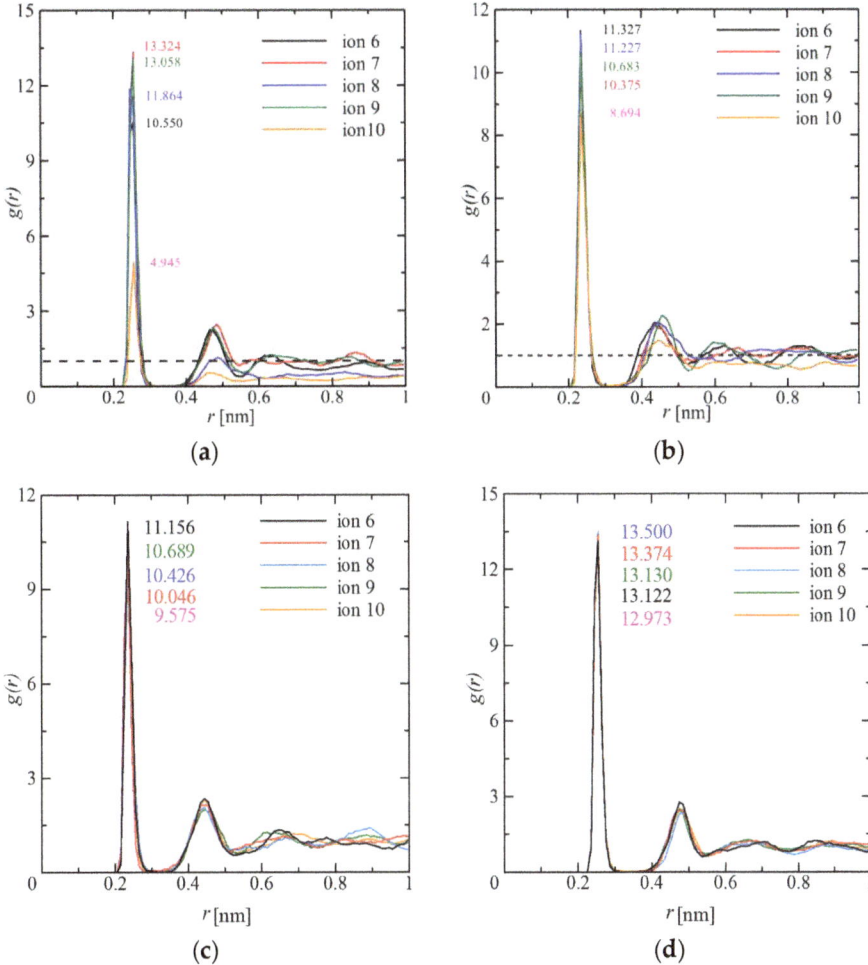

Figure 3. Radial distribution functions of water molecules around cations; (**a**) Sodium ions for 1500–1550 ps, (**b**) Calcium ions for 1500–1550 ps, (**c**) Sodium ions for 1950–2000 ps, (**d**) Calcium ions for 1950–2000 ps.

4.3. Tetrahedricity Parameter

The average value of the tetrahedricity parameter gradually decreased with time for all the solutions (See Table 5). This shows that the tetrahedral structure of the water molecules was forming. The decreasing rate of the average values in the cases of solutions with calcium chloride was lower than that in the cases of solutions with sodium chloride. This is because the hydration shells of calcium ions, which are stronger than the hydration shells of sodium ions, disturb the tetrahedral structure.

Table 5. Average values of the tetrahedricity parameter.

Solution	NaCl	NaCl + HPLC6	CaCl$_2$	CaCl$_2$ + HPLC6
1000–1050 ps	0.041	0.044	0.040	0.039
1500–1550 ps	0.038	0.039	0.039	0.040
1950–2000 ps	0.038	0.039	0.039	0.037

Figure 4a,b show the probability distributions of the tetrahedricity parameter, M_T, for salt solutions and mixed solutions in the period of 1500–1550 ps. The most probable values of M_T (the values of M_T for their highest probabilities) in the case of mixed solutions are higher than those in the case of salt solutions. This indicates that the HPLC6 inhibits water molecules forming a tetrahedral structure not only near the HPLC6 itself but also far from it.

Figure 4c,d show the distributions of the tetrahedricity parameter for the salt solutions and mixed solutions in the period of 1950–2000 ps. The distribution in Figure 4c is similar to that in Figure 4a, while the distribution in Figure 4d is not similar to that in Figure 4b. This is because the time changes in the positions of the calcium ions were predominant in the mixed solution in this period (See Figure 2d).

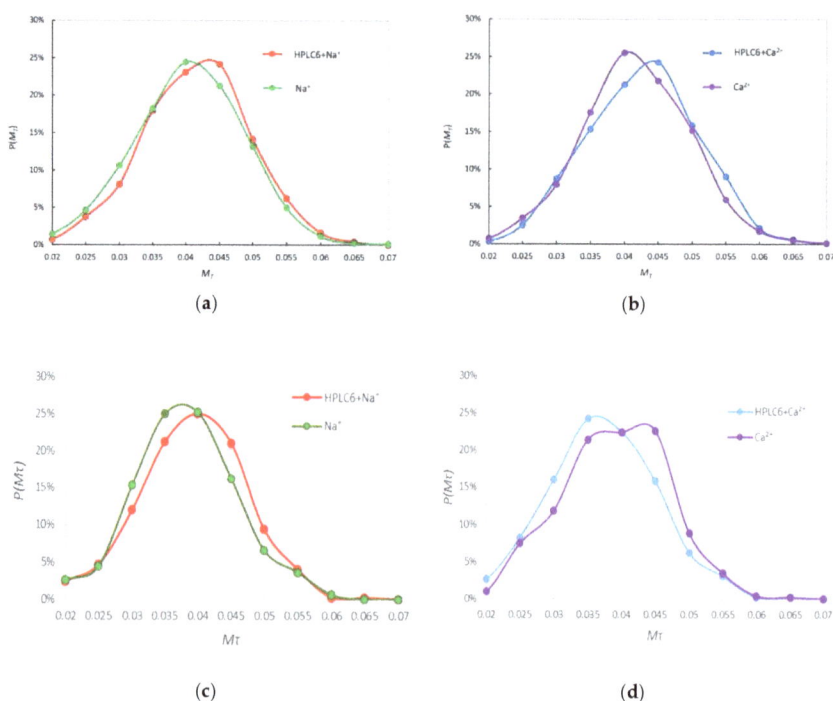

Figure 4. Distributions of the tetrahedricity parameter; (**a**) in the case with NaCl for 1500–1550 ps, (**b**) in the case with CaCl$_2$ for 1500–1550 ps, (**c**) in the case with NaCl for 1950–2000 ps, (**d**) in the case with CaCl$_2$ for 1950–2000 ps.

4.4. Average Number of Hydrogen Bonds

We evaluated the number of hydrogen bonds for a water molecule. The criteria for the hydrogen bond are as follows (The criterion for the angle is stricter than that proposed by Luzar and Chandler [41]):

(A)   The distance between oxygen atoms is lower than 3.5 Å.

(B)   The angle between the line connecting two oxygen atoms and the line connecting one of the two oxygen atoms and a covalent-bonded hydrogen atom is within $\pm 15°$.

We divided the computational domain into 9 layers in the x-direction, and examined the effects of ions and HPLC6 on the average number of hydrogen bonds per water molecule in each layer (See Figure 1). The thickness of each layer was 0.748 nm. In Ref. [16], the computational domain was divided into the regions of bulk ice, interface, and bulk water. The layer thickness is approximately half the interface width. Thus, the 1st, 2nd, and 3rd layers are in the bulk ice region, the 4th, 5th, 8th, and 9th layers are in the interface region, and the 6th and 7th layers are in the bulk water region.

Figure 5 shows the average number of hydrogen bonds for each layer for the periods of 1000–1050 ps and 1950–2000 ps. The first three layers were in the ice layer, while the other six layers were in the liquid region throughout the computation. Most of the residues of HPLC6 were located in the eighth layer. It is found from Figure 5a that the average numbers of hydrogen bonds for the single-solute solutions are lower than the average numbers for pure water in the layers in the liquid phase. In particular, the average number is lowest in the eighth and ninth layers for the HLPC6 solution (shown with the figure caption 'poly'). This is the result of hydrogen bonds between water molecules and the hydrophilic residues of HPLC6. Moreover, the hydrophobic hydration due to the hydrophobic residues kept water molecules away from HPLC6.

The average number of hydrogen bonds in the 4th, 5th, and 6th layers indicates lower values for salt solutions compared with the number for pure water. (The salt solutions are shown with the figure caption 'Na$^+$' and 'Ca^{2+}'. Note that the anion Cl$^-$ also contributed.) This is because water molecules are more attracted by the hydration shells of ions, particularly the cations in these layers. The number of hydrogen bonds decreased more than 4% in the 5th and 6th layers of the mixed solution of HPLC6 and CaCl$_2$ compared with the number of bonds in pure water, although the concentration of HPLC6 and CaCl$_2$ was 8.9 mM and 0.1 M, respectively. The number of calcium ions in the 5th layers increased from 1 (ion no. 3 in Figure 2d) to 3 (ion no. 3, 4 and 7) during the period 700–800 ps in the mixed solution. It can be surmised that the strong hydration shells of calcium ions prevent water molecules from forming the hydrogen bonds in these two layers.

In Figure 5b, the number of hydrogen bonds of salt solutions in the 7th and 8th layers is lower than that in the other cases in the period of 1950–2000 ps. This is because the time changes in the positions of some sodium ions (3 and 6 in Figure 2a) and calcium ions (1 and 7 in Figure 2b) in these layers are more noticeable than the time changes in the positions of the other ions.

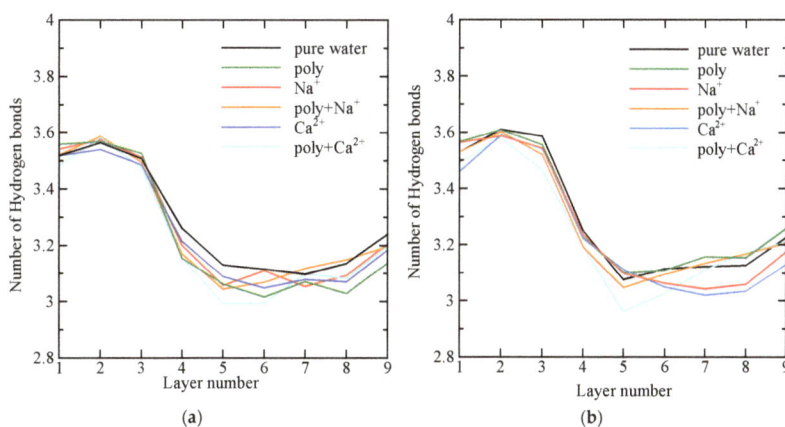

Figure 5. Number of hydrogen bonds in the layers; (**a**) for 1000–1050 ps, (**b**) for 1950–2000 ps.

4.5. Hydrogen Bond Correlation Function (HBCF)

Figure 6 shows the hydrogen bond correlation function (HBCF) as a function of time for the fifth and eighth layers. It is found from Figure 6a that the HBCF for the 5th layer of the NaCl solution and mixed solution indicates lower values than that for the pure water and HPLC6 solution at any time in the period of 1000–1050 ps. This indicates that the hydrogen bond among water molecules was disturbed by the ions and their motion in the 5th layer. A similar result was obtained for the CaCl$_2$ solution [37].

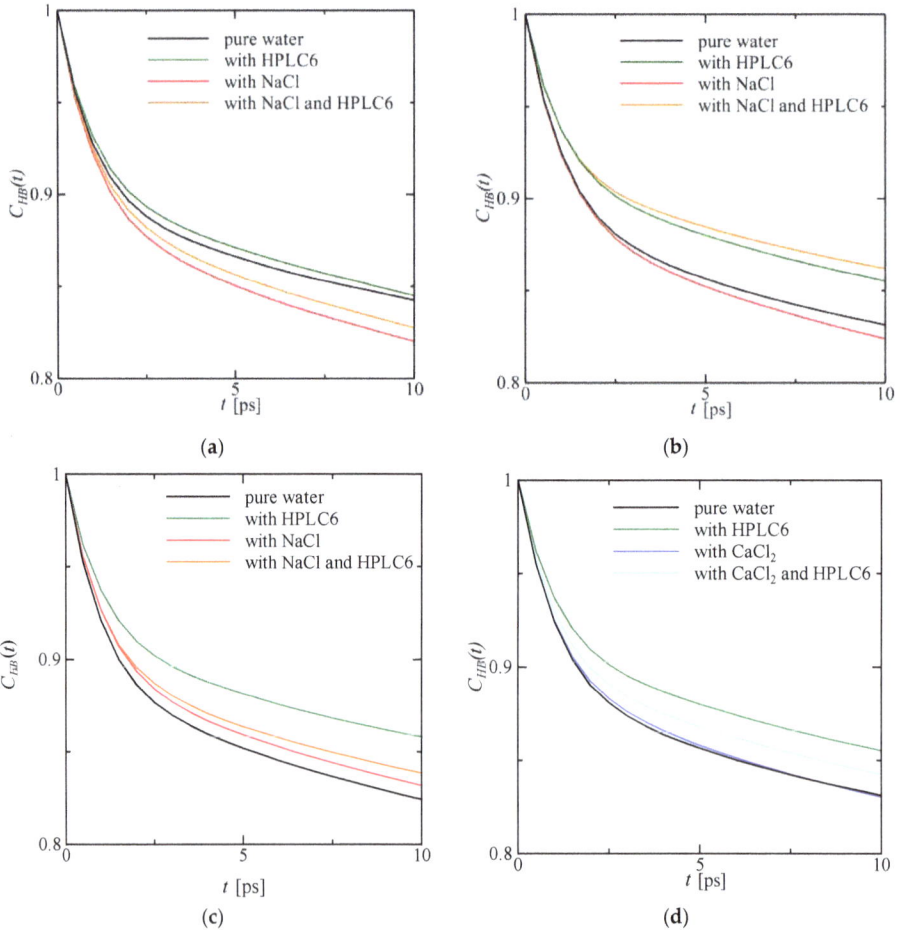

Figure 6. Hydrogen bond correlation functions in two layers; (**a**) 5th layer for 1000–1050 ps, (**b**) 8th layer for 1000–1050 ps, (**c**) 5th layer for 1950–2000 ps, (**d**) 8th layer for 1000–1050 ps.

It is found from Figure 6b that the HBCF for the eighth layer of the HPLC6 solution and mixed solution indicates higher values than that for the pure water and NaCl solution at any time in the period of 1000–1050 ps. This clearly indicates that the hydrogen bond among water molecules near the HPLC6 was enhanced by the HPLC6 molecule in the 8th layer. This may result in the formation of hydrophobic hydration shells around the alanine residues of HPLC6 in this layer. The effect of ions on the hydrogen bonding is not noticeable in this layer in this period probably because the hydration shells around the ions have not yet been formed well.

Figure 6c shows the HBCF of the solutions for the 5th layer in the period of 1950–2000 ps. The difference between the correlation function in the case of the HPLC6 solution and that in the other cases became noticeable compared with the difference shown in Figure 4a. This is probably because the space for forming the hydrogen bond among water molecules became small as a result of the growth of hydrophobic hydration shells around the HPLC6.

Figure 6d shows the HBCF of the solutions with $CaCl_2$ for the 8th layer in the period of 1000–1050 ps. The calcium ions in this layer did not disturb the hydrogen bond in the case of the salt solution. This is because only one calcium ion (6 in Figure 2b) was in this layer in this period. On the other hand, the ions in the case of the mixed solution disturbed the hydrogen bond in the layer in the period. This is probably because the effect of one calcium ion (1 in Figure 2d) in the 9th layer, which approached and entered into the 8th layer just after the period, was remarkable. This is consistent with the synergistic effects of ions and HPLC6.

4.6. Interaction between Cations and Amino Acid Residues

4.6.1. Surface-Bound Water Molecules

We allocated the HPLC6 model so that the four threonine residues were closest to the pyramidal face. Also, three alanine residues (the 6th, 17th and 28th) were close to the face. Thus, the spatial arrangement of HPLC6 was similar to that in Ref. [15].

We monitored the motion of many water molecules between the atom groups of the threonine residues and the ice surface. We discovered that at least three water molecules were nearly stationary, while the other water molecules nearby moved more rapidly in three locations; (1) near the hydroxyl group of the 24th threonine residue; (2) near the methyl group of the threonine residue; and (3) near the hydroxyl group of the 35th threonine residue. We believe that these nearly stationary water molecules are surface-bound water molecules.

4.6.2. Diffusion Coefficient

Table 6 shows the ratios of the diffusion coefficient of a cation nearest to a specific amino-acid residue to the diffusion coefficient of water. The arginine, lysine and glutamic-acid residues are located opposite to the four threonine residues with respect to the HPLC6 axis. Thus, these residues were inside the water region because the four threonine residues faced the interface.

This table also shows the ratios obtained by Nohara and Hagiwara [26] and Baştuğ and Kuyucak [42]. Ref. [26] dealt with the ions in supercooled water between ice layers with the prism faces, and used approximately one-third of the number of water molecules. Ref. [42] dealt with the ions in water at 273 K without any ice, and used a different potential function for water and ions.

Table 6. Ratio of diffusion coefficients of cations to the diffusion coefficient of water.

Na⁺			Ca²⁺			
Near Arginine	Nohara	Baştuğ	Near Lysine	Near Glutamic Acid	Nohara	Baştuğ
0.18	0.46	0.42	0.19	1.06	0.28	0.31

It is found from this table that the ratio for the sodium ion near the arginine residue (R in Table 3) is much lower than the ratio in [26,42]. This shows that the diffusion of the sodium ion in the vicinity of the arginine residue was attenuated. This is probably a result of electrostatic attraction between the oxygen atoms of water molecules in the hydration shell of the sodium ion and the C = NH group of the arginine residue because this group is positively electrified in water and the oxygen atoms are negatively electrified. Similarly, the ratio for the calcium ion near the lysine residue (K in Table 3) is lower than the ratio in [26,42]. The attenuation of diffusion of the calcium ion in the vicinity of the lysine residue is probably a result of electrostatic attraction between the oxygen atoms of water molecules in the hydration shells of the calcium ion and the amino group of the lysine residue. This is because this amino group is also positively electrified in water. On the other hand, the ratio for the calcium ion near the glutamic acid residue (E in Table 3) is much higher than the ratio in [26,42]. This shows that the diffusion of the calcium ion in the vicinity of the glutamic acid residue was enhanced. This is probably a result of electrostatic repulsion between the oxygen atoms of water molecules in the hydration shells of the calcium ion and the carboxyl group of the glutamic acid residue because the carboxyl group is negatively electrified in water. The side chains of all the other residues of HPLC6 except for the aspartic acid residue are neutrally electrified. It can be concluded that the diffusion of cations in the vicinity of HPLC6 depends on the electrified amino acid residues of HPLC6.

4.6.3. Hydrogen Bond Correlation Function (HBCF)

Figure 7 shows the hydrogen bond correlation function (HBCF) as a function of time for water molecules in the first hydration shell of the cations in the vicinity of the specific amino acid residues. The HBCF decreases with an increase in time. To discuss the robustness of the hydration shell of the cations, we compare the values of the HBCF at 10 ps. It is found from this figure and Figure 6 that the values for the cations near the electrified residues are lower than those for the cations near alanine residues and the values of water molecules in Figure 6. This shows that the first hydration shell of the cations in the vicinity of the electrified residues does not become robust and the water molecules in the first hydration shell can easily leave the shell. This weakening of the hydration shell of the cations enhances the motions of ions and water molecules. The motion enhancement of ions and water molecules can cause the attenuation of the incorporation of water molecules into the ice lattice at the interface. Also, the enhancement of ion motion is consistent with the decrease in the ion concentration of the mixed solution [19] described in the Introduction. Thus, the motion enhancement of ions and water molecules is a reason for the synergistic effect.

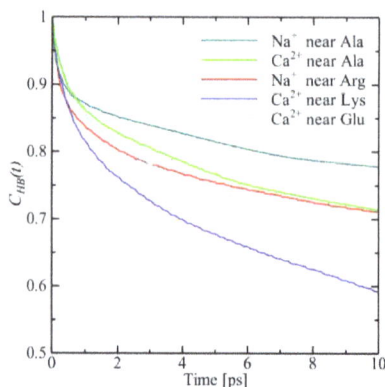

Figure 7. Hydrogen bond correlation functions of the cations.

The higher values of HBCF for the cations near alanine residues show that the first hydration shells of the cations in the vicinity of these residues remain robust. The robustness of the hydrophobic hydration shells around these residues is also expected. These robust shells also contribute to the synergistic effect. The discussion in this section is therefore consistent with the discussion in Section 4.4.

5. Conclusions

A molecular dynamics simulation was carried out on a mixture of water, a model of HPLC6 (winter flounder antifreeze protein), sodium ions, calcium ions, chloride ions and a thin ice layer. The main conclusions obtained are the following:

(1). The hydrogen bond among water molecules is inhibited by the cations even in the case of the mixed solutions. This is due to the hydration shells of the moving ions. This inhibition is more noticeable with an increase in the number density of ions and the valence of cations.

(2). The effective space for water molecules to form the hydrogen bonds becomes smaller in a region far from the HPLC6 as a result of the hydration shells of calcium ions in this region. This is also consistent with the synergistic effect of the mixed solutions of HPLC6 and $CaCl_2$.

(3). The negatively or positively electrified residues of HPLC6 respectively enhance or attenuate the diffusion of the cations in water.

(4). The hydrophobic hydration shells around the alanine residues of HPLC6 remain robust, as does the first hydration shells of the cations in the vicinity of these residues. On the other hand, the first hydration shells of the cations in the vicinity of the electrified residues do not become robust. The weakening of the hydration shells of the cations enhances the motions of ions and water molecules. The motion enhancement of ions and water molecules can cause the attenuation of the incorporation of water molecules into the ice lattice at the interface.

(5). The inhibition of the hydrogen bond due to the hydration shells of the moving ions in conclusion (1) and the weakening of the hydration shells of cations as a result of the electrified amino-acid residues nearby in conclusion (4) are the possible mechanisms of the synergistic effects suggested by Evans et al. [17].

Author Contributions: T.Y. and Y.H. conceived and designed the simulation. T.Y., T.K. and K.N. performed the simulation. T.Y., T.K., K.N. and Y.H. analyzed the data. T.Y. and Y.H. wrote the paper.

Funding: This research was funded by the Japan Society for the Promotion of Science through the Grant-in-Aid (KAKENHI) for Scientific Research (A) (15H02220).

Acknowledgments: The authors thank Hiroki Nada at the National Institute of Advanced Industrial Science and Technology Japan for his valuable comments and discussion.

Conflicts of Interest: The authors declare no conflict of interest.

References

1. Ustun, N.S.; Turhan, S. Antifreeze proteins: Characteristics, function, mechanism of action, sources and application to foods. *J. Food Process. Preserv.* **2015**, *39*, 3189–3197. [CrossRef]
2. Li, B.; Sun, D.-W. Novel methods for rapid freezing and thawing of foods—A review. *J. Food Eng.* **2002**, *54*, 175–182. [CrossRef]
3. Amir, G.; Rubinsky, B.; Basheer, S.Y.; Horowitz, L.; Jonathan, L.; Feinberg, M.S.; Smolinsky, A.K.; Lavee, L. Improved viability and reduced apoptosis in sub-zero 21-hour preservation of transplanted rat hearts using anti-freeze proteins. *J. Heart Lung Transpl.* **2005**, *24*, 1915–1929. [CrossRef] [PubMed]
4. Shitzer, A. Cryosurgery: Analysis and experimentation of cryoprobes in phase changing media. *J. Heat Transf.* **2011**, *133*. [CrossRef]
5. Grandum, S.; Yabe, A.; Tanaka, M.; Takemura, F.; Nakagomi, K. Characteristics of ice slurry containing antifreeze protein for ice storage applications. *J. Thermophys. Heat Transf.* **1997**, *11*, 461–466. [CrossRef]
6. Yang, D.S.C.; Sax, M.; Chakrabatty, A.; Hew, C.L. Crystal structure of an antifreeze polypeptide and its mechanistic implications. *Nature* **1988**, *333*, 232–237. [CrossRef] [PubMed]
7. Chao, H.; Houston, M.E., Jr.; Hodges, R.S.; Kay, C.M.; Sykes, B.D.; Loewen, M.C.; Davies, P.L.; Sönnichsen, F.D. A diminished role for hydrogen bonds in antifreeze protein binding to ice. *Biochemistry* **1997**, *36*, 14652–14660. [CrossRef] [PubMed]
8. Sicheri, F.; Yang, D.S.C. Ice-binding structure and mechanism of an antifreeze protein from winter flounder. *Nature* **1995**, *375*, 427–431. [CrossRef] [PubMed]

9. Haymet, A.D.J.; Ward, L.G.; Harding, M.M. Winter flounder "antifreeze" proteins: Synthesis and ice growth inhibition of analogues that probe the relative importance of hydrophobic and hydrogen-bonding interactions. *J. Am. Chem. Soc.* **1999**, *121*, 941–948. [CrossRef]
10. Baardsnes, J.; Kondejewski, L.H.; Hodges, R.S.; Chao, H.; Kay, C.; Davies, P.L. New ice-binding face for type I antifreeze protein. *FEBS Lett.* **1999**, *463*, 87–91. [CrossRef]
11. Davies, P.L.; Baardsnes, J.; Kuiper, M.J.; Walker, V.K. Structure and function of antifreeze proteins. *Philos. Trans. R. Soc. Lond. B* **2002**, *357*, 927–935. [CrossRef] [PubMed]
12. Jorov, A.; Zhorov, B.S.; Yang, D.S.C. Theoretical study of interaction of winter flounder antifreeze protein with ice. *Protein Sci.* **2004**, *13*, 1524–1537. [CrossRef] [PubMed]
13. Dalal, P.; Knickelbein, J.; Haymet, A.D.J.; Sönnichsen, F.D.; Madura, J. Hydrogen bond analysis of type 1 antifreeze protein in water and the ice/water interface. *PhysChemComm* **2001**, *4*, 32–36. [CrossRef]
14. Wierzbicki, A.; Dalal, P.; Cheatham, T.E., III; Knickelbein, J.E.; Haymet, A.D.J.; Madura, J.D. Antifreeze proteins at the ice/water interface: Three calculated discriminating properties for orientation of type I proteins. *Biophys. J.* **2007**, *93*, 1442–1451. [CrossRef] [PubMed]
15. Nada, H.; Furukawa, Y. Growth inhibition mechanism of an ice-water interface by a mutant of winter flounder antifreeze protein: A molecular dynamics study. *J. Phys. Chem. B* **2008**, *112*, 7111–7119. [CrossRef] [PubMed]
16. Todde, G.; Hovmöller, S.; Laaksonen, A. Influence of antifreeze proteins on the ice/water interface. *J. Phys. Chem. B* **2015**, *119*, 3407–3413. [CrossRef] [PubMed]
17. Evans, R.P.; Hobbs, R.S.; Goddard, S.V.; Fletcher, G.L. The importance of dissolved salts to the in vivo efficacy of antifreeze proteins. *Comp. Biochem. Physiol. Part A* **2007**, *148*, 556–561. [CrossRef] [PubMed]
18. Kristiansen, E.; Pedersen, S.A.; Zachariassen, K.E. Salt-induced enhancement of antifreeze protein activity: A salting-out effect. *Cryobiology* **2008**, *57*, 122–129. [CrossRef] [PubMed]
19. Hagiwara, Y.; Aomatsu, H. Supercooling enhancement by adding antifreeze protein and ions to water in a narrow space. *Int. J. Heat Mass Transf.* **2015**, *86*, 55–64. [CrossRef]
20. Hayakari, K.; Hagiwara, Y. Effects of ions on winter flounder antifreeze protein and water molecules near an ice/water interface. *Mol. Simul.* **2012**, *38*, 26–37. [CrossRef]
21. Nada, H.; Furukawa, Y. Anisotropy in growth kinetics at interfaces between proton-disordered hexagonal ice and water: A molecular dynamics study using the six-site model of H_2O. *J. Cryst. Growth* **2005**, *283*, 242–256. [CrossRef]
22. Yokoyama, T.; Hagiwara, Y. Molecular dynamics simulation for the mixture of water and an ice nucleus. *Mol. Simul.* **2003**, *29*, 235–248. [CrossRef]
23. Iwasaki, K.; Hagiwara, Y. Inhibition of ice nucleus growth in water by alanine dipeptide. *Mol. Simul.* **2004**, *30*, 487–500. [CrossRef]
24. Nobekawa, T.; Taniguchi, H.; Hagiwara, Y. Interaction between a twelve-residue segment of antifreeze protein type I, or its mutants, and water molecules. *Mol. Simul.* **2008**, *34*, 309–325. [CrossRef]
25. Nobekawa, T.; Hagiwara, Y. Interaction among the twelve-residue segment of antifreeze protein type I, or its mutants, water and a hexagonal ice crystal. *Mol. Simul.* **2008**, *34*, 591–610. [CrossRef]
26. Nohara, Y.; Hagiwara, Y. Diffusion of cations in salt solutions between ice walls. *Mol. Simul.* **2015**, *41*, 980–985. [CrossRef]
27. Gear, C.W. *Numerical Initial Value Problems in Ordinary Differential Equations*; Prentice-Hall: Upper Saddle River, NJ, USA, 1971.
28. Nosé, S. A molecular dynamics method for simulations in the canonical ensemble. *Mol. Phys.* **1984**, *52*, 255–268. [CrossRef]
29. Grandum, S.; Yabe, A.; Nakagomi, K.; Tanaka, M.; Takemura, F.; Kobayashi, Y.; Frivik, P.-E. Analysis of ice crystal growth for a crystal surface containing adsorbed antifreeze proteins. *J. Cryst. Growth* **1999**, *205*, 382–390. [CrossRef]
30. Abascal, J.L.F.; Sanz, E.; Fernández, R.G.; Vega, C. A potential model for the study of ices and amorphous water: TIP4P/Ice. *J. Chem. Phys.* **2005**, *122*. [CrossRef] [PubMed]
31. Jorgensen, W.J.; Chandrasekhar, J.; Madura, J.D.; Impey, R.W.; Klein, M.L. Comparison of simple potential functions for simulating liquid water. *J. Chem. Phys.* **1983**, *79*, 926–935. [CrossRef]
32. Frenkel, D.; Smit, B. *Understanding Molecular Simulation*; Academic Press: San Diego, CA, USA, 1996.

33. Ryckaert, J.P.; Ciccotti, G.; Berendsen, H.J.C. Numerical integration of the Cartesian equations of motion of a system with constraints: Molecular dynamics of *n*-alkanes. *J. Comput. Phys.* **1977**, *23*, 327–341. [CrossRef]

34. Cheng, A.; Merz, K.M., Jr. Ice-binding mechanism of winter flounder antifreeze proteins. *Biophys. J.* **1997**, *73*, 2851–2873. [CrossRef]

35. Jorgensen, W.L.; Tirado-Rives, J. The OPLS potential functions for proteins: Energy minimizations for crystals of cyclic peptides and crambin. *J. Am. Chem. Soc.* **1988**, *110*, 1657–1666. [CrossRef] [PubMed]

36. Chowdhuri, S.; Chandra, A. Molecular dynamics simulations of aqueous NaCl and KCl solutions: Effects of ion concentration on the single-particle pair, and collective dynamical properties of ions and water molecules. *J. Chem. Phys.* **2001**, *115*, 3732–3741. [CrossRef]

37. Chialvo, A.A. The structure of $CaCl_2$ aqueous solutions over a wide range of concentration. Interpretation of diffraction experiments via molecular simulation. *J. Chem. Phys.* **2003**, *119*, 8052–8061. [CrossRef]

38. Allen, M.P.; Tildesley, D.J. *Computer Simulation of Liquid*; Oxford Science: Oxford, UK, 1987.

39. Habasaki, J.; Ngai, K.L. Heterogeneous dynamics of ionic liquids from molecular dynamics simulations. *J. Chem. Phys.* **2008**, *129*, 194501. [CrossRef] [PubMed]

40. Holzmann, J.; Ludwig, R.; Geiger, A.; Paschek, D. Pressure and salt effects in simulated water: Two sides of the same coin? *Angew. Chem. Int. Ed.* **2007**, *46*, 8907–8911. [CrossRef] [PubMed]

41. Luzar, A.; Chandler, D. Hydrogen-bond kinetics in liquid water. *Nature* **1996**, *379*, 55–57. [CrossRef]

42. Baştuğ, T.; Kuyucak, S. Temperature dependence of the transport coefficients of ions from molecular dynamics simulations. *Chem. Phys. Lett.* **2005**, *408*, 84–88. [CrossRef]

© 2018 by the authors. Licensee MDPI, Basel, Switzerland. This article is an open access article distributed under the terms and conditions of the Creative Commons Attribution (CC BY) license (http://creativecommons.org/licenses/by/4.0/).

MDPI

St. Alban-Anlage 66

4052 Basel

Switzerland

Tel. +41 61 683 77 34

Fax +41 61 302 89 18

www.mdpi.com

Crystals Editorial Office

E-mail: crystals@mdpi.com

www.mdpi.com/journal/crystals